Introduction to Naval Architecture

Dedicated to

**Will, George, Phoebe, Millie,
Amelie and Maelys**

Introduction to Naval Architecture

Fifth Edition

Eric C. Tupper
BSc, CEng, RCNC, FRINA, WhSch

AMSTERDAM • BOSTON • HEIDELBERG • LONDON
NEW YORK • OXFORD • PARIS • SAN DIEGO
SAN FRANCISCO • SINGAPORE • SYDNEY • TOKYO
Butterworth-Heinemann is an imprint of Elsevier

Butterworth-Heinemann is an imprint of Elsevier.
The Boulevard, Langford Lane, Kidlington, Oxford, OX5 1GB
225 Wyman Street, Waltham, MA 02451, USA

First published as *Naval Architecture for Marine Engineers*, 1975
Reprinted 1978, 1981
Second edition published as *Muckle's Naval Architecture*, 1987
Third edition published as *Introduction to Naval Architecture*, 1996
Revised reprint 2000
Fourth edition 2004
Fifth edition 2013

British Library Cataloguing in Publication Data
A catalogue record for this book is available from the British Library

Library of Congress Cataloguing in Publication Data
A catalogue record for this book is available from the Library of Congress

ISBN: 978-0-08-098237-3

For information on all Elsevier Butterworth-Heinemann publications visit our website at http://books.elsevier.com

Typeset by MPS Limited, Chennai, India
www.adi-mps.com

Printed and bound in Great Britain

13 14 15 16 10 9 8 7 6 5

Contents

This page intentionally left blank

Preface to the Fifth Edition

GENERAL

Modern ships are large and complex and, unlike other forms of transport, many have to provide accommodation for human habitation over extended periods. They operate in a changeable environment which can be very hostile and their responses to that environment can be highly non-linear. To design and build efficient and effective ships can be very demanding.

For many years the naval architect made many simplifying assumptions in order to quantify the ship characteristics of stability, strength, speed and so on. Ships evolved slowly as making significant changes from existing practice was risky. Calculations had to be done manually using simple devices such as slide rules. With modern research, combined with massive increases in computing power, problems can be tackled more thoroughly. For instance, a ship's responses as an elastic body in irregular waves can be computed taking account of interactions between motions and of non-linearities. This requires a much deeper understanding of the subject and it is not possible to cover all aspects of the subject in depth in one volume.

This book provides a broad appreciation of the science and art of naval architecture explaining the subject in physical rather than in mathematical terms. This provides a clearer understanding of the underlying principles involved and should be adequate for many associated with maritime matters. For the naval architect proper, it provides the framework upon which to build the deeper understanding needed by a designer, builder or researcher.

ARRANGEMENT OF THE BOOK

Naval architecture involves two distinctly different tasks – the analysis of individual design characteristics and the synthesis of a design from those elements. Analysis calls upon procedures and theories developed over the years. It can be regarded as the science/engineering of the subject. Design is more an 'art' – both in an aesthetic sense and in arriving at a well balanced design meeting often conflicting requirements.

Analysis is relatively straightforward, although by no means simple. The basics are common to all types and sizes of ship and craft. Synthesis is more difficult both to achieve and to explain. It requires experience and flair. It is in the synthesis (design) stage that quite different vessels – in function, appearance and size – emerge. As an example, consider supporting a load on a circular support on a structure which is predominately composed of transverse and longitudinal bulkheads. Once a structural solution is sketched it can be checked to ensure it is adequately strong. Whilst there may be uncertainty in ascertaining the loads, the 'art' is in envisaging an efficient method of transmitting the loads from the circular support to the surrounding structure.

Textbooks can either deal with the naval architectural elements first and then show how they are applied during design or begin with the synthesis process showing what supporting naval architectural elements are needed and then considering these elements in turn. This book adopts the first approach. The earliest chapters deal with terminology, definitions and notation providing the means of communicating in a precise, unambiguous way. The next few chapters deal with 'analysis' and the later chapters with 'synthesis'. Where it is deemed appropriate, more detailed matter is placed in appendices to leave the main text more free flowing.

Advanced methods such as Computer Aided Design/Computer Aided Manufacture (CAD/CAM) systems are outlined, in general terms. The underlying naval architectural principles remain the same and it is these fundamentals this book covers. The practitioner should master them before embarking on more advanced methods. Often the advanced methods such as the latest the International Maritime Organisation (IMO) probabilistic methods of damage stability assessment are less transparent to the user. For this reason this book includes the traditional approaches in areas where it is felt they demonstrate more clearly the principles involved. Also many existing ships will have been designed using these earlier methods. It is the methods of application rather than the principles which have changed.

This edition takes account of the views of users of earlier editions on the balance and coverage of the subject. It reflects the continuing developments in technology, changes in international regulations and recent research.

SI units are used almost universally and are used in the main text. To assist those wishing to use data in Imperial units conversion tables are given in an appendix. So is the Froude notation because of its importance in early work on ship resistance.

Appendix E presents a range of questions based on each chapter of the book for use by students and lecturers.

References and websites are given where they provide information on recent developments or more detailed information on topics. Many of the reference documents can be accessed from websites and this is indicated. The Internet contains much other useful information but may not be on a site for long. Students can use these but they require some overarching knowledge of the subject before they can be used intelligently. This book aims to provide that understanding.

Acknowledgements

This current treatise has been developed from Muckle's Naval Architecture through several editions. Many figures are taken from the publications of the Royal Institution of Naval Architects. These are reproduced by kind permission of the Institution and a note is included in the captions. I am grateful for the help of SENER Ingeniera *y* Sistemas SA in providing information on their FORAN CAD/CAM system and to QinetiQ GRC for comments on their CAD system – PARAMARINE. I would like to thank my friend Robert Curry for his help in matters relating to ship structures, the IMO and operations in ice. In ship design, besides drawing upon my personal design experience I have drawn upon the extensive work of David Andrews and his team at University College, London.

I am grateful to my son, Simon, for his assistance in producing some of the illustrations.

This page intentionally left blank

Chapter 1

Introduction

GENERAL

Modern large ships are perhaps the most complex of modern engineering projects and represent the largest man-made mobile structures. Ships still carry over 90% of world trade and still carry large numbers of people on pleasure cruises and ferries in all areas of the globe. Ships, and other marine structures, are needed to exploit the riches of the deep. Their design, build, maintenance and operation, in all of which naval architects play a major role or exert considerable influence, are fascinating activities.

Although one of the oldest forms of transport, ships, their equipment and their function, are subject to constant evolution due to changes in world trade and technology and by pressure of economics. Other changes are driven by social changes and, in particular, by the public's desire for greater safety and for more protection of the environment.

A feature of many new designs is the variation in form of ships intended for relatively conventional tasks. This is for reasons of efficiency and has been made possible by the advanced analysis methods available, backed up by model experiment when needed. These enable unorthodox configurations to be adopted with confidence.

NAVAL ARCHITECTURE AND THE NAVAL ARCHITECT

What is naval architecture and what is required of a naval architect? In essence, one can say that naval architecture is the science of making a ship 'fit for purpose' and a naval architect is an engineer competent in naval architecture. A fuller answer on the nature of naval architecture is to be found in Ferreiro (2007). In summary, he defines naval architecture as:

> The branch of engineering concerned with the application of ship theory within the design and construction process, with the purpose of predicting the characteristics and performance of the ship before it is built.

Introduction to Naval Architecture. DOI: http://dx.doi.org/10.1016/B978-0-08-098237-3.00001-1

He defines ship theory as

The science explaining the physical behaviour of a ship, through the use of fundamental mathematics or empirically derived data.

The Ship

'Ship' should be interpreted broadly to mean any structure floating in water. It is usually self-propelled but may rely on tugs for movement. Others rely on the wind. Marine structures, such as harbour installations, are the province of the civil engineer.

The purpose of a merchant ship is to carry goods, perhaps people, safely across water. That of a warship is the support of government policy. In ordering a new merchant ship, the owner will have in mind a certain volume of cargo to be carried on voyages between certain ports with an average journey time. Each requirement will have an impact upon the ship design. For instance:

- The type of cargo may be able to be carried in bulk or may require packaging; it may be hazardous or it may require a special on-board environment.
- The volume of cargo will be the major factor in determining the size of the ship. There may be a need to move the cargo in discreet units of a specified size and weight.
- Ports, plus any rivers and canals to be negotiated, may place restrictions on the overall dimensions of the vessel. Depending on the port facilities, the ship may have to have more, or less, cargo handling equipment on board. The routes used also dictate the ocean areas to be traversed and hence the sea and weather conditions likely to be encountered.
- Schedules dictate the speed and hence the installed power. They may indicate desirable intervals between maintenance periods.

Fit for Purpose

To be fit for purpose, a ship must be able to operate safely and reliably. It must:

- float upright with enough watertight volume above the waterline to cope with waves and accidental flooding.
- have adequate stability to cope with operational upsetting moments and to withstand a specified degree of flooding following damage. It must not be so stable that motions become unpleasant.
- be able to maintain the desired average speed in the sea conditions it is likely to meet.
- be strong enough to withstand the loads it will experience in service.

- be capable of moving in a controlled way in response to movements of control surfaces, to follow a given course or manoeuvre in confined waters.

The ship must do all this economically with the minimum crew. This book deals with these various matters and brings them together in discussing the design process and the different ship types that emerge from that process. The design should be flexible because ship use is likely to change over the long life expected of ships.

Variety

Naval architecture is a fascinating and demanding discipline. It is fascinating because of the variety of floating structures and the many compromises necessary to achieve the most effective design. It is demanding because a ship is a very large capital investment. It must be safe for the people on board and the environment. Unlike many other forms of transport, the naval architect does not have the benefit of prototypes.

There are fishing vessels ranging from small local boats operating by day to ocean-going ships with facilities to deep freeze their catches. There are vessels for exploitation of undersea energy sources, gas and oil and extraction of minerals. There are oil tankers, ranging from small coastal vessels to giant supertankers. Other huge ships carry bulk cargoes such as grain, coal or iron ore. Ferries carry passengers between ports which may be only a few kilometres or hundreds apart. There are tugs for shepherding ships in port or for trans-ocean towing. Then there are dredgers, lighters and pilot boats without which a port could not function. Warships range from huge aircraft carriers through cruisers and destroyers to frigates, patrol boats, mine countermeasure vessels and submarines.

Increasingly naval architects are involved in the design of small craft such as yachts and motor cruisers. This reflects partly the much greater number of small craft, partly the increased regulation to which they are subject requiring a professional input and partly the increasingly advanced methods used in their design and new materials in their construction. In spite of the increasingly scientific approach, the design of small craft still involves a great deal of 'art'. Many are beautiful with graceful lines and lavishly appointed interiors. The craftsmanship needed for their construction is of the highest order.

Many naval architects are involved in offshore engineering – finding and exploiting oil, gas and mineral deposits. Their expertise has been needed for the design of the rigs and the many supporting vessels, including manned and unmanned submersibles used for maintenance of underwater installations. This involvement will continue as the riches of the ocean and ocean bed are exploited in the future and attention focuses on the polar regions.

Ships come in a variety of hull forms. Much of this book is devoted to single hull, displacement forms which rely upon displacing water to support their full weight. In some applications, particularly for fast ferries, multiple hulls are preferred because they provide large deck areas and good stability without excessive length. In planing craft, high speeds may be achieved by using dynamic forces to support part of the weight when under way. Surface effect ships use air cushions to support the weight of the craft, lifting it clear of the water and providing an amphibious capability. Hydrofoil craft rely on hydrodynamic forces on submerged foils under the craft to lift the main part of the hull above the waves.

Variety is not limited to appearance and function. Different materials are used— steel, wood, aluminium, reinforced plastics of various types and concrete. The propulsion system used to drive the craft through the water may be the wind but for most large craft is some form of mechanical propulsion. The driving power may be generated by diesels, steam or gas turbine, some form of fuel cell or a combination of these. Power will be transmitted to the propulsion device through mechanical or hydraulic gearing or by using electric generators and motors as intermediaries. The propulsor itself is usually some form of propeller, perhaps ducted, but may be a water or air jet. There are many other systems on board, such as means of manoeuvring the ship, electric power generation, hydraulic power for winches and other cargo handling systems.

Growing concern as regards pollution of the environment – the atmosphere and the oceans – is having an increasing impact on ship design and operations. ‘Greener’ forms of propulsion are being developed with greater emphasis on efficiency to reduce usage of fuel.

A ship can be a veritable floating township of several thousand people remaining at sea for several weeks. It needs electrics, air conditioning, sewage treatment plant, galleys, bakeries, shops, restaurants, cinemas and other leisure facilities. All these and the general layout must be arranged so that the ship can carry out its intended tasks efficiently. The naval architect has the problems of the land architect but, in addition, a ship must float, move, be capable of surviving in a very rough environment and withstand a reasonable level of damage. It is the naval architect who ‘orchestrates’ the design, calling upon the expertise of many other professions in achieving the best compromise between many, often conflicting, requirements. Naval architecture is a demanding profession because a ship is a major capital investment taking many years to create and expected to remain in service for 25 years or more. It is usually part of a larger transport system and must be properly integrated with the other elements of the overall system. A prime example of this is the container ship. Goods are placed in containers at the factory. These containers are of standard dimensions and are taken by road, or rail, to a port with specialised handling equipment where they are loaded on board. At the port of destination, they are off-loaded on to land transport. The use of containers means that ships need to spend far less time in port loading and unloading and the cargoes are more secure. Port fees are reduced and the ship is used more productively.

SAFETY

Most important is the safety of crew, ship and, increasingly, the environment. The design must be safe for normal operations and not be unduly vulnerable to mishandling or accident. No ship can be absolutely safe and a designer must take conscious decisions as to the level of risk judged acceptable in the full range of scenarios in which the ship can expect to find itself. There will always be a possibility that the design conditions will be exceeded. The risk of this and the potential consequences must be assessed and only accepted if they are judged unavoidable or the level of risk is acceptable. Acceptable, that is, to the owner, operator and the general public and not least to the designer who has ultimate responsibility. Even where errors on the part of others have caused an accident, the designer should have considered such a possibility and taken steps to minimise the consequences. For instance, in the event of collision, the ship must have a good chance of surviving or of remaining afloat long enough for passengers to be taken off safely. This brings with it the need for a whole range of life-saving equipment.

Naval architects must work closely with those who build, maintain and operate the ships they design. This need for teamwork and the need for each player to understand the others' needs and problems are the themes of a book published by The Nautical Institute in 1999.

THE IMPACT OF TECHNOLOGY AND COMPUTERS

Over the last half-century, changing technology has had a tremendous impact upon how ships are designed, built, operated and maintained. The following are examples:

- Satellites enable ships to locate their position accurately using *global positioning systems*. They can pick up distress signals and locate the casualty for rescue organisations. They can measure sea conditions over wide areas and facilitate the routeing of ships to avoid worst storms.
- Modern materials require much less maintenance, reducing operating costs and manpower demands. New hull treatments permit much longer intervals between dockings and reduce pollution of the ocean.
- Modern equipment is generally more reliable. Modularisation and repair by replacement policies reduce downtime and the number of repair staff needed on board. Electronically controlled operating and surveillance systems enable fewer operators to cope with large main propulsion systems and a wide range of ship's services.

The biggest impact has been the influence of the computer which has made a vital contribution to many of the changes referred to above. But it is in the sequence of design, build, maintaining and running of ships that their influence has been greatest for the naval architect. In some cases, these processes have changed almost beyond recognition, although the underlying principles and objectives remain the same.

In Design

Computer-aided design (CAD) systems enable concept designs to be produced more rapidly, in greater detail and with greater accuracy. Once the concept design is agreed, the same CAD system can be used for the contract design phase. Three-dimensional graphics and virtual reality techniques can be used to interact with others more efficiently. CAD systems are integrated suites of related programs and can accommodate advanced programs for structural strength evaluations, motion predictions and so on.

In Production

Once construction is approved, data can be passed to the chosen shipyard in digital form. This reduces the risk of misinterpretation of design intent. Provided the designer's CAD and builder's *computer-aided manufacture* systems are compatible, the builder's task in producing information for the production process can be reduced. The builder can develop the database as the design is developed to provide all the details for manufacture and, later, to pass on to those who have to operate and maintain the equipment and systems.

In Operation

CAD systems make it easier to pass information on to those who are to operate and maintain the ship. Hydrostatic, hold and tank capacity, stability and strength data can be fed into the ship's own software systems to assist the captain in loading and operating the ship safely. Listings of equipment and fittings, with code numbers, will ensure that any replacements and spares will meet the *form, fit and function* requirements. Computer-based decision-aiding systems can help the master by providing guidance on the loading sequences to eliminate the possibility of jeopardising the stability or strength. In warships they can assist the captain when under enemy attack by suggesting the optimum actions to take in defending the ship. It is emphasised that they are only used in an advisory capacity in these roles. They do not reduce the master's or captain's responsibility.

Computer-based simulators can assist in training navigators, machinery controllers and so on. These simulators can be produced to various levels of realism, depending upon the need. The computer can provide external stimuli, through goggles or screens, which the operator can expect to experience in practice. For instance, a navigational simulator can provide pictures of a harbour and its approaches including other ships.

A computer is only a tool, albeit a very powerful one, and it must be used intelligently. It is an aid to the human, although artificial intelligence techniques can be used to help a relatively inexperienced person. So-called *expert systems* can store information on how a number of very experienced engineers would view a certain problem in a variety of circumstances. Thus, a less-experienced person (at least in that particular type of vessel or situation) can be guided into what might be termed good practice.

This is not to say that the tasks of the designer, builder or operator have been made easy. Some of the more humdrum activities such as tedious manual calculations of volumes and weights have been removed, but more knowledge is needed to carry out the total task. Whereas in the past, a simple longitudinal hull strength calculation, using a standard wave, was all that was possible, a much more complex assessment is now demanded.

SUMMARY

Naval architecture can be an interesting and rewarding profession. A feel for the variety to be found in the profession can be obtained by reading the memoirs of eminent naval architects such as Marshall Meek (2003) and Ken Rawson (2006) or a book such as that by Fred Walker (2010).

Ships vary greatly in size and use, and their associated capabilities and qualities will be discussed in more detail as the fundamentals of the subject are covered later in the book. A general text such as this can provide only a general understanding of the problems involved and the methods used in tackling them. To apply the methods, a student may need more detail and this can be found in specialist textbooks dealing with specific subjects such as powering and strength. Individual scientific papers can provide more information on more advanced techniques.

This page intentionally left blank

Chapter 2

Definition and Regulation

INTRODUCTION

In any engineering discipline, and naval architecture is no exception, precision is necessary in defining the products created. Communications with others must be clear and unambiguous. An internationally recognised terminology has grown up over the years to aid this definition. Some of the terms will be unfamiliar to those coming new to shipping. Others will be familiar from everyday usage but may have a special and very precise meaning in naval architecture. A reference to the right hand side of a ship can be ambiguous; which side is meant depends upon where the person addressed is standing and in which direction he/she is facing. To refer to the starboard side of the ship allows no such misunderstanding.

A notation has also become generally accepted (the use of V and L to denote speed and length are examples) for representing physical entities in text – as a form of shorthand – and in equations. Appendix A outlines the notation commonly used and the units used for different quantities.

The meaning of some terms used here will become apparent later.

DEFINITION

Defining the Hull Form

A ship's hull form determines many of its main attributes, stability characteristics and resistance, and therefore the power needed for a given speed, seaworthiness, manoeuvrability and load-carrying capacity. It is important, therefore, that hull shape be defined accurately. The basic descriptors such as length and beam used must be defined. Not all authorities use the same definitions, and it is important that the reader of a document, or user of a computer program, checks the definitions adopted. Those used in this chapter are those generally used in the United Kingdom and most are internationally accepted.

Introduction to Naval Architecture. DOI: http://dx.doi.org/10.1016/B978-0-08-098237-3.00002-3

The Geometry of the Hull

A ship's hull is three dimensional and is usually symmetrical about a fore and aft plane. Throughout this book, a symmetrical hull form is assumed unless otherwise stated. The hull envelope shape is defined by its intersection with three sets of mutually orthogonal planes. The horizontal planes are known as *waterplanes* and the lines of intersection are known as *waterlines*. The planes parallel to the middle line plane cut the hull in *bow and buttock lines*, the middle line plane itself defining the *profile*. The intersections of the transverse, that is the athwartships, planes define the *transverse sections*.

Two sets of main hull dimensions are used. *Moulded* dimensions are those between the inner surfaces of the hull envelope. Dimensions measured to the outside of the plating are meant if there is no qualifying adjective. Moulded dimensions are used to find the internal volumes of the hull – a rough indicator of earning capacity.

Five different lengths used in naval architecture are defined below. The first three lengths are those most commonly used to define the ship form and are as follows (Figure 2.1).

- *The length between perpendiculars* (LBP or L_{PP}) is the distance measured along the summer load waterplane (the design waterplane for warships) from the after to the fore perpendicular. The *after perpendicular* is commonly taken as the line passing through the rudder stock. The *fore perpendicular* is the vertical line through the intersection of the forward side of the stem with the summer load waterline.

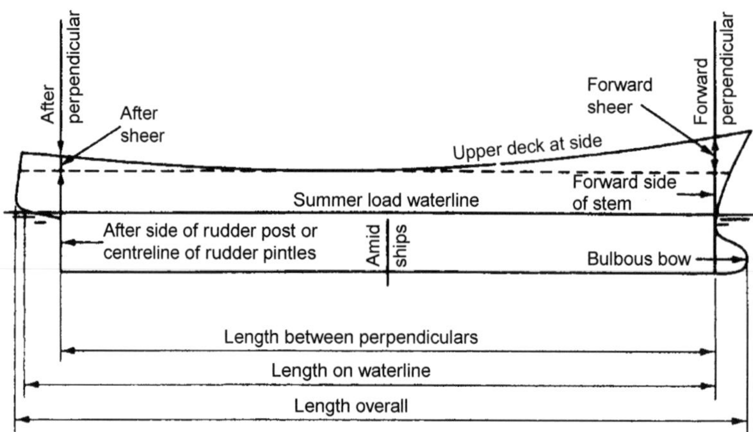

FIGURE 2.1 Principal dimensions.

- *The length overall* (LOA or L_{OA}) is the distance between the extreme points forward and aft measured parallel to the summer (or design) waterline. Forward the point may be on the raked stem or on a bulbous bow.
- *The length on the waterline* (LWL or L_{WL}) is the length on the waterline, at which the ship happens to be floating between the intersections of the bow and after end with the waterline. If not otherwise stated the summer load (or design) waterline is to be understood.
- *The scantling length* is used in classification society rules to determine the required scantlings. The scantling length is the LBP but is not to be less than 96% of LWL and need not be more than 97% of LWL
- *Subdivision length.* A length used in damage stability calculations carried out in accordance with International Maritime Organisation (IMO) standards. It is basically the ship length embracing the buoyant hull and the reserve of buoyancy.

The mid-point between the perpendiculars is called *amidships* or *midships*. The section of the ship at this point by a plane normal to both the summer waterplane and the centreline plane of the ship is called the *midship section*. It may not be the largest section of the ship – in general this will be a section somewhat aft of amidships. The breadth of the ship at any point is called the *beam*. Unless otherwise qualified, the beam quoted is usually that amidships at the summer waterplane. Referring to Figure 2.2, the *moulded beam* is the greatest distance between the inside of plating on the two sides of the ship at the greatest width at the section chosen. The *breadth extreme* is measured to the outside of plating but will also take account of any overhangs or flare.

The ship *depth* (Figure 2.2) varies along the length but is usually quoted for amidships. The *moulded depth* is measured from the underside of the deck plating at the ship's side to the top of the inner flat keel plate. Unless otherwise specified, the depth is to the uppermost continuous deck. Where a rounded deck edge, gunwhale, is fitted the convention used is indicated in Figure 2.2.

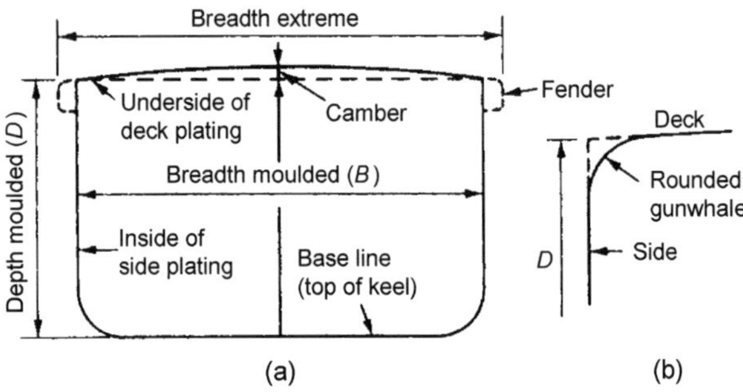

FIGURE 2.2 Breadth measurements.

Sheer (Figure 2.1) is a measure of how much a deck rises towards the stem and stern. For any position along the length of the ship, it is defined as the height of the deck at side at that position above the deck at side amidships.

Camber or *round of beam* is defined as the rise of the deck in going from the side to the centre as shown in Figure 2.3. Decks are cambered to enable water to run off them more easily. For ease of construction, and reduce cost, camber is applied usually only to weather decks, and straight line camber often replaces the older parabolic curve.

The bottom of a ship, in the midships region, is usually flat but not necessarily horizontal. If the line of bottom is extended out to intersect the moulded breadth line (Figure 2.3), the height of this intersection above the keel is called the *rise of floor* or *deadrise*. Most ships have a flat keel and the extent to which this extends athwartships is termed the *flat of keel* or *flat of bottom*.

In some ships, the sides are not vertical at amidships. If the upper deck beam is less than that at the waterline, it is said to have *tumble home*, the value being half the difference in beams. If the upper deck has a greater beam the ship is said to have *flare*. Most ships have flare at a distance from amidships particularly towards the bow.

The *draught* of the ship at any point along its length is the distance from the keel to the waterline. If a *moulded draught* is quoted, it is measured from the inside of the keel plating. For navigation purposes, it is important to know the maximum draught. This will be taken to the bottom of any projection below keel such as a bulbous bow, propeller or sonar dome. If a waterline is not quoted, the design waterline is usually intended. To aid the captain, draught marks are placed near the bow and stern and remote reading devices

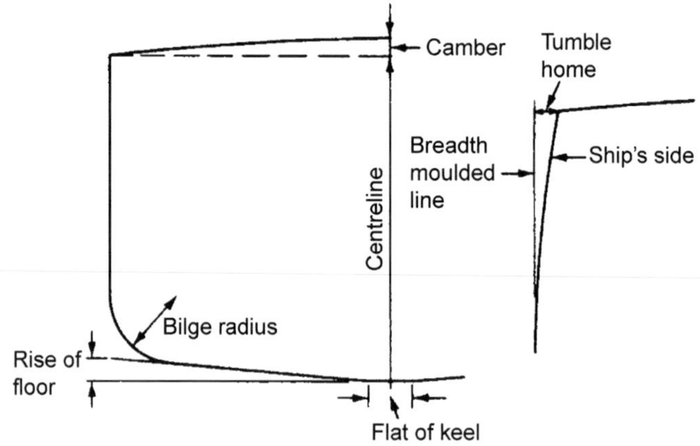

FIGURE 2.3 Section measurements.

for draught are often provided. The difference between the draughts forward and aft is referred to as the *trim*. Trim is said to be *by the bow* or *by the stern* depending upon whether the draught is greater forward or aft, respectively. Often draughts are quoted for the two perpendiculars. Being a flexible structure, a ship will usually bend slightly fore and aft, the curvature varying with the loading. The ship is said to *hog* or *sag* when the curvature is concave down or up respectively. The amount of hog or sag is the difference between the actual draught amidships and the mean of the draughts at the fore and after perpendiculars. In calculations, the curvature of the ship due to hog or sag is usually taken to be parabolic in shape.

Air draught is the vertical distance from the summer waterline to the highest point in the ship, usually the top of a mast. This dimension is important for ships that need to go under bridges in navigating rivers or canals or when entering port. In some cases, the topmost section of the mast can be struck (lowered) to enable the ship to pass under an obstruction.

Freeboard is the difference between the depth at side and the draught. That is, it is the height of the deck at side above the waterline. The freeboard is usually greater at the bow and stern than at amidships due to sheer. This helps create a drier ship in waves. Freeboard is important in determining safety and stability at large angles.

In defining the relative position of items, the following terms are used:

- *Ahead* and *astern (aft)* are used for an item either forward of or behind another.
- *In way of* to denote proximity to. (Note: It does not denote an obstruction.)
- *Abreast* to denote to the side of.

Representing the Hull Form

The hull form is portrayed graphically by the *lines plan* or *sheer plan* (Figure 2.4). This shows the various curves of intersection between the hull

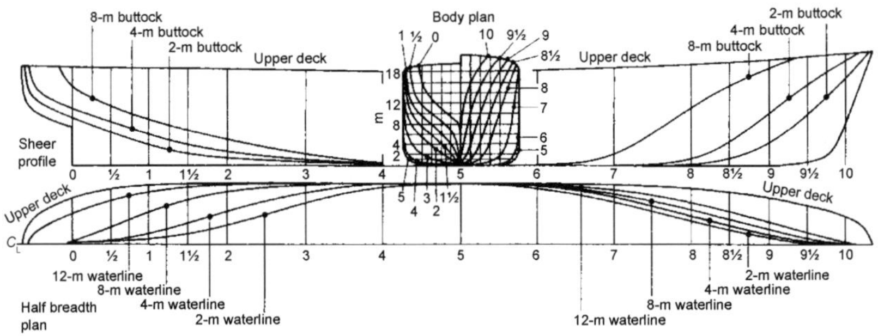

FIGURE 2.4 Lines plan.

and the three sets of orthogonal planes. For symmetrical ships, only one half is shown. The curves showing the intersections of vertical fore and aft planes with the hull are grouped in the *sheer profile*, the waterlines in the *half breadth plan*, and the sections by transverse planes in the *body plan*. In merchant ships, the transverse sections are numbered from aft to forward. In warships, they are numbered from forward to aft, although the forward half of the ship is still, by tradition, shown on the right hand side of the body plan. The distances of the various intersection points from the middle line plane are called *offsets*.

Clearly the three sets of curves making up the lines plan are interrelated as they represent the same three-dimensional body. This interdependency is used in manually fairing the hull form, each set being faired in turn and the changes in the other two noted. At the end of an iterative process, the three sets will be mutually compatible. Fairing is now achieved by computer using mathematical fairing techniques, such as the use of curved surface patches, to define the hull. It is often generated directly from the early design processes in the computer. The old manual process is easier to describe and illustrates the intent. Manual fairing was done first in the design office on a reduced scale drawing. To aid production, the lines were then *laid off* full scale on the floor of a building known as the mould loft. That is, they were scribed on the floor. They were then re-faired. Some shipyards used a reduced scale, say one-tenth, for use in the building process. Nowadays, data are passed to a shipyard in digital form where it is then fed into a computer-aided manufacturing system.

In some ships, particularly large carriers of bulk cargo, the transverse cross section is constant for some fore and aft distance near amidships. This portion is known as the *parallel middle body*.

Appendages of the main hull, such as shaft bossings, or sonar domes, are faired separately.

Hull Characteristics

Having defined the hull form, it is possible to derive a number of characteristics which have significance in determining the general performance of the ship. As a floating body, a ship in equilibrium will displace its own weight of water, as explained later. Thus the volume of the hull below the design load waterline must represent a weight of water equal to the weight of the ship at its designed load. This *displacement*, as it is called, can be defined as:

$$\Delta = \rho g \nabla$$

where ρ is the density of the water in which the ship is floating, g is the acceleration due to gravity and ∇ is the underwater volume.

For flotation, stability and hydrodynamic performance generally, it is this displacement, expressed either as a volume or a force that is of interest.

For rule purposes Classification Societies may use a *moulded displacement* which is the displacement within the moulded lines of the ship between perpendiculars.

It is useful to have a feel for the fineness of the hull form. This is provided by a number of *form coefficients* or *coefficients of fineness*. These are defined as follows, where ∇ is the volume of displacement:

- Block coefficient $(C_B) = \dfrac{\nabla}{L_{PP}\,BT}$

  where:

  L_{PP} is length between perpendiculars,
  B is the extreme breadth underwater and
  T is the mean draught.

- Coefficient of fineness of waterplane $(C_{WP}) = \dfrac{A_W}{L_{WL}B}$

  where:

  A_W is waterplane area,
  L_{WL} is the waterline length and
  B is the extreme breadth of the waterline.

- Midship section coefficient $(C_M) = \dfrac{A_M}{BT}$

  where:

  A_M is the midship section area and
  B is the extreme underwater breadth amidships.

- Longitudinal prismatic coefficient $(C_p) = \dfrac{\nabla}{A_M L_{pp}}$

  It will be noted that $C_M \times C_p = C_B$.

- Vertical prismatic coefficient $(C_{VP}) = \dfrac{\nabla}{A_W T}$

These coefficients are ratios of the volume of displacement to various circumscribing rectangular or prismatic blocks, or of an area to the circumscribing rectangle. Use has been made of displacement and not the moulded dimensions because the coefficients are used in the early design stages and are more directly related to most aspects of ship performance. Practice varies, however, and moulded dimensions may be needed in applying some classification societies' rules.

Some typical values (for rough guidance only) are presented in Table 2.1.

The values of these coefficients can provide useful information about the ship form. For instance, the low values of block coefficient for cargo liners would be used for high-speed refrigerated ships. The low block coefficient value for icebreakers reflects the hull form forward which is shaped to help the ship drive itself up on to the ice and break it. The great variation in size

TABLE 2.1 Typical Values of Coefficients of Fineness

Type of Vessel	Block Coefficient	Prismatic Coefficient	Midship Area Coefficient
Crude oil carrier	0.82–0.86	0.82–0.90	0.98–0.99
Product carrier	0.78–0.83	0.80–0.85	0.96–0.98
Dry bulk carrier	0.75–0.84	0.76–0.85	0.97–0.98
Cargo ship	0.60–0.75	0.61–0.76	0.97–0.98
Passenger ship	0.58–0.62	0.60–0.67	0.90–0.95
Container ship	0.60–0.64	0.60–0.68	0.97–0.98
Ferries	0.55–0.60	0.62–0.68	0.90–0.95
Frigate	0.45–0.48	0.60–0.64	0.75–0.78
Tug	0.54–0.58	0.62–0.64	0.90–0.92
Yacht	0.15–0.20	0.50–0.54	0.30–0.35
Icebreaker	0.60–0.70		

and speed of modern ship types means that the coefficients of fineness can also vary greatly. The coefficients of a similar ship in terms of use, size and speed should be used for early work on a new design.

The block coefficient indicates whether the form is *full* or *fine* and whether the waterlines will have large angles of inclination to the middle line plane at the ends. Large values will lead to greater wavemaking resistance at speed. A slow ship can afford a relatively high block coefficient as its resistance is predominately frictional. A high value is good for cargo carrying and is often obtained by using a length of parallel middle body, perhaps 15–20% of the total length.

The angle a waterline makes with the centreline at the bow is termed the *angle of entry* and influences resistance. As speed increases, a designer will reduce the length of parallel middle body to give a lower prismatic coefficient, keeping the same midship area coefficient. As speed increases still further, the midship area coefficient will be reduced, usually by introducing a rise of floor. A low value of midship section coefficient indicates a high rise of floor with rounded bilges. It will be associated with a higher prismatic coefficient. Finer ships will tend to have their main machinery spaces nearer midships to get the benefit of the fuller sections. There must be a compromise between this and the desire to keep the shaft length as short as possible.

A large value of vertical prismatic will indicate body sections of U form; a low value will indicate V sections. These features affect the seakeeping performance including slamming.

DISPLACEMENT AND TONNAGE

Displacement

A ship's *displacement* significantly influences its behaviour at sea. Displacement is a force but the term *mass displacement* can also be used. Displacement varies in service from *light displacement* to *fully loaded displacement*. For most performance studies, the latter is usually assumed.

Deadweight

Although influencing its behaviour in service, displacement is not a direct measure of a ship's carrying capacity, i.e. its earning power. To measure capacity, *deadweight* and *tonnage* are used.

Deadweight, or *deadmass*, is the difference between the load displacement up to the minimum permitted freeboard and the lightweight or light displacement. Lightweight is the weight of the hull and machinery, so the deadweight includes the cargo, fuel, water, crew and effects. The term *cargo deadweight* is used for the cargo alone. A table of deadweight against draught, for fresh and salt water, is provided to a ship's master in the form of a *deadweight scale*. This may be in the form of a diagram, a set of tables or, more likely these days, as software.

Tonnage

Ton in this usage originally derived from *tun*, which was a wine cask. The number of tuns a ship could carry was a measure of its capacity. Thus tonnage is a volume measure, not a weight measure, and for many years the standard ton was taken as 100 cubic feet (2.83 m^3). Because of the way it is now calculated, the unit of tonnage can no longer be regarded as a figure with a fixed volume of 2.83 m^3. Two 'tonnages' are of interest to the international community, namely:

- the *gross* tonnage based on the volume of all enclosed spaces and representing the overall size of a vessel and
- the *net* tonnage based on the volume of cargo spaces plus the volume of passenger spaces multiplied by a coefficient so representing its carrying capacity.

The former can also be regarded as a measure of the difficulty of handling and berthing and the latter of earning ability. At one time, differences between systems adopted by different countries, in making allowances such as for machinery spaces, caused many anomalies. Sister ships could have different tonnages merely because they were registered in different countries—they 'flew different flags'. To remove these anomalies, and establish an internationally approved system, the IMO produced the

International Convention on Tonnage Measurement of Ships. The tonnages are defined as follows:

$$\text{Gross tonnage (GT)} = K_1 V$$

$$\text{Net tonnage (NT)} = K_2 V_c \left(\frac{4T}{3D}\right)^2 + K_3 \left(N_1 + \frac{N_2}{10}\right)$$

where:

V = total volume of all enclosed spaces of the ship in cubic metres
$K_1 = 0.2 + 0.02 \log_{10} V$
V_c = total volume of cargo spaces in cubic meters
$K_2 = 0.2 + 0.02 \log_{10} V_c$

$$K_3 = 1.25 \frac{\text{GT} + 10{,}000}{10{,}000}$$

D = moulded depth amidships in metres
T = moulded draught amidships in metres
N_1 = number of passengers in cabins with not more than eight berths
N_2 = number of other passengers

$N_1 + N_2$ = total number of passengers the ship is permitted to carry.
In using these formulae:

- When $N_1 + N_2$ is less than 13, N_1 and N_2 are to be taken as zero.
- The factor $(4\,T/3D)^2$ is not to be taken as greater than unity and the term $K_2V_c\,(4\,T/3D)^2$ is not to be taken as less than 0.25 GT.
- NT is not to be less than 0.30 GT.
- All volumes included in the calculation are measured to the inner side of the shell or structural boundary plating, whether or not insulation is fitted, in ships constructed of metal. Volumes of appendages are included but spaces open to the sea are excluded.

Because of the modifying *K* factors GT and NT are no longer a direct measure of volume and they are stated as dimensionless numbers. The word ton is no longer used.

Special tonnages are calculated for ships operating through the Suez and Panama Canals. They are shown on separate certificates and charges for the use of the canals are based on them.

Although displacement and tonnage are different, they can be roughly related for a given ship types. Approximations for displacement are as follows:

General cargo ship = 2.0 × GT = 1.4–1.6 × Dwt.
Passenger ship = 1.1 × GT.

Container ship = 1.4 × Dwt.
Bulk carrier = 1.2–1.3 × Dwt.

FREEBOARD AND LOAD LINES

Load lines are associated with the name of Samuel Plimsoll who introduced a bill to Parliament to limit the draught to which a ship could be loaded to provide some minimum watertight volume of ship above the waterline. This led to a *statutory freeboard* and provided an insurance against a merchant ship being lost. Freeboard is measured downwards from the *freeboard deck* which is the uppermost complete deck exposed to the weather and sea, the deck and the hull below it having permanent means of watertight closure. A lower deck than this can be used as the freeboard deck provided it is permanent and continuous fore and aft and athwartships.

A basic freeboard is given in the Load Line Regulations, the value depending upon ship length and whether it carries liquid cargoes only in bulk. This basic freeboard has to be modified for the block coefficient, length to depth ratio, the sheer of the freeboard deck and the extent of superstructure. The rules governing this are somewhat complex but the intention is to provide a simple visual check that a laden ship has sufficient reserve of buoyancy for its intended service.

When all corrections have been made to the basic freeboard, the figure arrived at is termed the *Summer freeboard*. This distance is measured down from a line denoting the top of the freeboard deck at side and a second line is painted on the side with its top edge passing through the centre of a circle (Figure 2.5). The initials of the associated classification society are marked along the horizontal line.

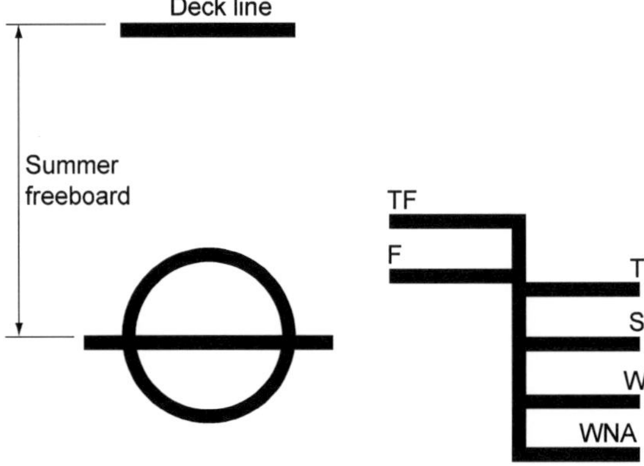

FIGURE 2.5 Load line markings.

To allow for different water densities and the severity of conditions likely to be met in different seasons and areas of the world, a series of extra lines are painted on the ship's side. Relative to the Summer freeboard, for a Summer draught of T, the other freeboards are as follows:

- The Winter freeboard is $T/48$ greater.
- The Winter North Atlantic freeboard is 50 mm greater still.
- The Tropical freeboard is $T/48$ less.
- The Fresh Water freeboard is $\Delta/40\,t$ cm less, where Δ is the displacement in tonne and t is the tonnes/cm immersion.
- The Tropical Fresh Water freeboard is $T/48$ less than the Fresh Water freeboard.

Safety of Life at Sea

As might be expected, ships carrying passengers are subject to particularly stringent rules. A vessel is defined as a passenger ship when it is designed to carry more than 12 passengers. It is issued with a *Passenger Certificate* when it has been checked for compliance with the regulations. Various maritime nations had rules for passenger ships before 1912 but it was the loss of the *Titanic* in that year that focused international concern on the matter. An international conference was held in 1914 but it was not until 1932 that the International Convention for the Safety of Life at Sea (SOLAS) was signed by the major nations. The Convention has been reviewed at later IMO (see below) conferences in the light of experience. The Convention covers a wide range of topics including watertight subdivision, damaged stability, fire, life-saving appliances, radio equipment, navigation and machinery and electrical installations.

INTERNATIONAL AND NATIONAL REGULATORY BODIES

General

Maritime trade is international in character and it is desirable that it be carried on safely with minimum impact upon the environment. For economic reasons it is necessary to create a 'level playing field'. This requires those involved to agree standards and then set up mechanisms for monitoring their achievement. These topics are discussed in more detail in Kristiansen (2005).

In this section, the historical development of the organisations and more general aspects of their work are presented. How they impact on ship design, construction and operation is discussed later in this book.

As Kristiansen points out, safety is regulated by laws and regulations originating in different legal sources, including international, national and case law; the United Nations Law of the Seas (UNCLOS); European Union Directives; the IMO; the Rules of the Classification Societies and the

International Association of Classification Societies; and Port State Control (PSC) Memorandum of Understanding (MOU) guidelines.

For merchant ships, the most influential body is the IMO. Whilst the IMO debates and agrees standards recommended for use of the international community, they are not an enforcing agency. For a given ship, legislation is the responsibility of the government of the country in which the ship is registered. Enforcement of that legislation rests with the registering nation with some powers vested in the countries whose ports the ship uses. Generally warships are designed to rather higher standards than those laid down by IMO, reflecting the fact that they are designed to withstand enemy action.

International Law

The legal right of countries to carry out lawful business freely across the oceans – the principle of the freedom of the seas – is an important one. The international basis for this is set out in UNCLOS, the main features of which are that ships can sail without restriction in all waters; the country of registration (i.e. the Flag State) has the principal jurisdiction over the ship; and other countries have limited jurisdiction over the ship.

For a country with a coastline, there are specific areas of interest:

- the exclusive economic zone stretching 200 nautical miles (nm) from its coast, within which it has very limited jurisdiction over other countries' ships – mainly related to pollution;
- its territorial waters stretching 12 nm from the coast, where it has full jurisdiction and
- its own ports.

The International Maritime Organisation

General

In the mid-nineteenth century, shipping nations got together to develop international regulations to be followed to improve safety at sea. Following some major losses at sea, it was proposed that a permanent international body should be established to promote maritime safety more effectively, but the setting up of that body had to await the establishment of the United Nations after the Second World War. In 1948, an international conference in Geneva adopted a convention formally establishing the *Inter-Governmental Maritime Consultative Organization*, the name being changed to IMO (www.imo.org) in 1982. The Convention entered into force in 1958. The Organisation met the following year and in 1960 IMO adopted a new version of the *International Convention for the Safety of Life at Sea*, which is the most important of all treaties dealing with maritime safety. The freeboard to which a ship is loaded makes a significant contribution to its safety and is defined in the Convention. IMO then turned its attention to such matters as the facilitation of international maritime traffic, load lines and the carriage of dangerous goods, while the system of measuring the tonnage of ships was revised. Although much of the legislation is in reaction to problems encountered, the organisation is increasingly adopting a proactive policy.

The *International Convention on Load Lines* (1930) was based on the principle of reserve buoyancy, although it was recognised then that the freeboard should also ensure adequate stability and avoid excessive stress on the ship's hull as a result of overloading. The 1966 Load Lines Convention determined the freeboard of a ship by subdivision and damage stability calculations. The regulations take into account the potential hazards present in different geographic zones and different seasons of the year. An annex deals with doors, freeing ports, hatchways and other items. The main aim is to ensure the watertight integrity of ships' hulls below the freeboard deck and the weathertight integrity on and above the freeboard deck. All assigned load lines must be marked amidships on each side of the ship, together with the deck line. Ships intended for the carriage of timber deck cargo are assigned a smaller freeboard.

A Protocol entering into force in 2000 harmonised the Convention's survey and certification requirements with those contained in the SOLAS and MARPOL conventions. This Protocol revised parts of the Load Lines Convention and introduced the tacit amendment procedure, under which amendments adopted enter into force 6 months after acceptance unless they are rejected by one-third of Parties. Usually, the period from adoption to acceptance is 2 years. Amendments to the 1988 Load Lines Protocol include a number of important revisions, in particular to regulations concerning strength and intact stability of ships.

Although safety is IMO's most important responsibility, pollution became a major problem as the amount of oil being transported by sea and the size of oil tankers grew. The most important of the pollution-related measures is the *International Convention for the Prevention of Pollution from Ships, 1973*, as modified by the Protocol of 1978 (*MARPOL* 73/78). This covers accidental and operational oil pollution as well as pollution by chemicals, goods in packaged form, sewage, garbage and air pollution.

In the 1970s, a *Global Maritime Distress and Safety System* was initiated, with the establishment of the *International Mobile Satellite Organization*, greatly improving the provision of radio and other messages to ships. It became fully operational in 1999. Now a ship that is in distress anywhere in the world is virtually guaranteed assistance, even if the ship's crew do not have time to radio for help, as the message will be transmitted automatically. In some remote ocean areas, it may take some time for assistance to arrive.

Two IMO initiatives relate to the human element in shipping. In 1998, the *International Safety Management Code,* and in 1997, the revised *International*

Convention on Standards of Training, Certification and Watchkeeping for seafarers entered into force. These greatly improved seafarer standards.

The 2000s saw a number of new conventions, including those on:

- *anti-fouling systems* (AFS 2001);
- *ballast water management* to prevent the spread of alien species (BWM 2004);
- *ship recycling* (Hong Kong International Convention for the Safe and Environmentally Sound Recycling of Ships, 2009) and
- *International Ship and Port Facility Security Code* – a new, comprehensive security regime for international shipping, including the *Convention for the Suppression of Unlawful Acts Against the Safety of Maritime Navigation.* Inter alia this introduces the right of a State Party to board a ship flying the flag of another State Party.

Structure of IMO

The Organisation consists chiefly of an Assembly and Council backed by five main Committees which are in turn supported by sub-committees. The Assembly consists of all Member States and not only meets once every 2 years in regular sessions but may also be convened in an extraordinary session if necessary. It approves the work programme, votes the budget and elects the Council.

The Council is elected by the Assembly for 2-year terms beginning after each regular session of the Assembly. It is the Executive organ and responsible, under the Assembly, for supervising the work of the Organisation. Between Assemblies, the Council generally performs all the functions of the Assembly.

The Maritime Safety Committee

This is the highest technical body of the Organisation and consists of all Member States. Its functions are to 'consider any matter within the scope of the Organisation concerned with aids to navigation, construction and equipment of vessels, manning from a safety standpoint, rules for the prevention of collisions, handling of dangerous cargoes, maritime safety procedures and requirements, hydrographic information, log books and navigational records, marine casualty investigations, salvage and rescue and any other matters directly affecting maritime safety'. An expanded Maritime Safety Committee adopts amendments to conventions such as SOLAS and includes all Member States as well as those countries which are Party to conventions such as SOLAS.

The Marine Environment Protection Committee

This committee comprises all Member States and is concerned with prevention and control of pollution from ships. In particular, it is concerned with the adoption and amendment of conventions and other regulations and measures to ensure their enforcement.

Conventions

By the time IMO came into existence, several important international conventions existed, including the International Convention for the Safety of Life at Sea of 1948, the International Convention for the Prevention of Pollution of the Sea by Oil of 1954, and treaties dealing with load lines and the prevention of collisions at sea.

IMO became responsible for ensuring that the majority of these conventions were kept up to date and for developing new conventions as required. It is now responsible for nearly 50 international conventions and agreements and has adopted numerous protocols and amendments.

Each convention includes conditions which have to be met before it enters into force. Generally, the more important and complex the document, the more stringent the conditions for its entry into force. For example, for the International Convention for the Safety of Life at Sea, 1974, entry into force required acceptance by 25 States with merchant fleets comprising not less than 50% of the world's gross tonnage.

A Government's acceptance of a convention obliges it to take the measures required by the convention. Often national law has to be enacted or changed to enforce the provisions of the convention. Special facilities may have to be provided, an inspectorate appointed, and adequate notice must be given to ship owners, shipbuilders and so on.

Much of the regulation requires certificates to show that the requirements of the various instruments have been met. In many cases, this involves a survey which may mean the ship being out of service for several days. To reduce the problems caused by different survey dates and periods between surveys, a harmonised system of ship survey and certification was introduced in 2000. This covers survey and certification requirements of the conventions on SOLAS, load lines, pollution and a number of codes covering the carriage of dangerous substances. Briefly the harmonised system provides a 1-year standard survey interval, some flexibility in timing of surveys, dispensations to suit the operational programme of the ship and maximum validity periods for Certificates of 5 years for cargo ships and 1 year for passenger ships. The main changes to the SOLAS and Load Line Conventions are that annual inspections are made mandatory for cargo ships with unscheduled inspections discontinued.

SOLAS and the Collision Regulations require ships to comply with rules on design, construction and equipment. SOLAS coverage includes life-saving equipment, both the survival craft (lifeboats and life rafts) and personal (life jackets and immersion suits). Numbers of such items are stated on the Safety Certificate.

Enforcing Regulations

Since IMO has no power to enforce their rules, regulations and resolutions, each participating country needs to apply international rules through laws passed by its own governing body (Parliament). A Government Department is given responsibility for drafting the regulations, getting them agreed and seeing that they are followed by ship owners/operators. This is known as ***Flag State Control***. Although the governments that ratify conventions are responsible for their implementation in ships which fly their flag, it becomes the responsibility of owners to ensure that their ships meet IMO standards. The International Safety Management Code is meant to ensure they do, by requiring them to produce documents specifying that their ships do meet the requirements.

Flag States

The national body appointed by a country's government to implement its laws acts as Flag State on behalf of that country. When it is satisfied with the technical documentation submitted for, and inspections of a ship, the Flag State will register it and issue the necessary safety-related certificates. Some Flag States accept foreign vessels as well as their own and have become 'international' or 'offshore' registers. The standards of some of these offshore registries have been questioned in the past and they became known as *Flags of Convenience* as there was a suspicion that they had a more lenient approach to registration. The Flag State may delegate its responsibilities to an independent certifying authority – typically a classification society that is legally recognised by the Flag State.

Within the United Kingdom, the government department concerned is the *Department for Transport* and within that the *Maritime and Coastguard Agency (MCA)*.

Port State Control (PSC)

Whilst a certificate of registration should denote the same standard whichever state has issued it, a Port State can challenge a vessel if it suspects it does not comply with the certificate. The legal basis for PSC in Europe is the Paris MOU on PSC signed by 19 European countries and Canada in 1982. Similar arrangements cover other countries. The Port State can carry out inspections, on a sampling basis, of ships visiting its ports. It can require any deficiencies found to be put right in a defined period, detain a vessel until important deficiencies have been corrected and in an extreme case ban a repeated offender.

The Maritime Coastguard Agency, UK

In the United Kingdom, the MCA is the government organisation, within the Department for Trade, tasked with maritime matters and the registering of ships. Established in 1998 by merging the Coastguard and Marine Safety Agencies, it is responsible for providing a 24-h maritime search and rescue service, pollution prevention and response and the inspection and enforcement of standards of ships and the registration of ships and seafarers.

In doing this, it works in accord with and/or issues:

- Acts of Parliament, the highest level of UK law.
- Statutory Instruments (SIs) containing the majority of UK laws.
- Marine (M) Notices publicising to the shipping and fishing industries important safety, pollution prevention and other relevant information.

The MCA operates the *UK Ship Register* which is one the oldest and most prestigious shipping registers in the world. It is highly regarded and has a reputation for maintaining the highest international standards. It also issues guidance to surveyors and publications containing its interpretation of the relevant regulations with practical advice on design and testing requirements to be considered when approval for new shipboard arrangements is requested.

The MCA delegates authority to UK recognised classification societies to carry out certain technical and survey services on its behalf but PSC surveys are carried out only by MCA surveyors.

Some of the matters that are regulated in this way are touched upon in other chapters, including subdivision of ships and carriage of grain and dangerous cargoes. Tonnage measurement has been discussed already. The other major area of regulation is the freeboard demanded and this is covered by the *Load Line Regulations*.

Classification Societies

There are many classification societies which cooperate through the *International Association of Classification Societies* (IACS) (www.iacs.org.uk). A list of the main ones, with websites, is included under references. Much useful information can be gleaned from these web sites. The work of the classification societies is exemplified by Lloyd's Register (LR) of London, founded in 1760, and the oldest society. LR of Shipping, like Lloyd's Insurance Corporation, began life in Edward Lloyd's coffee shop in London, although the two organisations have now quite different functions. In the second half of the eighteenth century, marine insurers used meetings in the coffee house to develop a system for the independent technical assessment of the ships presented to them for insurance cover. In 1760, a Committee was formed for this purpose. An attempt was made to 'classify' the condition of each ship on an annual basis. The condition of the hull was classified A, E, I, O or U, according to the excellence of its construction and its adjudged continuing soundness (or otherwise). Equipment was classified G, M or B: simply, good, middling or bad. In time, G, M and B were replaced by 1, 2 or 3, which is the origin of the expression 'A1', meaning 'first or highest class'.

With time, the practice of assigning different classifications has been superseded, with some exceptions. Today a vessel either meets the relevant Class Society's Rules or it does not. As a consequence it is either 'in' or 'out' of 'class'. However, each of the classification societies has developed a series of notations that may be granted to a vessel to indicate that it is in compliance with some additional mandatory or voluntary criteria that may be either specific to that vessel type or that are in excess of the standard classification requirements, e.g. the ice class notation. Today LR classes some 6700 ships totalling about 96 million in gross tonnage.

When a ship is built to class, it must meet the requirements laid down by the classifying society as regards design and build. Materials, structure, machinery and equipment must be of the required quality. Construction is surveyed to ensure proper standards of workmanship are adhered to. Later in life, to retain its class, a ship must be surveyed at regular intervals. The scope and depth of these surveys reflect the age and service of the ship. Thus, standards of safety, quality and reliability are set and maintained. Some societies have developed schemes for monitoring the condition of ships to assist in their inspection and maintenance. A database is created using a digitised representation of the vessel. Results of class surveys and owners' inspections are input to the database which can be accessed on board ship or ashore.

Classification applies to ships and floating structures extending to machinery and equipment such as propulsion systems, liquefied gas containment systems and so on.

For many years, the Rules of classification societies were in tabular form basing the scantlings required for different types of ship on their dimensions and tonnage. These gave way to rational design standards, and now computer-based assessment tools allow a designer to optimise the design with minimum scantlings and making it easier to produce. For a ship to be designed directly using analysis requires an extensive specification on how the analyses are to be carried out and the acceptance criteria to apply. Sophisticated analysis tools are needed to establish the loads to which the ship will be subject.

Classification societies are becoming increasingly involved in the classification of naval vessels. Typically, they cover the ship and ship systems, including stability, watertight integrity, structural strength, propulsion, fire safety and life saving. They do not cover the weapon systems themselves but do cover the supporting systems. A warship has to be 'fit for service' as does any ship. The technical requirements to make them fit for service will differ, as would the requirements for a tanker from those for a passenger ship. In the case of the warship, the need to take punishment as a result of enemy action, including shock and blast, will lead to a more rugged design. There will be more damage scenarios to be considered with redundancy built into systems so that they are more likely to remain functional after damage.

The involvement of classification societies with naval craft has a number of advantages. It means warships will meet at least the internationally agreed safety standards to which merchant ships are subject. The navy concerned benefits from the worldwide organisation of surveyors to ensure equipment, materials or even complete ships are of the right quality.

Classification societies are international in character and independent of government but they often have delegated powers from governments to carry out many of the statutory functions mentioned earlier. They carry out surveys and certification and carry out statutory surveys covering the international conventions on load lines, cargo ship construction, safety equipment, pollution prevention, grain loading and so on, and issue International Load Line Certificates, Passenger Ship Safety Certificates and so on. The actual registering of ships is carried out by the government organisation concerned. Naturally, owners find it easier to arrange registration of their ships with a government, and to get insurance cover, if the ship is built and maintained in accordance with the rules of a classification society.

The International Association of Classification Societies

Cooperation between Classification Societies goes back to the International Load Line Convention of 1930 and its recommendations that societies should collaborate to secure *"as much uniformity as possible in the application of the standards of strength upon which freeboard is based..."*.

A conference of major societies in 1939 agreed on further cooperation and a second major class society conference in 1955 led to the creation of Working Parties on specific topics. In 1968, IACS was formed by seven leading societies. The value of their combined level of technical knowledge and experience was quickly recognised, and in 1969, IACS was given consultative status with the IMO. It remains the only non-governmental organisation with Observer status which is able to develop and apply Rules.

Compliance with the IACS Quality System Certification Scheme (details on the IACS website) is mandatory for IACS Membership. Now the main societies forming IACS collectively class over 90% of all commercial tonnage involved in international trade.

IACS is governed by a Council, on which each Member is represented. Under the Council is a General Policy Group which develops and implements actions giving effect to the policies, directions and long-term plans of the Council.

Unified Requirements

Unified requirements (URs) are minimum technical requirements adopted by the IACS Members and incorporated in their Rules and practices. Each IACS Member is free to adopt more stringent requirements.

Common Rules

Common Rules are IACS URs covering broad areas of classification requirements which, once adopted by the IACS Council, are applied by all Members. The IACS Common Structural Rules are a comprehensive set of minimum requirements for the classification of the hull structures of bulk carriers and double-hull oil tankers.

Unified Interpretations

These provide uniform interpretations of IMO instruments on those matters which in the Convention are left to the satisfaction of the Flag Administration or where more precise wording is found to be necessary.

STANDARDS

A ship, its build and the equipment fitted in it are subject to an array of standards. Some are internationally recognised; others are national in scope and others will relate to the products and activities of an individual company. In many cases, it is necessary to show compliance with a standard, within stipulated limits, by means of physical testing. Standards are necessary to ensure products, and services meet the desired standards of quality, environmental friendliness, safety, reliability, efficiency and inter-changeability at an economical cost.

The *International Organization for Standardization (ISO)*, formed in 1946, is the world's largest developer and publisher of International Standards. It is a network of the national standards institutes of 160 countries and has more than 18,500 International Standards. Standards are developed by Working Groups formed of national representative experts from industry on the particular subject of the standard. Most of ISO standards are specific to a particular product, material or process. However, ISO 9001 (quality) and ISO 14001 (environment) are 'generic management system standards'. That is, the same standard can be applied to any organisation, large or small.

National Standards are developed under a similar process. British Standards Institution is the United Kingdom's National Standards Body (NSB) and it was the world's first such body. It represents UK interests across all of the European and international standards organisations.

In the United States, the Department of Defense (DOD) has MIL (Military) standards which evolved during the Second World War from the need to ensure proper performance, maintainability and inter-changeability of military equipment. The number of standards reached nearly 30,000 by 1990 and the DOD decided to encourage the use of industry standards, such as ISO 9000 series for quality assurance. MIL standards are still widely used, including in NATO. The United States also has the ASTM standards that are industry developed standards for testing materials.

Companies may have developed their own standards of procedure and quality over the years and may choose to retain them – perhaps because they are more stringent than related national documents. In general, however, companies use widely accepted specifications to ensure their customers understand what they are getting.

Whilst internationally agreed standards are vital, it is inevitable that such standards will:

- be the essential minimum as they have to be agreed by many countries and
- take time to agree after the need first becomes apparent.

The 'best practice' will normally require standards higher than these minima.

Standards are optional and only become requirements when they are referred to as requirements in classification society Rules, in statutory flag state regulations or ship/equipment contracts.

Precision

Specifications used must be precise in that they must be:

- quoted accurately including the version applicable and
- unambiguous without the need for interpretation. For instance, if a pressure test is invoked it must state the test medium, the test pressure, the test duration, what (if any) drop off in pressure is acceptable with time after initial application and so on.

Impact of Rules and Regulations on Design

A ship designer must satisfy not only the owner's stated requirements but also the IMO regulations and classification society rules. The first will define the type of ship and its characteristics such as size and speed. The second broadly ensures that the ship will be safe and acceptable in ports throughout the world. They control such features as subdivision, stability, fire protection, pollution prevention and manning standards. The third sets out the 'engineering' rules by which the ship can be designed to meet the demands placed on it. They will reflect the properties of the materials used in construction and the loadings the ship is likely to experience in the intended service.

There are three basic forms the rules of a classification society may take:

- Prescriptive standards describing exactly what is required, reflecting that society's long experience and the gradual trends in technological development. They enable a design to be produced quickly and do not require the designer to have advanced structural design knowledge. They are not well suited to novel design configurations or to incorporating new, rapidly changing, technological developments. Because of this, the performance standard approach is increasingly favoured.

- Performance standards which are flexible in that they set out aims to be achieved but leave the designer free to decide how to meet them within the overall constraints of the rules. They set standards and criteria to which the design must conform to provide the degree of safety and reliability demanded.
- The safety case approach which considers the totality of risks the ship is subject to within a variety of operational scenarios and predictable incidents. A *formal safety assessment* (FSA) involves identifying hazards, assessing the risks/effects associated with each hazard, considering alternative strategies and making decisions so as to reduce the risks and their consequences to acceptable levels. Put another way the designer thinks what might go wrong, the consequences if it does go wrong, the implications for the design of reducing, or avoiding, the risk and making a conscious decision on how to manage the situation. Thus a designer might decide that although an event is of very low probability, its repercussions are so serious that something must be done to reduce the hazard.

Although attractive in principle, FSA is an expensive approach and is likely to be used for individual projects only if they are high profile ones. It can be used, however, as the basis for developing future classification and convention requirements. One problem, particularly for radically new concepts is foreseeing what might happen and under what circumstances.

It is clear that probability theory is going to play an increasing part in design safety assessments and development. To quote two examples:

- When considering longitudinal strength, the designer must assess the probability of the ship meeting various sea conditions, the need to operate or merely survive in these conditions, the probability of the structure having various levels of built-in stress and the probable state of the structure in terms of loss of plate thickness due to corrosion.
- In considering collision at sea, consideration must be given to the density of traffic in the areas in which the ship is to operate. Then there are the probabilities that the ship will be struck at a certain point along its length by a ship of a certain size and speed, that the collision will cause damage over a certain length of hull and the state of watertight doors and other openings. Then some allowance must be made for the actions of the crew in containing the incident.

Statistics are being gathered to help quantify these probabilities but many still require considerable judgement on the part of the designer.

SUMMARY

It is important that a ship's principal geometric features are defined and characterised in an unambiguous way. It will be shown later how these parameters can be calculated and used. The international community has adopted a common system of notation and units to aid in communicating what can be quite complex concepts. Whilst displacement and deadweight are measures of weight, gross and net tonnage are measures of volume. A number of international and national bodies are involved in creating and implementing a range of measures aimed at improving ship safety and the protection of the environment. Important amongst these are the IMO and the classification societies. It is interesting to see how they have developed and operate.

Whilst all legal requirements must be met, engineers have a much broader responsibility to the public and the profession and must do their best, using all available knowledge. It is important that engineers keep abreast of developments in their field through continuing professional development and that they abide by the code of practice for their profession.

Chapter 3

Ship Form Calculations

INTRODUCTION

A ship's three-dimensional hull form can be represented by a series of curves which are the intersections of the hull with three sets of mutually orthogonal planes. Naval architects need to find the areas and volumes enclosed by such curves and surfaces and their centres of area or volume. To find the centres of areas and volumes it is necessary to obtain their first moments about chosen axes. For some calculations the moments of inertia of areas are needed, which are their second moments of the area, again about chosen axes. These properties could be calculated mathematically, by integration, if the form could be expressed in mathematical terms. This is not easy to do precisely and approximate methods of integration are usually adopted, certainly for manual calculations. These methods rely upon representing the actual hull curves by ones which are defined by simple mathematical equations. In the simplest case a series of straight lines are used.

APPROXIMATE INTEGRATION

One could draw the shape, the area of which is required, on squared paper and count the squares included within it. If mounted on a uniform card the figure could be balanced on a pin to obtain the position of its centre of gravity – its centroid of area. Such methods would be very tedious but illustrate the principle of what is being attempted. To obtain an area it is divided into a number of sections by a set of parallel lines which are usually, but not necessarily, equally spaced.

Trapezoidal Rule

If the points at which the parallel lines intersect the area perimeter are joined by straight lines, the area can be represented approximately by the summation

Introduction to Naval Architecture. DOI: http://dx.doi.org/10.1016/B978-0-08-098237-3.00003-5

of the set of trapezia so formed. This is illustrated in Figure 3.1. The area of the shaded trapezium is:

$$A_n = \frac{1}{2}h_n(y_n + y_{n+1})$$

where h_n is the distance between the ordinates y_n and y_{n+1}.

Any area can be divided into two, each with part of its boundary a straight line. Such a line can be chosen as the axis about which moments are taken. This simplifies the representation of the problem as in Figure 3.2 which uses equally spaced lines, called *ordinates*. The device is apt for ships which are usually symmetrical about their middle line planes, and areas such as those of waterplanes can be treated as two halves.

Referring to Figure 3.2, the curve ABC has been replaced by two straight lines, AB and BC with ordinates y_0, y_1 and y_2 distance h apart. The area is the sum of the two trapezia so formed:

$$\text{Area} = \frac{h(y_0 + y_1)}{2} + \frac{h(y_1 + y_2)}{2} = \frac{h(y_0 + 2y_1 + y_2)}{2}$$

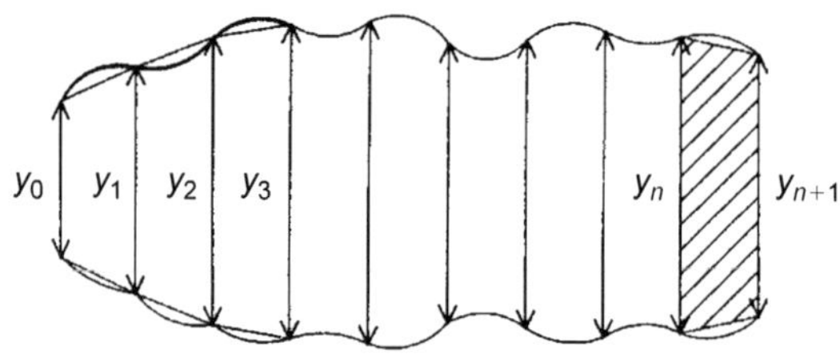

FIGURE 3.1 Trapezoidal rule.

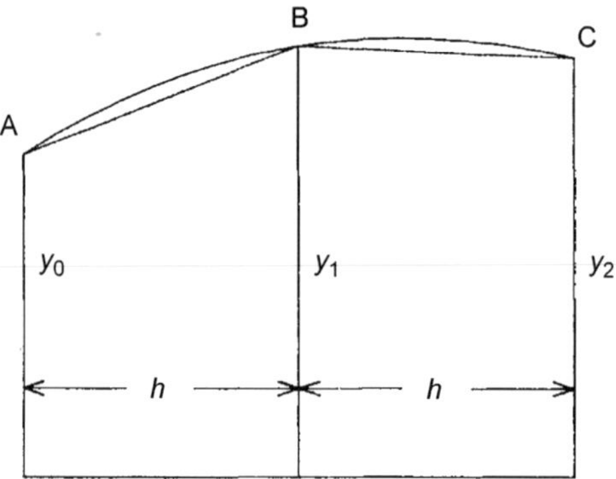

FIGURE 3.2 Area with three ordinates.

The accuracy with which the area under the actual curve is calculated will depend upon how closely the straight lines mimic the curve. The accuracy of representation can be increased by using more ordinates and a smaller interval h. Generalising, for $n+1$ ordinates the area will be given by:

$$\text{Area} = \frac{h(y_0 + 2y_1 + 2y_2 + \cdots + 2y_{n-1} + y_n)}{2}$$

To calculate the volume of a three-dimensional solid the values of its cross-sectional areas at equally spaced intervals can be calculated as above. These areas can then be used as the new ordinates in a *curve of areas* which can be integrated to obtain the volume.

Simpson's First Rule

The trapezoidal rule, using straight lines to replace the actual ship curves, has limited accuracy. Many naval architectural calculations are carried out using what are known as Simpson's rules. In these the actual curve is represented by a mathematical equation of the form:

$$y = a_0 + ax_1 + a_2x^2 + a_3x^3$$

The curve, shown in Figure 3.3, is represented by three equally spaced ordinates y_0, y_1 and y_2. It is convenient to choose the origin to be at the base of y_1 to simplify the algebra but the results would be the same wherever the origin is taken. The curve extends from $x=-h$ to $x=+h$ and the area under it is:

$$\begin{aligned} A &= \int_{-h}^{+h} (a_0 + a_1x + a_2x^2 + a_3x^3)\mathrm{d}x \\ &= \left[a_0x + a_1x^2/2 + a_2x^3/3 + a_3x^4/4\right]_{-h}^{+h} \\ &= 2a_0h + 2a_2h^3/3 \end{aligned}$$

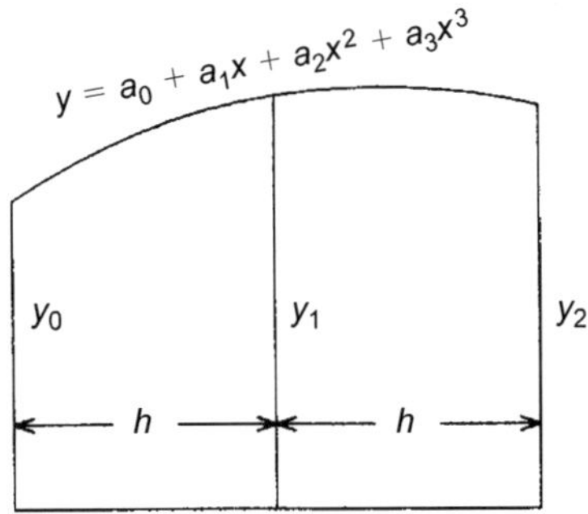

FIGURE 3.3 Simpson's first rule.

Now the three ordinates defining the curve are:

$$y_0 = a_0 - a_1 h + a_2 h^2 - a_3 h^3$$
$$y_1 = a_0$$
$$y_2 = a_0 + a_1 h + a_2 h^2 + a_3 h^3$$

It would be convenient to be able to express the area of the figure as a simple sum of the ordinates each multiplied by some factor to be determined. Assuming that A can be represented by:

$$A = Fy_0 + Gy_1 + Hy_2$$

then:

$$\begin{aligned} A &= (F+G+H)a_0 - (F-H)a_1 h + (F+H)a_2 h^2 - (F-H)a_3 h^3 \\ &= 2a_0 h + 2a_2 h^3/3 \end{aligned}$$

These equations are satisfied by:

$$F = H = h/3 \quad \text{and} \quad G = 4h/3$$

Hence:

$$A = \frac{h}{3}(y_0 + 4y_1 + y_2)$$

This is known as *Simpson's First Rule* or *3 Ordinate Rule*.

This rule can be generalised to any figure defined by an odd number of evenly spaced ordinates, by applying the First Rule to ordinates 0–2, 2–4, 4–6 and so on, and then summing the results. This provides the rule for $n + 1$ ordinates:

$$A = \frac{h}{3}(y_0 + 4y_1 + 2y_2 + 4y_3 + 2y_4 + 4y_5 + \cdots + 4y_{n-1} + y_n)$$

In many cases it is sufficiently accurate for a merchant ship hull to use 10 divisions with 11 ordinates but it is worth checking by eye whether the ordinates appear to define the actual curves reasonably accurately. Warship hulls have greater curvature and are usually represented by 20 divisions with 21 ordinates. The areas at the bow and stern are more curved and greater accuracy can be obtained by introducing intermediate ordinates, as in Figure 3.4.

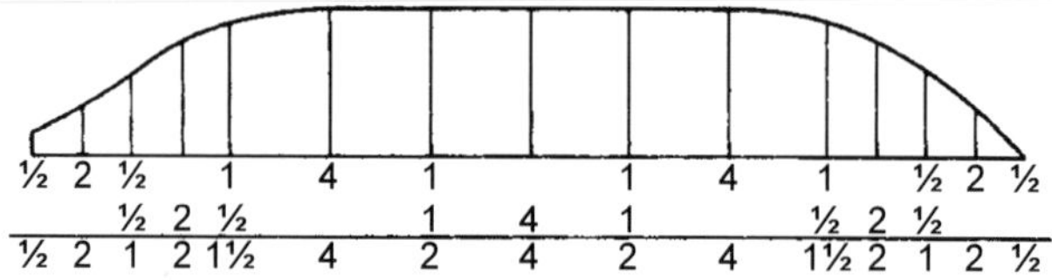

FIGURE 3.4 Intermediate ordinates with multipliers.

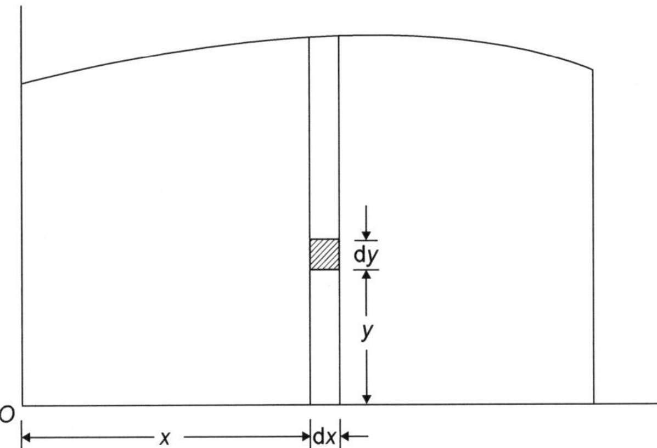

FIGURE 3.5 Moments of area.

The figure gives the Simpson multipliers to be used for each consecutive area defined by three ordinates.

The total area is given by:

$$A = \frac{h}{3}\left(\tfrac{1}{2}y_0 + 2y_1 + y_2 + 2y_3 + 1\tfrac{1}{2}y_4 + 4y_5 + 2y_6 + 4y_7 + 2y_8 + 4y_9 + 1\tfrac{1}{2}y_{10} + 2y_{11} + y_{12} + 2y_{13} + \tfrac{1}{2}y_{14}\right)$$

where y_1, y_3, y_{11} and y_{13} are the extra ordinates.

The method outlined above for calculating areas can be applied to evaluating any integral. Thus it can be applied to the first and second moments of area. Referring to Figure 3.5, the moments will be given by:

$$\text{First moment} = \iint x \,\mathrm{d}x\,\mathrm{d}y = \int xy\,\mathrm{d}x \text{ about the } y\text{-axis}$$

$$= \iint y \,\mathrm{d}x\,\mathrm{d}y = \int \frac{1}{2}y^2\,\mathrm{d}x \text{ about the } x\text{-axis}$$

$$\text{Second moment} = \iint x^2 \,\mathrm{d}x\,\mathrm{d}y = \int x^2\,y\,\mathrm{d}x \text{ about the } y\text{-axis} = I_y$$

$$= \iint y^2 \,\mathrm{d}x\,\mathrm{d}y = \int \frac{1}{3}y^3\,\mathrm{d}x \text{ about the } x\text{-axis} = I_x$$

The calculations can be set out in tabular form.

Worked Example 3.1

Calculate the area between the curve, defined by the ordinates below, and the *x*-axis. Calculate the first and second moments of area about the *x*- and *y*-axes and the position of the centroid of area.

x	0	1	2	3	4	5	6	7	8
y	1	1.2	1.5	1.6	1.5	1.3	1.1	0.9	0.6

Solution

There are nine ordinates spaced one unit apart. The results can be calculated in a tabular fashion as in Table 3.1.

Noting that in this case $h = 1$

$$\text{Area} = \frac{29.8}{3} = 9.93\ \text{m}^2$$

$$\text{First moment about } y\text{-axis} = \frac{111.2}{3} = 37.07\ \text{m}^3$$

$$\text{Centroid from } y\text{-axis} = \frac{37.07}{9.93} = 3.73\ \text{m}$$

$$\text{First moment about } x\text{-axis} = 0.5 \times \frac{38.78}{3} = 6.463\ \text{m}^3$$

$$\text{Centroid from } x\text{-axis} = \frac{6.463}{9.93} = 0.65\ \text{m}$$

$$\text{Second moment about } y\text{-axis} = \frac{546.4}{3} = 182.13\ \text{m}^4$$

$$\text{Second moment about } x\text{-axis} = \frac{1}{3} \times \frac{52.378}{3} = 5.82\ \text{m}^4$$

Table 3.1 Tabular Calculations

x	y	SM	$F(A)$	xy	$F(M_y)$	x^2y	$F(I_y)$	y^2	$F(M_x)$	y^3	$F(I_x)$
0	1.0	1	1.0	0	0	0	0	1.0	1.0	1.0	1.0
1	1.2	4	4.8	1.2	4.8	1.2	4.8	1.44	5.76	1.728	6.912
2	1.5	2	3.0	3.0	6.0	6.0	12.0	2.25	4.50	3.375	6.750
3	1.6	4	6.4	4.8	19.2	14.4	57.6	2.56	10.24	4.096	16.384
4	1.5	2	3.0	6.0	12.0	24.0	48.0	2.25	4.50	3.375	6.750
5	1.3	4	5.2	6.5	26.0	32.5	130.0	1.69	6.76	2.197	8.788
6	1.1	2	2.2	6.6	13.2	39.6	79.2	1.21	2.42	1.331	2.662
7	0.9	4	3.6	6.3	25.2	44.1	176.4	0.81	3.24	0.729	2.916
8	0.6	1	0.6	4.8	4.8	38.4	38.4	0.36	0.36	0.216	0.216
Total			29.8		111.2		546.4		38.78		52.378

The second moment of an area is always least about an axis through its centroid. If the second moment of an area, A, about an axis x from its centroid is I_x and I_{xx} is that about a parallel axis through the centroid:

$$I_{xx} = I_x - Ax^2$$

In the Worked Example 3.1, the second moments about axes through the centroid and parallel to the x- and y-axes are respectively:

$$I_{xx} = 5.82 - 9.93(0.65)^2 = 1.62 \text{ m}^4$$

$$I_{yy} = 182.13 - 9.93(3.73)^2 = 43.97 \text{ m}^4$$

Where there are large numbers of ordinates the arithmetic in the table can be simplified by halving each Simpson multiplier and then doubling the final summations so that:

$$A = \frac{2h}{3}\left(\tfrac{1}{2}y_0 + 2y_1 + y_2 + \cdots + 2y_n + \tfrac{1}{2}y_{n+1}\right)$$

Application to Waterplane Calculations

Most of the waterplanes the naval architect is concerned with are symmetrical about the x-axis so the calculations can be carried out for one half and doubled for the complete waterplane. This is done in Worked Example 3.2.

Worked Example 3.2

The summer waterplane of a ship is defined by a series of half ordinates (metres) at 14.1 m separation, as follows:

Station	1	2	3	4	5	6	7	8	9	10	11
Half ordinates	0.10	5.20	9.84	12.80	14.04	14.40	14.20	13.70	12.60	10.06	1.30

Calculate the area of the waterplane, the position of its centroid of area and its second moments of area.

Solution

A table can be constructed (as shown in Table 3.2). In Table 3.2, $F(A)$ represents SM $\times y$; $F(M) =$ SM $\times$ lever $\times y$; $F(I)$ long = SM $\times$ lever $\times$ lever $\times y$ and $F(I)$ trans = SM $\times y^3$. From the summations in the table:

The area of the waterplane = $2/3 \times 14.1 \times 327.4 = 3077$ m^2

The centroid of area is aft of amidships by $14.1 \times 107.84/327.4 = 4.64$ m

(Note that there is no need to calculate the moment in absolute terms)

The longitudinal second moment of area about amidships

$= 2/3 \times 14.1 \times 14.1 \times 14.1 \times 1896$

$= 3{,}543{,}000$ m^4

Table 3.2 Tabular Calculations

Station	Half, y Ordinates	SM	F(A)	Lever	F(M)	Lever	F(I)long	yyy	F(I) trans
1	0.10	1	0.10	5	0.50	5	2.50	0	0
2	5.20	4	20.80	4	83.20	4	332.80	141	562
3	9.84	2	19.68	3	59.04	3	177.12	953	1906
4	12.80	4	51.20	2	102.40	2	204.80	2097	8389
5	14.04	2	28.08	1	28.08	1	28.08	2768	5535
6	14.40	4	57.60	0	0.00	0	0.00	2986	11,944
7	14.20	2	28.40	−1	−28.40	−1	28.40	2863	5727
8	13.70	4	54.80	−2	−109.60	−2	219.20	2571	10,285
9	12.60	2	25.20	−3	−75.60	−3	226.80	2000	4001
10	10.06	4	40.24	−4	−160.96	−4	643.84	1018	4072
11	1.30	1	1.30	−5	−6.50	−5	32.50	2	2
Total			327.40		−107.84		1896.04		52,423

The minimum longitudinal second moment will be about the centroid of area and given by:

$$I_L = 3,543,000 - 3077(4.64)^2 = 3,477,000 \text{ m}^4$$

The transverse second moment $= 2/3 \times 1/3 \times 14.1 \times 52423$
$= 164,300 \text{ m}^4$

Other Simpson's Rules

Other rules can be deduced for figures defined by different numbers of evenly spaced ordinates, that for four ordinates becoming:

$$A = \frac{3h}{8}(y_0 + 3y_1 + 3y_2 + y_3)$$

This is known as *Simpson's Second Rule* and its proof would be similar to that of the First Rule. It can be extended to cover 7, 10, 13 and so on ordinates, becoming:

$$A = \frac{3h}{8}(y_0 + 3y_1 + 3y_2 + 2y_3 + 3y_4 + \cdots + 3y_{n-1} + y_n)$$

A special case is where the area between two ordinates is required when three are known. If, for instance, the area between ordinates y_0 and y_1 of Figure 3.3 is needed:

$$A_1 = \frac{h}{12}(5y_0 + 8y_1 - y_2)$$

This is called *Simpson's 5, 8 minus 1 Rule* and it will be noted that if it is applied to both halves of the curve then the total area becomes:

$$A = (h/12)[(5y_0 + 8y_1 - y_2) + (-y_0 + 8y_1 + 5\,y_2)]$$
$$= (h/12)(4y_0 + 16y_1 + 4y_2)$$

That is it becomes:

$$A = \frac{h}{3}(y_0 + 4y_1 + y_2)$$

which is in accord with Simpson's First Rule.

Unlike others of Simpson's rules the 5, 8, −1 Rule cannot be applied to moments. A corresponding rule for moments, derived in the same way as those for areas, is known as *Simpson's 3, 10 minus 1 Rule* and gives the moment of the area bounded by y_0 and y_1 about y_0, as:

$$M = \frac{h^2}{24}(3y_0 + 10y_1 - y_2)$$

If in doubt about the multiplier to be used, a simple check can be applied by considering the area or moment of a simple rectangle.

Tchebycheff's Rules

In arriving at Simpson's rules, equally spaced ordinates were used and varying multipliers for the ordinates deduced. The equations concerned can be solved to find the spacing needed for ordinates if the multipliers are to be unity. For simplicity of the mathematics involved, the curve is assumed to be centred upon the origin, $x = 0$, with the ordinates arranged symmetrically about the origin. Thus for an odd number of ordinates the middle one will be at the origin. Rules so derived are known as *Tchebycheff's rules* and they can be represented by the equation:

$$A = \frac{\text{Span of curve on } x\text{-axis} \times \text{Sum of ordinates}}{\text{Number of ordinates}}$$

Thus for a curve spanning two units, $2h$, and defined by three ordinates:

$$A = \frac{2h}{3}(y_0 + y_1 + y_2)$$

The spacings required of the ordinates are given in Table 3.3.

TABLE 3.3 Tchebycheff Ordinate Spacings

Number of Ordinates	Spacing Each Side of Origin ÷ The Half Length				
2	0.5773				
3	0	0.7071			
4	0.1876	0.7947			
5	0	0.3745	0.8325		
6	0.2666	0.4225	0.8662		
7	0	0.3239	0.5297	0.8839	
8	0.1026	0.4062	0.5938	0.8974	
9	0	0.1679	0.5288	0.6010	0.9116
10	0.0838	0.3127	0.5000	0.6873	0.9162

TABLE 3.4 Gauss Rule Ordinate Spacing and Multipliers

Number of Ordinates	Spacing Each Side of Midordinate	Multiplier as a Factor of the Half Length
2	0.57735	0.50000
3	0.00000	0.44444
	0.77460	0.27778
4	0.33998	0.32607
	0.86114	0.17393
5	0.00000	0.28445
	0.53847	0.23931
	0.90618	0.11846

Integral = sum of products × total base length.

Gauss Rules

From the point of view of manual calculations Simpson's rules have the advantage of using equally spaced ordinates; Tchebycheff's rules the advantage of constant multipliers. A third set of rules, known as the Gauss rules have unequal ordinate spacing and multipliers as shown in Table 3.4.

The Gauss rules are more accurate than Simpson's or Tchebycheff's rules but were not often used in manual calculations because of their relative complexity. In computer programs this is not such a problem and they can be used to obtain the greater accuracy.

Relative Accuracy of Rules

An analysis has shown that:

- odd-ordinate Simpson's rules are preferred as they are only marginally less accurate than the next higher even number rule;
- even-ordinate Tchebycheff's rules are preferred as they are as accurate as the next highest odd-ordinate rule;
- a Tchebycheff rule with an even number of ordinates is rather more accurate than the next highest odd number Simpson's rule. That is, the four-ordinate Tchebycheff's rule is more accurate than the five-ordinate Simpson's rule;
- the five-ordinate Gauss rule gives an accuracy comparable with that of the nine-ordinate Simpson's or Tchebycheff's rule.

Polar Coordinates

The rules discussed above have been illustrated by figures defined by a set of parallel ordinates and this is most convenient for waterplanes. For transverse sections of a ship a problem can arise at the turn of bilge unless closely spaced ordinates are used in that area. An alternative is to adopt polar coordinates radiating from some convenient pole, O, on the centreline (Figure 3.6).

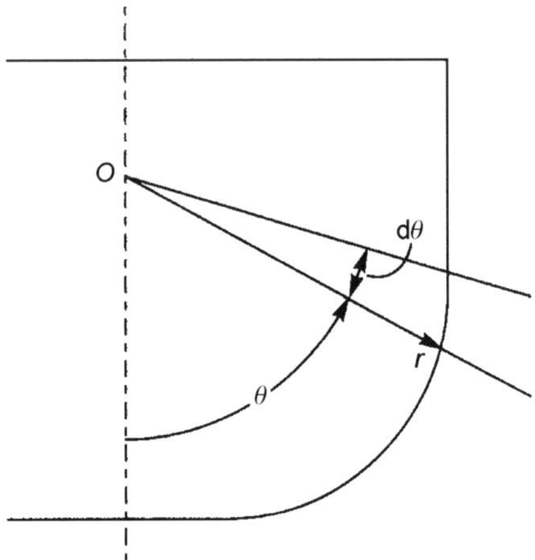

FIGURE 3.6 Polar coordinates.

$$\text{Area of the half section} = \frac{1}{2}\int_0^{180} r^2 \mathrm{d}\theta$$

If the section shape is defined by a number of radial ordinates at equal angular intervals the area can be determined using one of the approximate integration methods. Since the deck edge is a point of discontinuity one of the radii should pass through it. This can be arranged by careful selection of O for each transverse section.

SPREADSHEETS

The type of calculations discussed above, lend themselves to the use of computer spreadsheets using a program such as Microsoft Excel. A spreadsheet can be produced for the calculations in Table 3.1. This is done in Table 3.5. The first four columns present the ordinate number and the values of x, y and Simpson's multiplier. Assuming the x values are in cells B3–B11, the y values in C3–C11 and the SM values in D3–D11, then:

- the figure to go in cell E3 is obtained by an instruction of the form '= C3*D3' without the quotes, and so on for the rest of column E;
- the figure to go in cell F3 is obtained by an instruction of the form '= B3*C3' without the quotes, and so on for the rest of column F;
- the figure to go in cell G3 is obtained by an instruction of the form '= D3*F3' without the quotes, and so on for the rest of column G;
- the figure to go in cell H3 is obtained by an instruction of the form '= B3*B3*C3' without the quotes, and so on for the rest of column H;
- the figure to go in cell I3 is obtained by an instruction of the form '= D3*H3' without the quotes, and so on for the rest of column I;
- the figure to go in cell J3 is obtained by an instruction of the form '= C3*C3' without the quotes, and so on for the rest of column J;
- the figure to go in cell K3 is obtained by an instruction of the form '= D3*J3' without the quotes, and so on for the rest of column K;
- the figure to go in cell L3 is obtained by an instruction of the form '= C3*C3*C3' without the quotes, and so on for the rest of column L;
- the figure to go in cell M3 is obtained by an instruction of the form '= D3*L3' without the quotes, and so on for the rest of column M.

The summation can be done for columns E, G, I, K and M by using the instruction '= SUM(E3:E11)' and so on, or the Excel $\sum$ function can be used.

Then the area is obtained by (1/3)[SUM(E3:E11)]; the first moment about the y-axis by (1/3)[SUM(G3:G11)]; the centroid of area from the y-axis by moment/area or, in this case by [SUM(G3:G11)]/[SUM(E3:E11)]. It should be noted that the ordinate spacing in this case is unity. Had it been h, say, then the area would be given by (h/3)[SUM(E3:E11)] and so on.

TABLE 3.5 Spreadsheet

	A	B	C	D	E	F	G	H	I	J	K	L	M
1	Ordinates	x	y	SM	$F(A)$	xy	$F(M_y)$	xxy	$F(I_y)$	yy	$F(M_x)$	yyy	$F(I_x)$
2													
3	1	0.000	1.000	1.000	1.000	0.000	0.000	0.000	0.000	1.000	1.000	1.000	1.000
4	2	1.000	1.200	4.000	4.800	1.200	4.800	1.200	4.800	1.440	5.760	1.728	6.912
5	3	2.000	1.500	2.000	3.000	3.000	6.000	6.000	12.000	2.250	4.500	3.375	6.750
6	4	3.000	1.600	4.000	6.400	4.800	19.200	14.400	57.600	2.560	10.240	4.096	16.384
7	5	4.000	1.500	2.000	3.000	6.000	12.000	24.000	48.000	2.250	4.500	3.375	6.750
8	6	5.000	1.300	4.000	5.200	6.500	26.000	32.500	130.000	1.690	6.760	2.197	8.788
9	7	6.000	1.100	2.000	2.200	6.600	13.200	39.600	79.200	1.210	2.420	1.331	2.662
10	8	7.000	0.900	4.000	3.600	6.300	25.200	44.100	176.400	0.810	3.240	0.729	2.916
11	9	8.000	0.600	1.000	0.600	4.800	4.800	38.400	38.400	0.360	0.360	0.216	0.216
12													
13													
14	Total				29.800		111.200		546.400		38.780		52.378

More complex functions can be built into the tables as the complexity of the calculations increases. There are many shortcuts that can be used by those familiar with the software and students will be aware of these through other applications. The great value of the spreadsheet is that templates can be created for common calculations and thoroughly checked. Then, in subsequent use, possible errors are restricted to the inputting of the basic data.

Excel has been used extensively for the tabular calculations in Appendix B.

SUMMARY

Areas and volumes enclosed by typical ship curves and surfaces, together with their first and second moments, can be calculated by a number of approximate methods. Computer spreadsheets can be used to assist in these calculations. The methods can be applied quite widely in engineering applications other than naval architecture and they provide the means of evaluating many of the integrals called up by the theory outlined in later chapters.

Chapter 4

Flotation

INTRODUCTION

We now consider the concepts of equilibrium and buoyancy, how to establish the draughts at which a ship will float in water and how these draughts will be affected by additions or movements of weight.

EQUILIBRIUM

Equilibrium of a Body Floating in Still Water

A body floating freely in still water experiences a downward force due to gravity. If the body has a mass *m*, this force will be *mg* and is known as its *weight*. If the body is in equilibrium there must be a force of the same magnitude and in the same line of action as the weight but opposing it. Otherwise the body would move. This opposing force is generated by the hydrostatic pressures acting on the body Figure 4.1. These act normal to the body's surface and can be resolved into vertical and horizontal components. The sum of the vertical components is known as the *buoyancy force*

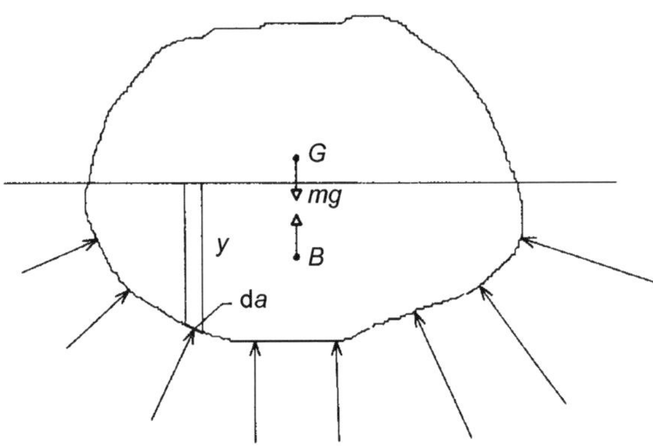

FIGURE 4.1 Floating body.

Introduction to Naval Architecture. DOI: http://dx.doi.org/10.1016/B978-0-08-098237-3.00004-7

and must equal the body's weight otherwise it would rise or sink. The horizontal components must cancel out or the body would move sideways. The two forces must act in the same vertical line or the body would be subject to a moment and rotate. The gravitational force mg can be imagined as concentrated at a point G which is the centre of mass, commonly known as the *centre of gravity*. Similarly the opposing force can be imagined to be concentrated at a point B.

The pressure acting on a small element of the surface, $\mathrm{d}a$, a depth y below the surface is:

$$\text{Pressure} = \text{density} \times \text{gravitational acceleration} \times \text{depth} = \rho g y$$

and the normal force on an element of area $\mathrm{d}a = \rho g y \, \mathrm{d}a$

If φ is the angle of inclination of the body's surface to the horizontal then the vertical component of force is:

$$(\rho g y \, \mathrm{d}a)\cos\varphi = \rho g(\text{volume of vertical element})$$

Summating over the whole volume the total vertical force is:

$$\rho g \nabla$$

where ∇ is the immersed volume of the body.

This is the weight of the displaced water. This vertical force 'buoys up' the body and is known as the *buoyancy force* or simply *buoyancy*. The point, B, through which it acts is the centroid of volume of the displaced water and is known as the *centre of buoyancy*.

Since the buoyancy force is equal to the weight of the body, $m = \rho\nabla$.

In other words the mass of a floating body equals the mass of the water displaced by that body.

If the density of a body is greater than that of the water, the weight of water it can displace is less than its own weight and it will sink to the bottom. If held by a spring balance its apparent weight would be reduced by the weight of water it displaced – due to the water pressures acting upon it. This leads to *Archimedes' Principle* which states that when a solid is immersed in a fluid it experiences an upthrust equal to the weight of the fluid displaced. This explains why divers find it easier to lift heavy items underwater.

To illustrate the principle, consider a rowing boat in a swimming pool. If a large lump of iron in the boat is dropped over the side it will sink to the bottom of the pool. But will the depth of the water, as measured at the side of the pool, increase or decrease? It will in fact decrease. Initially the boat displaces water equal to the weight of itself and the lump of iron. After jettisoning the iron it displaces only its own weight. The iron is displacing its own volume, however, which is less than the volume of water equal to its

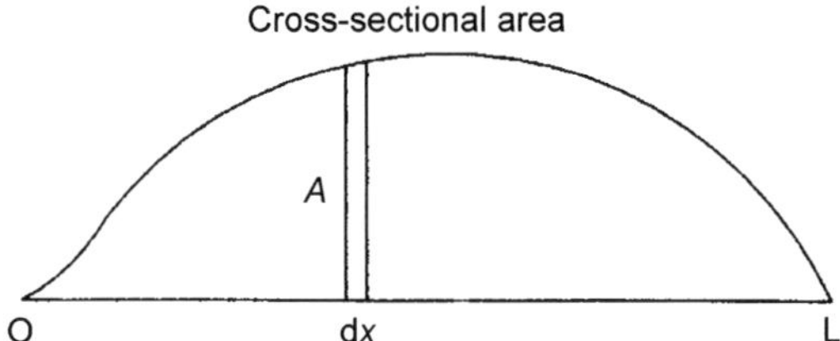

FIGURE 4.2 Cross-sectional area curve.

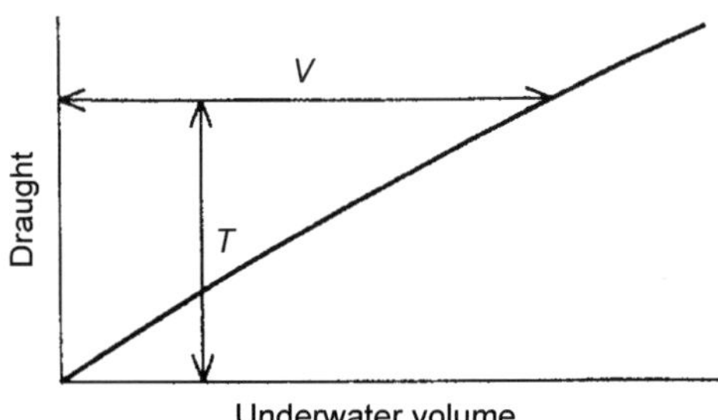

FIGURE 4.3 Volume curve.

weight. Hence the total volume of water displaced by the boat and iron decreases and the water level in the pool drops.

Underwater Volume

Once a ship form is defined the underwater volume can be calculated. If the immersed areas of a number of sections throughout the length of a ship are calculated, a sectional area curve can be drawn as in Figure 4.2. The underwater volume, or *volume of displacement*, is given by the area under the curve and is represented by:

$$\nabla = \int A \, \mathrm{d}x$$

If immersed cross-sectional areas are calculated to a number of waterlines parallel to the design waterline, then the volume up to each can be determined and plotted against draught as in Figure 4.3. The volume corresponding to any given draught T can be read off provided the waterline at T is parallel to those used in deriving the curve.

One method of finding the underwater volume is to use *Bonjean* curves. These are curves of immersed cross-sectional areas plotted against draught for each transverse section. They are usually drawn on the ship profile as in

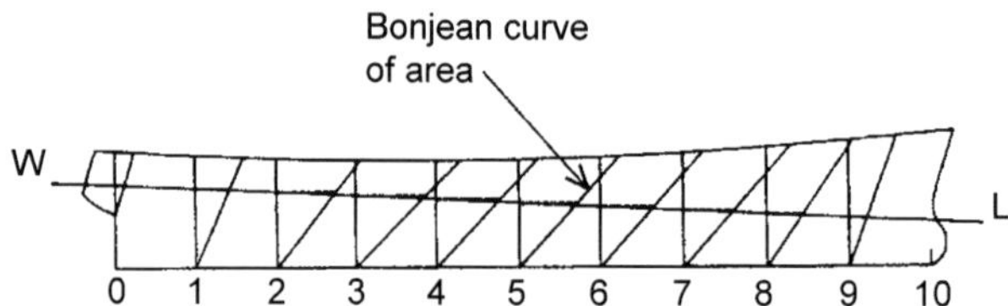

FIGURE 4.4 Bonjean curves.

Figure 4.4. Suppose the ship is floating at waterline WL. The immersed areas for this waterline are obtained by drawing horizontal lines from the intercept of the waterline with the middle line of a section to the Bonjean curve for that section. Having the areas for all the sections, the underwater volume and its longitudinal centre of buoyancy can be calculated.

When displacement was calculated manually, it was customary to use a *displacement sheet*. The displacement up to the design waterline was determined by using Simpson's rules applied to half ordinates measured at waterlines and at sections. The calculations were done in two ways. First the areas of sections were calculated and integrated in the fore and aft direction to give volume. Then areas of waterplanes were calculated and integrated vertically to give volume. The two volume values had to be the same if the arithmetic was correct. The displacement sheet was also used to calculate the vertical and longitudinal positions of the centre of buoyancy. This text has concentrated on calculating the characteristics of a floating body. It is helpful to have these concepts developed in more detail using numerical examples and this is done in Appendix B. The calculation lends itself very well to the use of Excel spreadsheets.

Reserve of Buoyancy

An intact ship, floating freely, will have watertight spaces above the waterline. Should water enter the ship for some reason the ship will sink lower in the water (and generally trim and heel) until the weight of water that has entered is balanced by the additional buoyancy generated by the extra watertight volume. The total buoyancy of the extra volume available represents a *reserve of buoyancy*. It is usually expressed as a percentage of the intact displacement and gives some measure of a ship's ability to sustain damage.

THE METACENTRE

The metacentre related to heeling is known as the *transverse metacentre* (there is another related to trimming, known as the *longitudinal metacentre*). Its position can be found by considering small inclinations of a ship about its

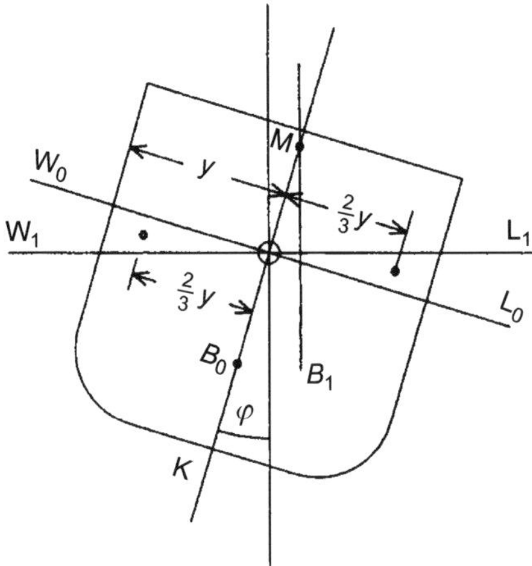

FIGURE 4.5 Transverse metacentre.

centreline, Figure 4.5. For small angles, say 2° or 3°, the upright and inclined waterlines will intersect at O on the centreline. The volumes of the emerged and immersed wedges must be equal for constant displacement.

For small angles the emerged and immersed wedges at any section, W_0OW_1 and L_0OL_1, are approximately triangular. If y is the half ordinate of the original waterline at the cross section the emerged or immersed section area is:

$$\tfrac{1}{2}y \times y \tan \varphi = \tfrac{1}{2}y^2\varphi$$

for small angles, and the total volume of each wedge is:

$$\int \tfrac{1}{2}y^2\varphi \, \mathrm{d}x$$

integrated along the length of the ship.

This volume is effectively moved from one side to the other and for triangular sections the transverse movement will be $4y/3$ giving a total transverse shift of buoyancy of:

$$\int \tfrac{1}{2}y^2\varphi \, \mathrm{d}x \times 4y/3 = \varphi \int 2y^3/3 \; \mathrm{d}x$$

since φ is constant.

The expression within the integral sign is the second moment of area, or the moment of inertia, of a waterplane about its centreline. It may be denoted by I, whence the transverse movement of buoyancy is:

$$I\varphi \quad \text{and} \quad \nabla \times B_0B_1 = I\varphi$$

so that $B_0B_1 = I\varphi/\nabla$, where ∇ is the total volume of displacement.

Referring to Figure 4.5 for small angles:

$$B_0B_1 = B_0M_0\varphi \quad \text{and} \quad B_0M_0 = I/\nabla$$

Thus the height of the metacentre above the centre of buoyancy is found by dividing the second moment of area (inertia) of the waterplane about its centreline by the volume of displacement. The height of the centre of buoyancy above the keel, *KB*, is the height of the centroid of the underwater volume above the keel, and hence the height of the metacentre above the keel is:

$$KM = KB + BM$$

The difference between *KM* and *KG* gives the *metacentric height, GM*.

The Transverse Metacentre for Simple Geometric Forms

To illustrate this concept consider some simple geometric forms.

Vessel of Rectangular Cross Section

Consider the form in Figure 4.6 of breadth *B* and length *L* floating at draught *T*. If the cross section is uniform throughout its length, the volume of displacement = *LBT*.

The second moment of area of waterplane about the centreline = $LB^3/12$. Hence:

$$BM = \frac{LB^3}{12LBT} = B^2/12T$$

The height of the centre of buoyancy above keel, $KB = T/2$ and that of the metacentre above the keel is $KM = T/2 + B^2/12T$.

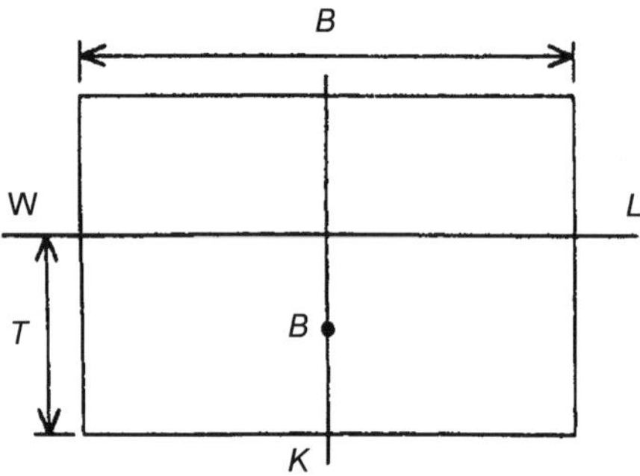

FIGURE 4.6 Rectangular section vessel.

TABLE 4.1 KM Values

T	$0.5T$	$18.75/T$	KM
1	0.5	18.75	19.25
2	1.0	9.37	10.37
3	1.5	6.25	7.75
4	2.0	4.69	6.69
5	2.5	3.75	6.25
6	3.0	3.12	6.12

The height of the metacentre depends upon the draught and beam but not the length. At small draught relative to beam, the second term predominates and at zero draught KM would be infinite.

Consider the case where B is 15 m for draughts varying from 1 to 6 m. Then:

$$KM = \frac{T}{2} + \frac{15^2}{12T} = 0.5T + \frac{18.75}{T}$$

KM values for various draughts are shown in Table 4.1 and KM and KB are plotted against draught in Figure 4.7. Such a diagram is called a *metacentric diagram*. KM is large at small draughts and falls rapidly with increasing draught. If the calculations were extended KM would reach a minimum value and then start to increase. The draught at which KM is minimum can be

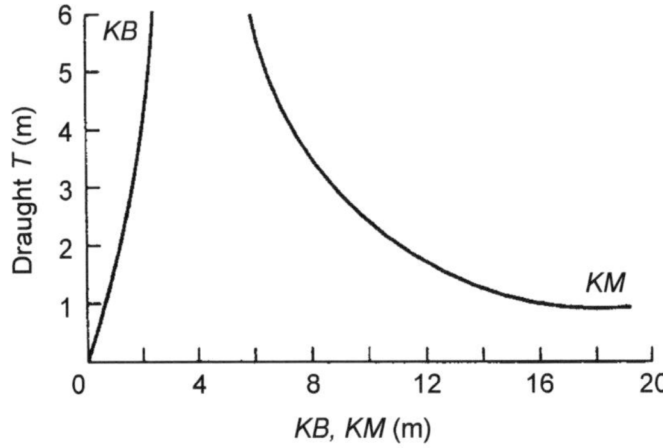

FIGURE 4.7 Metacentric diagram.

found by differentiating the equation for KM with respect to T and equating to zero. That is, KM is a minimum at T given by:

$$\mathrm{d}KM/dT = ½ - B^2/12T^2$$

In the example KM is minimum when the draught is 6.12 m.

Vessel of Constant Triangular Section

A vessel of triangular cross section floats apex down, the breadth at the top being B and the depth D. The breadth of the waterline at draught T is given by:

$$\begin{aligned} b &= (T/D) \times B, \textit{leading to}: \\ I &= (L/12) \times [(T/D) \times B]^3 \\ \nabla &= L \times (T/D) \times B \times T/2 \\ BM &= I/V = B^2T/6D^2 \\ KB &= 2T/3 \\ KM &= 2T/3 + B^2T/6D^2 \end{aligned}$$

The curves of KM and KB against draught are both straight lines with zero at zero draught.

Vessel of Circular Cross Section

A circular cylinder, of radius R and centre of section O, floats with its axis horizontal. For any waterline, above or below O, and for any inclination, the buoyancy force always acts through O. That is, KM is independent of draught and equal to R.

Metacentric Diagrams

The positions of B and M depend only upon the geometry of the ship and the draughts at which it is floating. They can therefore be determined without knowledge of the loading of the ship that causes it to float at those draughts, or of the density of the fluid in which it floats. A *metacentric diagram*, in which KB and KM are plotted against draught, is a convenient way of defining the positions of B and M for a range of waterplanes. Waterplanes parallel to the design or load waterplane are normally used.

Figure 4.8 shows a typical diagram. The vertical scale is used to represent the draught and a line is drawn at 45° to this. For a given draught T_1 a horizontal line is drawn intersecting this line in D_1 through which a vertical line is drawn. On this vertical line the positions of M_1 and B_1 are set out such that their distances from D_1 represent their distances above or below the waterline represented by D_1. Whilst B_1 will always lie below the 45° line, M_1 may lie above or below it. Since KB is roughly proportional to draught for normal ship forms the B curve will approximate a straight line passing through the origin. At moderate draughts the M curve will be such that KM

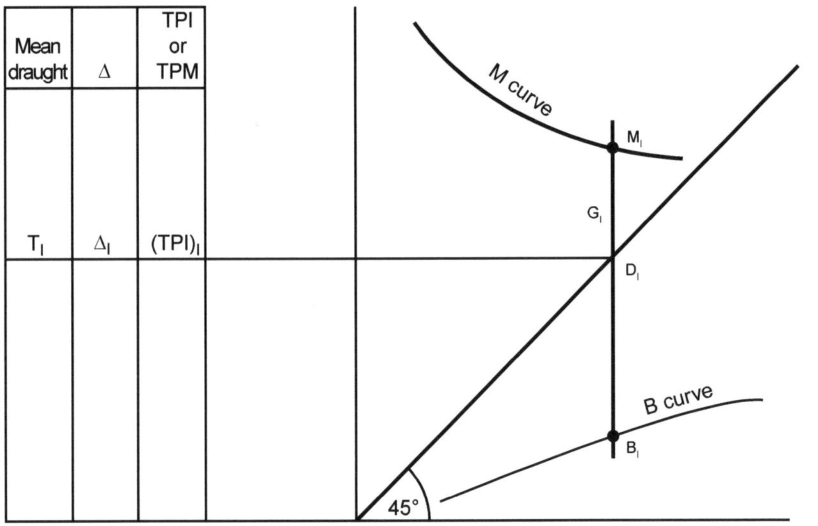

FIGURE 4.8 Metacentric diagram.

is reducing with draught. A table is included in the figure giving the values of draught, displacement and tonnes per unit immersion (TPI) for a range of draughts. These will be based on sea water of standard density. The positions of G can be shown for the ship conditions assumed for various draughts.

TRIM

Suppose a ship, floating at waterline W_0L_0 (Figure 4.9), is caused to trim through a small angle, θ, at constant displacement (by moving a weight longitudinally, say), to a new waterline W_1L_1 intersecting the original waterplane in a transverse axis through F.

The volumes of the immersed and emerged wedges must be equal so, for small θ:

$$\int 2y_f(x_f\theta)\mathrm{dx} = \int 2y_a(x_a\theta)dx$$

where y_f and y_a are the waterplane half breadths at distances x_f and x_a from F.

This is the condition that F is the centroid of the waterplane and F is known as the *centre of flotation*. For small trims at constant displacement a ship trims about a transverse axis through the centre of flotation. For most ships F is somewhat aft of amidships.

If a small weight is added to a ship it will sink and trim until the extra buoyancy generated equals the weight and the centre of buoyancy of the added buoyancy is vertically below the centre of gravity of the added weight.

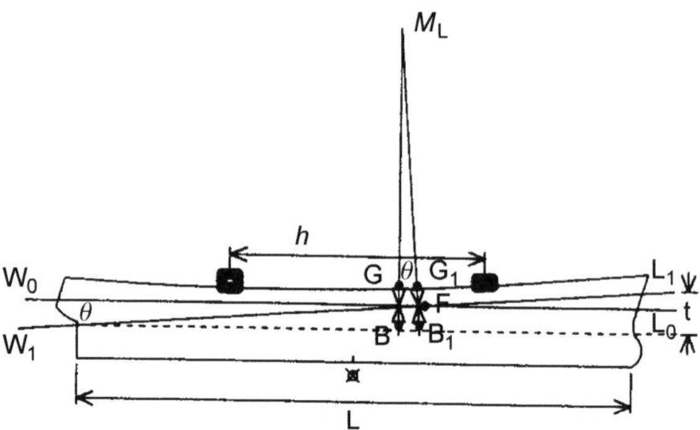

FIGURE 4.9 Trim changes.

If the weight is added in the vertical line of the centre of flotation then the ship sinks bodily with no trim as the centre of buoyancy of the added layer will be above F. Generalising this a small weight placed anywhere along the length can be regarded as being initially placed at F to cause sinkage and then moved to its actual position, causing trim. In other words, it can be regarded as a weight acting at F and a trimming moment about F.

Moment to Change Trim

If the ship in Figure 4.9 is trimmed by moving a weight, w, from its initial position to a new position h forward, the trimming moment will be wh. This will cause the centre of gravity of the ship to move from G to G_1 and the ship will trim causing B to move to B_1 such that:

$$GG_1 = wh/W$$

and B_1 is vertically below G_1.

The trim is the difference in draughts forward and aft. The change in trim angle can be taken as the change in that difference divided by the longitudinal distance between the points at which the draughts are measured. From Figure 4.9:

$$\tan\theta = t/L = GG_1/GM_L = wh/WGM_L$$

from which:

$$wh = t \times W \times GM_L/L$$

This is the moment that causes a trim t, so the moment to cause unit change of trim is:

$$WGM_L/L$$

The *moment to change trim* (MCT) 1 m is a convenient figure to quote to show how easy a ship is to trim. It should be noted that whilst most authorities use a metre as the unit change, some use the MCT by 1 cm. It is recommended, therefore, that the full title is given. e.g. MCT 1 m.

The value of MCT is very useful in calculating the draughts at which a ship will float for a given condition of loading. Suppose it has been ascertained that the weight of the ship is W and the centre of gravity is x forward of amidships and that at that weight with a waterline parallel to the design waterline it would float at a draught T with the centre of buoyancy y forward of amidships. There will be a moment $W(y-x)$ taking it away from a waterline parallel to the design one. The ship trims about the centre of flotation by an amount equal to the moment divided by the MCT for the waterplane. The draughts at any point along the length can be found by simple ratios.

Worked Example 4.1

A ship of mass 5000 t, 98 m long, floats at draughts of 5.5 m forward and 6.2 m aft, being measured at the extreme ends. The longitudinal metacentric height is 104 m and the centre of flotation is 2.1 m aft of amidships. Determine the MCT 1 cm and the new end draughts when a mass of 85 t, which is already on board, is moved 30 m forward.

Solution

$$\text{MCT 1 cm} = \frac{W \times GM_L}{100\,L}$$

$$= \frac{5000 \times 9.81 \times 104}{100 \times 98} \quad \text{where } g = 9.81\ m/s^2$$

$$= 520.5 \text{ MNm}$$

As the mass is already on board there will be no bodily sinkage. The change of trim is given by the trimming moment divided by MCT.

$$\text{Change in trim} = \frac{85 \times 9.81 \times 30}{520.5}$$

$$= 48.1 \text{ cm by the bow}$$

The changes in draught will be:

$$\text{Forward} = 48.1 \times \frac{(98/2) + 2.1}{98} = 25.1 \text{ cm}$$

$$\text{Aft} = 48.1 \times \frac{(98/2) - 2.1}{98} = 23.0 \text{ cm}$$

The new draughts become 5.751 m forward and 5.97 m aft.

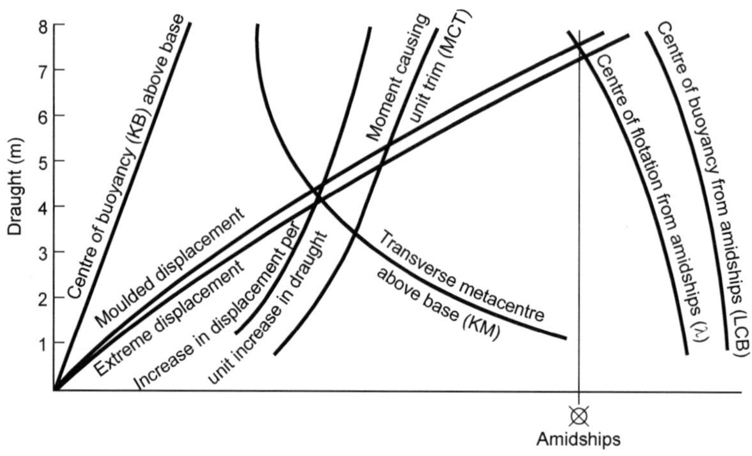

FIGURE 4.10 Hydrostatic curves.

HYDROSTATIC CURVES

It is customary to obtain the displacement and position of B, M and F for a range of waterplanes parallel to the design waterplane and plot them against draught measured vertically. Such sets of curves are called *hydrostatic curves*. Each curve will have its own scale along the horizontal axis.

The curves in Figure 4.10 show moulded and extreme displacement. The former was mentioned earlier. The latter, normally shown simply as the displacement curve, allows for the displacement to the outside of the plating and outside the perpendiculars, bossings, bulbous bows, etc. It is this displacement which is relevant to the discussion of flotation and stability. The additions to the moulded figure can have a measurable effect upon displacement and the position of B.

Tonnes per Unit Immersion

It will be noted that the curves include one for the increase in displacement for unit increase in draught. If a waterplane has an area A, then the increase in displaced volume for unit increase in draught at that waterplane is $1 \times A$. The increase in displacement will be $\rho g A$. For $\rho = 1025\ \text{kg/m}^3$ and $g = 9.81\ \text{m/s}^2$ increase in displacement per metre increase in draught is:

$$1025 \times 9.81 \times 1 \times A = 10{,}055\, A \text{ newtons}$$

It is usual to quote the increase in displacement in terms of the *extra tonnes per unit immersion*. As with MCT, it is necessary to know the unit of immersion used. It may be 1 m or 1 cm. The latter is to be preferred if the

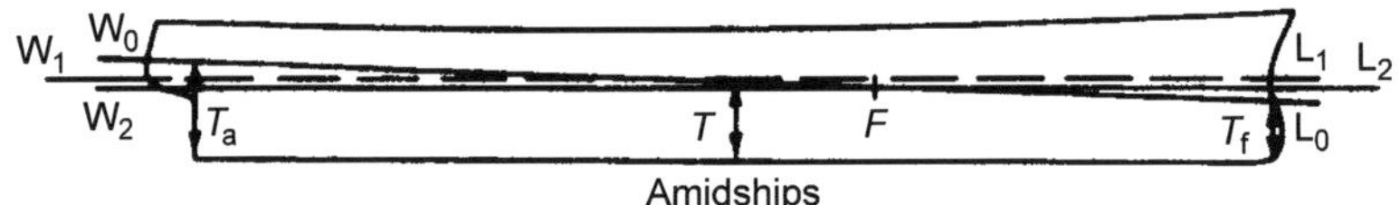

FIGURE 4.11 Determination of displacement.

area of the waterplane changes appreciably over 1 m. It is often, then, abbreviated to TPC. The increase in displacement per unit increase in draught is useful in calculations where weights are added to a ship provided the change in draught is small.

Hydrostatic curves are useful for working out the draughts and the value of *GM*, in various conditions of loading. This is done for all normal working conditions of the ship and the results supplied to the master.

PROBLEMS IN TRIM

Determination of Displacement from Observed Draughts

Suppose draughts at the perpendiculars are T_a and T_f as in Figure 4.11. The mean draught will be $T = (T_a + T_f)/2$ and a first approximation to the displacement could be obtained by reading off the corresponding displacement, Δ, from the hydrostatic curves. In general, W_0L_0 will not be parallel to the waterlines for which the hydrostatic curves were computed. If waterline W_1L_1, intersecting W_0L_0 at amidships, is parallel to the design waterline then the displacement read from the hydrostatics for draught T is in fact the displacement to W_1L_1. It has been seen that because ships are not symmetrical fore and aft they trim about F. As shown in Figure 4.11 (where F is shown ahead of amidships), the displacement to W_0L_0 is less than that to W_1L_1, the difference being the layer $W_1L_1L_2W_2$, where W_2L_2 is the waterline parallel to W_1L_1 through F on W_0L_0. If λ is the distance of F forward of amidships then the thickness of layer $= \lambda \times t/L$ where $t = T_a - T_f$.

If i is the increase in displacement per unit increase in draught:

$$\text{Displacement of layer} = \lambda \times ti/L \text{ and the actual displacement}$$
$$= \Delta - \lambda \times ti/L$$

Whether the correction to the displacement read off from the hydrostatics initially is positive or negative depends upon whether the ship is trimming by the bow or stern and the position of F relative to amidships. This can be determined using a simple sketch.

If the ship is floating in water of a different density to that for which the hydrostatics were calculated a further correction is needed in proportion to the two density values. At a given draught the displacement will be larger if the water in which ship is floating is greater than the standard.

This calculation for displacement has assumed that the keel is straight. It is likely to be curved, even in still water, so that a draught taken at amidships may not equal $(d_a + d_f)/2$ but have some value d_m giving a deflection of the hull, δ. If the ship sags the above calculation would underestimate the volume of displacement. If it hogs it would overestimate the volume. It is reasonable to assume the deflected profile of the ship is parabolic, so that the deflection at any point distant x from amidships is $\delta[1 - (2x/L)^2]$. Hence:

$$\text{Volume correction} = \int b\delta[1 - (2x/L)^2]\mathrm{d}x$$

where b is the waterline breadth at the point considered.

Unless an expression is available for b in terms of x this cannot be integrated mathematically and must be evaluated by approximate integration using the ordinates for the waterline.

Longitudinal Position of the Centre of Gravity

Suppose a ship is floating in equilibrium at a waterline W_0L_0 as in Figure 4.12 with the centre of gravity distant x from amidships, a distance yet to be determined. The centre of buoyancy B_0 must be directly beneath G. Now assume the ship brought to a waterline W_1L_1 parallel to those used for the hydrostatics, which cuts off the correct displacement. The position of the centre of buoyancy will be at B_1, distant y from amidships, a distance that can be read from the hydrostatics for waterline W_1L_1. It follows that if t was the trim, relative to W_1L_1, when the ship was at W_0L_0:

$$\Delta(y - x) = t \times (\text{moment to cause unit trim}) \text{ and:}$$

$$x = y - \frac{t \times \text{MCT}}{\Delta}$$

giving the longitudinal centre of gravity.

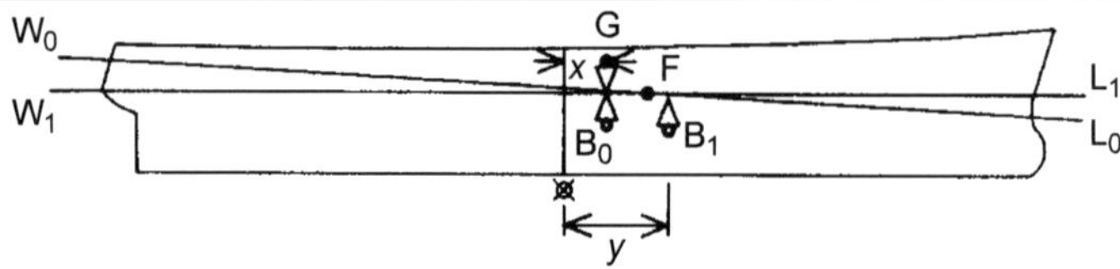

FIGURE 4.12 Determining longitudinal centre of gravity.

Direct Determination of Displacement and Position of G

The above methods for finding the displacement and longitudinal position of G are usually sufficiently accurate when trim is small. For more accuracy and larger trims the Bonjean curves can be used. If the end draughts, distance L apart, are observed then the draught at any particular section can be calculated, since:

$$T_x = T_a - (T_a - T_f)\frac{x}{L}$$

where x is the distance from where T_a is measured.

These draughts can be corrected for hog or sag. The calculated draughts at each section can be set upon the Bonjean curves and the immersed areas read off. The immersed volume and position of the centre of buoyancy can be found by approximate integration. For equilibrium, the centre of gravity and centre of buoyancy must be in the same vertical line and the position of the centre of gravity follows. Using the density of water in which the ship is floating, the displacement can be determined.

TRANSVERSE WEIGHT MOVEMENTS

In Figure 4.13 a ship is shown upright and at rest in still water. If a small weight w is shifted transversely through a distance h, the centre of gravity of the ship, originally at G_0, moves to G_1 such that $G_0G_1 = wh/W$. The ship will heel through an angle φ causing the centre of buoyancy to move to B_1 vertically below G_1 to restore equilibrium. It will be seen that:

$$\frac{G_0G_1}{GM} = \tan\varphi \quad \text{and} \quad \tan\varphi = \frac{wh}{W \times GM}$$

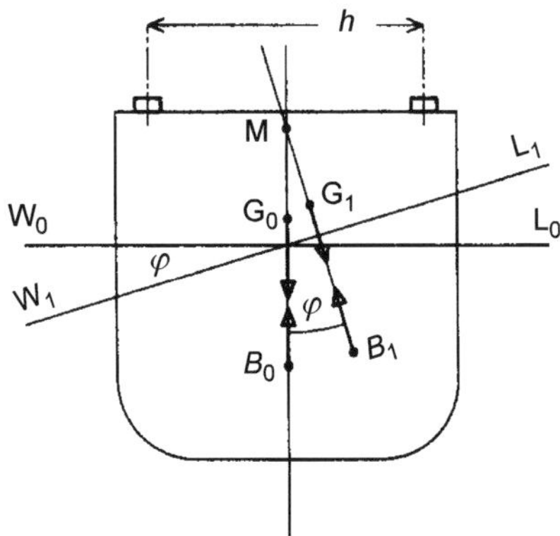

FIGURE 4.13 Transverse movement of weight.

This applies whilst the angle of inclination remains small enough for M to be regarded as a fixed point. For most conventional hull forms M can be taken as sensibly fixed up to angles of heel of about 10°.

SUMMARY

For a ship of known displacement and cg position, the draughts at which it will float can be found using its hydrostatic curves. These curves define a number of hydrostatic characteristics of the hull arising from its geometry. They can be used to determine the change in draught due to adding or removing weights and the change in trim and heel due to moving weights fore and aft or transversely. The metacentre is the point at which successive buoyancy forces intersect as a vessel is inclined. For small angles it will be a fixed point on the centreline. A more detailed discussion on this topic, with both worked and set examples, is to be found in Barrass and Derrett (2012).

Chapter 5

Stability

INTRODUCTION

A naval architect is concerned with a ship's:

- stability of attitude (heel and trim) when afloat;
- directional stability associated with manoeuvring.

Directional stability is dealt with in manoeuvring. Here we consider the stability of attitude.

It is relatively easy to define what is required but difficult to ensure the aims are achieved. The aims can be said to be that the ship should:

- float without excessive heel or trim in still water and, if disturbed from its position of equilibrium by some force, will return to its original attitude when the disturbance is removed;
- be able to withstand the forces and moments imposed on it in service without sinking and without excessive heel or trim;
- be able to withstand a reasonable amount of flooding as the result of damage without sinking whilst accepting that no ship can be completely safe.

For small disturbances in still water achieving the first aim is not difficult. The draughts can be determined from the conditions for equilibrium. The waterplane and underwater shapes are known and straightforward calculations can show whether the ship is stable. The problem is in deciding the standards of stability required to ensure the other aims are met. The problem of surviving normal in-service conditions is complicated by the following considerations:

- The ship condition, including displacement and amount of fluids in tanks, varies
- The sea state can vary from smooth to very rough
- Winds will vary in strength and direction
- Heeling and trimming moments may be introduced by ship operations
- Heel is caused as the ship turns.

Introduction to Naval Architecture. DOI: http://dx.doi.org/10.1016/B978-0-08-098237-3.00005-9

In waves the ship's underwater form and the waterline shape will be constantly changing. It will be acted on by wave and wind forces. Because many of these forces are cyclic in nature their periods relative to the natural frequencies of the ship's motion are important and resonance may greatly magnify the resulting motions.

In trying to meet the aim of surviving an accident, the above uncertainties apply to the conditions at the time of the incident. In the case of a collision there are extra considerations such as where and how extensive the damage; how deep the penetration; the structural damage resulting; the state of watertight boundaries and closures and the actions the crew take to deal with the situation. In the case of grounding uncertainties include the nature of the seabed; the state of the tide and the location and area of the ship's bottom affected.

Whilst various measures of a ship's stability can be assessed and standards set, these do not directly give the probability that the ship might be lost. The traditional criteria do, however, lead to designs that are safe in most operational scenarios.

THE APPROACH

A full investigation of a ship's stability is complex and, even with the power of modern computers, some assumptions and approximations must be made. In order to study stability methodically we consider:

- the intact stability of the ship in still water at small angles;
- the extension to stability at larger angles of inclination and the features of the stability curves obtained are compared with those for previous successful ships;
- the behaviour of the ship in wind and waves;
- the flooding, loss of stability and the risk of loss following an incident are then assessed. Based on seagoing experience, the position and extent of the damage are expressed in probabilistic terms as are the likely sea and wind conditions at the time.

STABILITY AT SMALL ANGLES

Transverse Stability and the Transverse Metacentre

The metacentre is key in the consideration of small angle stability. In Figure 5.1 a ship floating originally at waterline W_0L_0 is rotated through a small angle by an external force and then floats at waterline W_1L_1. Because the angle of heel is small the two waterlines are assumed to intersect on the ship's centreline.

The inclination does not affect the position of G, the ship's centre of gravity, provided no weights are free to move. The inclination does, however, affect the underwater shape and the centre of buoyancy moves from B_0 to B_1

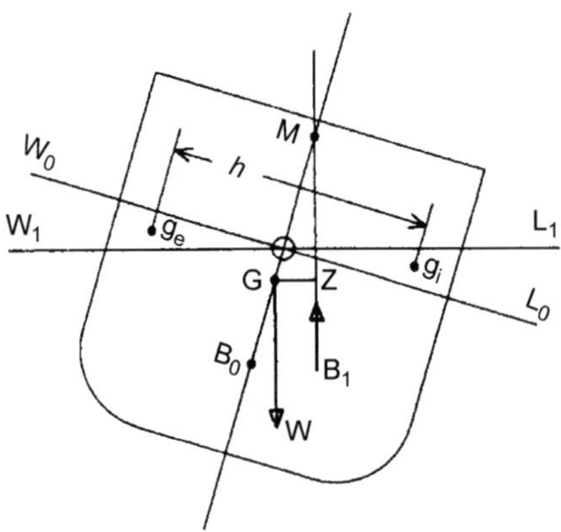

FIGURE 5.1 Small angle stability.

due to a volume, υ, represented by W_0OW_1, emerging from the water and an equal volume, represented by L_0OL_1, being immersed.

If g_e and g_i are the centroids of the emerged and immersed wedges and $g_eg_i = h$, then

$$B_0B_1 = \frac{\upsilon \times h}{\nabla}$$

where ∇ is the total volume of the ship.

In general a ship will trim slightly when it is inclined at constant displacement. Strictly B_0, B_1, g_e, etc. are the projections of the actual points on to a transverse plane. The line joining B_0 to B_1 will be parallel to that joining g_e and g_i.

The buoyancy acts vertically upwards (normal to the waterline W_1L_1) through B_1 and intersects the original vertical at M. M is known as the *transverse metacentre* or *metacentre*. The weight $W = mg$ acting downwards through G and the buoyancy force, of equal magnitude, acting upwards are not in the same line but form a couple $W \times GZ$, where GZ is the perpendicular on to B_1M drawn from G. If M is above G this couple will restore the body to its original position and in this condition the body is said to be in stable equilibrium or simply stable. $GZ = GM \sin \varphi$ and is called the *righting lever*. GM is known as the *metacentric height*. For a given position of G, as M can be taken as fixed for small inclinations, GM will be constant for any particular waterline. Although G (and hence GM) can vary with the loading of the ship even for a given displacement, BM will be constant for a given waterline.

In Chapter 4 it was shown that:

$$BM = I/(\text{volume of displacement})$$

In Figure 5.1 M is above G, giving positive stability, and *GM* is regarded as positive in this case. If, when inclined, the new position of the centre of buoyancy, B_1, is directly under G, the three points M, G and Z are coincident and there is no moment acting on the ship. When the disturbing force is removed the ship will remain in the inclined position. The ship is said to have neutral stability. Both *GM* and *GZ* are zero.

A third possibility is that, after inclination, the new centre of buoyancy will lie to the left of G. There is then a moment $W \times GZ$ which will take the ship further from the vertical. In this case the ship is said to be unstable and it may heel to a considerable angle or even capsize. For unstable equilibrium M is below G and both *GM* and *GZ* are negative.

The above considerations apply to what is called the *initial stability* of the ship, which is when the ship is upright or very nearly so. The degree of stability is directly proportional to the metacentric height and *GM* can be taken as the criterion of initial stability.

Longitudinal Stability

The principles involved are the same as those for transverse stability, but for longitudinal inclinations the stability depends upon the distance between the centre of gravity and the longitudinal metacentre. In this case the distance between the centre of buoyancy and the longitudinal metacentre will be governed by the second moment of area of the waterplane about a transverse axis passing through its centroid. For normal ship forms this quantity is many times that for the second moment of area about a longitudinal axis and BM_L is large compared with BM_T, often commensurate with the length of the ship. It is thus virtually impossible for an undamaged conventional ship to be unstable when inclined about a transverse axis:

$$KM_L = KB + BM_L = KB + I_L/\nabla$$

where I_L is the second moment of the waterplane area about a transverse axis through its centre of flotation.

Stability of a Fully Submerged Body

In this case there is no waterplane and B is a fixed point determined by the shape of the body. There is no metacentre as such and the forces of weight and buoyancy will always act vertically through G and B respectively at any angle. Stability will be the same for inclination about any axis. It will be positive if B is above G. As a submersible is an elastic body it will compress as the depth of submergence increases. Water is effectively incompressible so the buoyancy force reduces with increasing depth and the body experiences a net downward force causing it to sink further. Thus it is unstable in depth variation. The decrease in buoyancy must be compensated for by pumping out water or by forces generated by the hydroplanes. If the submarine moves into water of a different density there will again be an imbalance in forces due to the changed buoyancy force. There is no 'automatic' compensation such as a surface vessel experiences when the draught adjusts in response to density changes.

SPECIAL CASES IN STABILITY

There are some special situations which are interesting. They are:

- a wall-sided hull form;
- the effect of suspended weights;
- the presence of free surfaces inside the ship.

Wall-Sided Ship

A ship is said to be *wall-sided* when its sides are vertical in way of the waterline over its whole length (Figure 5.2). Any turn of bilge, flare, etc. must not be exposed by the inclination of the ship nor must the deck edge become immersed. Because the vessel is wall-sided, successive heeled waterplanes intersect on the ship's centreline. The emerged and immersed wedges will have sections which are right-angled triangles of equal area. Let the new position of the centre of buoyancy B_1 after inclination through φ be α and β relative to the centre of buoyancy position in the upright condition, measured parallel to the ship's principal axes. Using the notation shown in the figure:

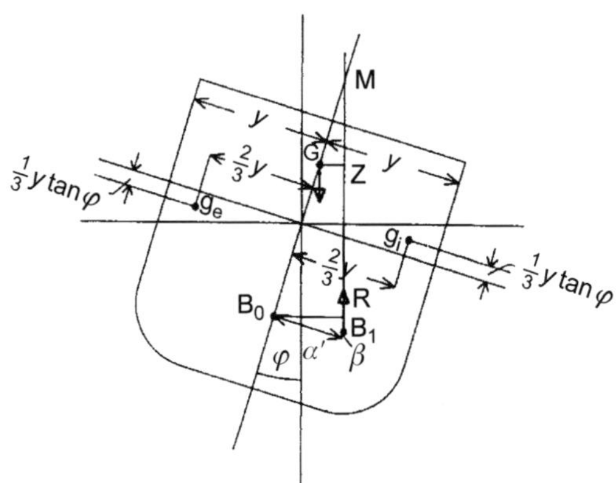

FIGURE 5.2 Wall-sided ship.

$$\text{Transverse moment of volume shift} = \int \frac{y}{2} \times y \tan\varphi \, dx \times \frac{4y}{3}$$
$$= \int \frac{2}{3} \times y^3 \tan\varphi \, dx$$
$$= \tan\varphi \int \frac{2}{3} y^3 \, dx$$
$$= I \tan\varphi$$

where I is the second moment of area of the waterplane about the centreline. Therefore:

$$\alpha = I \tan\varphi / V = B_0M \tan\varphi \quad \text{since } B_0M = \frac{I}{V}$$

Similarly the vertical moment of volume shift is

$$\int \frac{1}{2} y^2 \tan\varphi \times \frac{2}{3} y \tan\varphi \, dx = \int \frac{y^3}{3} \tan^2\varphi \, dx = \frac{1}{2} \tan^2\varphi$$

and

$$\beta = \frac{1}{2} I \tan^2\varphi / \nabla = \frac{1}{2} B_0M \tan^2\varphi$$

From the figure:

$$B_0R = \alpha \cos\varphi + \beta \sin\varphi$$
$$= B_0\text{M} \tan\varphi \cos\varphi + \frac{1}{2} B_0M \tan^2\varphi \sin\varphi$$
$$= \sin\varphi \, (B_0M + \frac{1}{2} B_0M \tan^2\varphi)$$

Now

$$GZ = B_0R - B_0G \sin\varphi$$
$$= \sin\varphi(B_0M - B_0G + \frac{1}{2} B_0M \tan^2\beta\varphi)$$
$$= \sin\varphi(GM + \frac{1}{2} B_0M \tan^2\varphi)$$

This is called the *wall-sided formula*. It is often reasonably accurate for full hull forms up to angles of about 10°. It can be regarded as a refinement of the simple expression $GZ = GM \sin\varphi$.

Influence on Stability of a Freely Hanging Weight

Consider a weight w suspended freely from a point h above its centroid. If the ship heels slowly the weight moves transversely and takes up a new position, vertically below the suspension point. As far as the ship is concerned the weight seems to be located at the suspension point. Compared to the situation with the weight fixed, the ship's centre of gravity will rise (and GM reduced) by GG_1, where

$$GG_1 = wh/W$$

where W is the total weight of the ship.

Weights free to move in this way should be avoided, but this is not always possible. Consider a weight being lifted by a shipboard crane. As soon as the weight is lifted off the deck its effect on stability is as though it were at the crane's jib. The resulting rise in G may cause a stability problem. If the weight is being lifted from a jetty there will be an increase in displacement and the ship will initially heel towards the jetty. This effect is important to the design of heavy lift ships.

Effect of Liquid Free Surfaces

A ship in service will usually have tanks which are partially filled with liquids. These may be the fuel and water tanks the ship is using for its own services or tanks carrying liquid cargoes. When the ship is heeled slowly through a small angle the liquid surface moves so as to remain horizontal. In this discussion a quasi-static condition is considered so that sloshing of the liquid is avoided. Different considerations apply to the dynamic conditions of a ship rolling where 'sloshing' forces may be significant to the structural strength of tanks.

For small angles, and assuming the liquid surface does not intersect the top or bottom of the tank, the volume of the wedge that moves is (Figure 5.3)

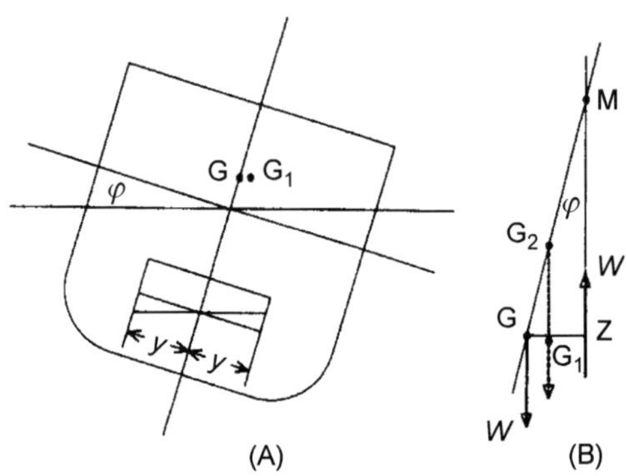

FIGURE 5.3 Fluid free surface.

$$\int \frac{1}{2} y^2 \varphi \, \mathrm{d}x, \text{ integrated over the length, } l, \text{ of the tank}$$

Assuming the wedges can be treated as triangles, the moment of transfer of volume is

$$\int \frac{1}{2} y^2 \varphi \, \mathrm{d}x \times \frac{4y}{3} = \varphi \int \frac{2}{3} y^3 \, \mathrm{d}x = \varphi I_1$$

where I_1 is the second moment of area of the liquid surface. The moment of mass moved $= \rho_f \varphi I_1$, where ρ_f is the density of the fluid in the tank. The centre of gravity of the ship will move because of this shift of mass to a position G_1 and

$$GG_1 = \rho_f g \varphi I_1 / W = \rho_f g \varphi I_1 / \rho g \nabla = \rho_f \varphi \nabla_1 / \rho \nabla$$

where ρ is the density of the water in which the ship is floating and V is the volume of displacement.

The effect on the transverse movement of the centre of gravity is to reduce GZ by the amount GG_1 as shown in Figure 5.3B. That is, there is an effective reduction in stability. Since $GZ = GM \sin \varphi$ for small angles, the influence of the shift of G to G_1 is equivalent to raising G to G_2 on the centreline so that $GG_1 = GG_2 \tan \varphi$ and the righting moment is given by

$$W(GM \sin \varphi - GG_2 \cos \varphi \tan \varphi) = W(GM - GG_2) \sin \varphi$$

Thus the effect of the free surface is equivalent to a rise of GG_2 of the centre of gravity, the 'loss' of GM being:

$$\text{Free surface effect } GG_2 = \rho_f I_1 / \rho \nabla$$

By analogy with the loss of stability due to the suspended weight, the water in the tank behaves as though its weight acts through some point above the centre of the tank, height I_1/v above the centroid of the fluid in the tank, where v is the volume of fluid. In effect the tank has its own 'metacentre' through which its fluid weight acts. The fluid weight is $\rho_f v$ and the centre of gravity of the ship will be effectively raised through GG_2, where

$$W \times GG_2 = \rho \nabla \times GG_2 = (\rho_f v)(I_1 / v) = \rho_f I_1$$

and

$$GG_2 = \rho_f I_1 / \rho \nabla \text{ as before}$$

This loss is the same whatever the height of the tank in the ship or its transverse position. If the loss is sufficiently large, the ship's metacentric height will become negative and it will heel over. The free surfaces of tanks must be kept to a minimum. One way of reducing them is to subdivide wide tanks into two or more narrow ones. Assuming a tank has a constant section and length, l, the second moment of area without division is $lB^3/12$. With a centre division the sum of the second moments of area of the two tanks is $(l/12)(B/2)^3 \times 2 = lB^3/48$. The introduction of a centre division has reduced the free surface effect to a quarter of its original value. Using two bulkheads to divide the tank into three equal width sections reduces the free surface to a ninth of its original value. Thus subdivision is very effective and it is common to subdivide the double bottom of ships. The main tanks of ships carrying liquid cargoes must be designed taking free surface effects into account and their breadths are reduced by providing centreline or wing bulkheads. Some particulate cargoes, such as grain, carried in bulk can move as the ship heels causing a similar effect to a free surface as discussed later.

Free surface effects must be taken into account in ship design and operators must be aware of their significance and use the tanks in ways intended. Both transverse and longitudinal stability will be affected. The importance of free surfaces to a ship's stability was shown by the loss of the *Herald of Free Enterprise* in 1987. In this case the free surface was that of water entering through the bow doors and covering the vehicle deck.

THE INCLINING EXPERIMENT

As the position of the centre of gravity is so important for stability it must be established accurately for the ship as completed. During design it is determined by calculation by considering all weights making up the ship – steel, outfit, fittings, machinery and systems – and assessing their individual centres of gravity. These data lead to the displacement and centre of gravity of the lightship. For particular conditions of loading the weights of all items to be carried must then be added at their appropriate centres of gravity to give the loaded displacement and centre of gravity. There are many fittings in a ship and it is difficult to account for all items accurately in such calculations, and the lightship weight and centre of gravity are measured experimentally.

The experiment, the *inclining experiment*, involves causing the ship to heel to small angles by moving known weights known distances transversely across the deck and observing the angles of inclination. The draughts of the ship are noted together with the water density. Ideally the experiment is conducted when the ship is complete, but this is not generally possible. There will usually be a number of items both to go on and to come off the ship (e.g. staging and tools). The weights and centres of gravity of these must be assessed and the condition of the ship as inclined corrected.

A typical setup is shown in Figure 5.4. Two sets of two weights, all of w, are placed on each side of the ship at about amidships, the port and starboard sets being h apart. Set 1 is moved a distance h to a position

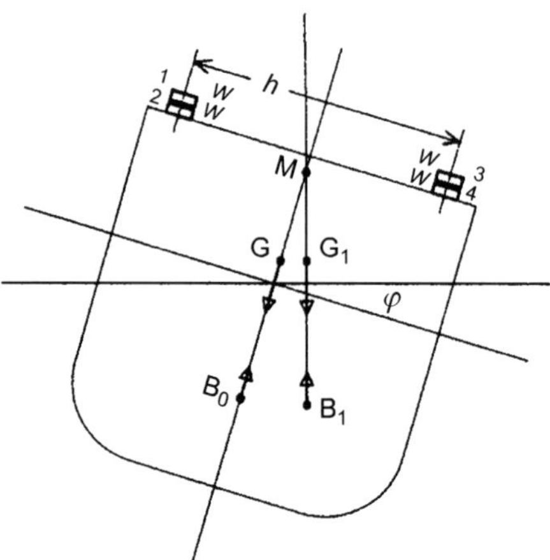

FIGURE 5.4 Inclining experiment.

alongside sets 3 and 4. G moves to G_1 as the ship inclines to a small angle and B moves to B_1. It follows that:

$$GG_1 = \frac{wh}{W} = GM \tan \varphi \quad \text{and} \quad GM = wh \cot \varphi / W$$

φ can be obtained by using two long pendulums, one forward and one aft, suspended from the deck into the holds. If d and l are the shift and length of a pendulum, respectively, $\tan \varphi = d/l$.

To improve the accuracy of the experiment, several shifts of weight are used. Thus, after set 1 has been moved, a typical sequence would be to move successively set 2, replace set 2 in original position followed by set 1. The sequence is repeated for sets 3 and 4. At each stage the angle of heel is noted and the results plotted to give a mean angle for unit applied moment. When the metacentric height has been obtained, the height of the centre of gravity is determined by subtracting GM from the value of KM given by the hydrostatics for the mean draught at which the ship was floating and allowing for the water density. This KG must be corrected for the weights to go on and come off. The longitudinal position of B, and hence G, can be found using the recorded draughts.

In this simple explanation equal weights and equal transfer distances are used. In practice the weights and distances may differ but the same principles apply.

To obtain accurate results a number of precautions have to be observed. First the experiment should be conducted in calm water with little wind. Inside a dock is good as this eliminates the effects of tides and currents. The ship must be floating freely when records are taken so mooring lines must be slack and the brow lifted clear. All weights must be secured and tanks must be empty or pressed full to avoid free surface effects. If this is not possible soundings of the tanks must be taken and allowance made. If the ship does not return to its original position when the inclining weights are restored it is an indication that some loose weight has moved, or that fluid has moved from one tank to another, possibly through a leaking valve. The number of people on board must be kept to a minimum, and those present must go to defined positions when readings are taken. The motions of the pendulum bobs are damped by immersion in a trough of water.

The draughts must be measured accurately at stem and stern, and at amidships, assuming such marks exist, in case the ship is hogging or sagging. The density of water is taken by hydrometer at several positions around the ship and at several depths to give a good average figure. If the ship has a large trim at the time of inclining it might not be adequate to use the hydrostatics to give the displacement and the longitudinal and vertical positions of B. In this case detailed calculations should be carried out to find these quantities for the inclining waterline.

Every passenger ship regardless of size, and every cargo ship of length 24 m and upwards, is required to be inclined upon its completion and its stability determined. Details of stability are supplied to the master. Subsequently when alterations are made to the ship which materially affect the stability information supplied to the master, amended stability information must be provided. *IACS; REC 031 (2004)* sets out a unified procedure for inclining a ship.

At periodical intervals not exceeding 5 years, a lightweight survey is carried out on all passenger ships to verify any changes in lightship displacement and longitudinal centre of gravity. The ship is re-inclined whenever, in comparison with the approved stability information, a deviation from the lightship displacement exceeding 2% or a deviation of the longitudinal centre of gravity exceeding 1% of length is found or anticipated.

Ships commonly grow in displacement with years in service and this growth usually leads to degradation in stability, as additions tend to be made high in the ship. If this is suspected to be significant a repeat inclining experiment is carried out during the life of the ship. For UK warships it is usual to allow for an increase in displacement of 0.65% per annum in service and an increase in *KG* of 0.45% per annum.

STABILITY AT LARGE ANGLES

We now consider stability for larger angles of heel.

Atwood's Formula

At larger angles of inclination *M* can no longer be regarded as a fixed point. The metacentric height is no longer a suitable measure of stability and the value of the *righting arm*, *GZ*, is used instead.

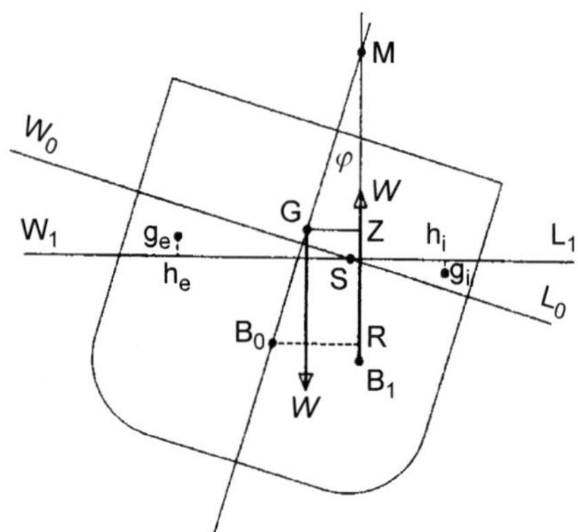

FIGURE 5.5 Atwood's formula.

Referring to Figure 5.5 assume the ship is in equilibrium under the action of its weight and buoyancy with W_0L_0 and W_1L_1 the waterlines when upright and when inclined through φ, respectively. These two waterlines will cut off the same volume of buoyancy but will not, in general, intersect on the centreline but at some point S.

A volume represented by W_0SW_1 has emerged and an equal volume, represented by L_0SL_1, has been immersed. Let this volume be v. Using the notation shown in the figure, the horizontal shift of the centre of buoyancy is given by

$$B_0R = v \times h_e h_i / \nabla \quad \text{and} \quad GZ = B_0R - B_0G \sin \varphi$$

This expression for GZ is often called *Atwood's formula*.

The wall-sided formula, derived earlier, can be regarded as a special case of Atwood's formula.

Angle of Loll

If the ship has a positive GM it will be in equilibrium when GZ is zero, i.e.:

$$0 = \sin \varphi(GM + \tfrac{1}{2}B_0M \tan^2 \varphi)$$

This equation is satisfied by two values of φ. The first is $\sin \varphi = 0$, or $\varphi = 0$. This is the case with the ship upright as is to be expected. The second value is given by

$$GM + \tfrac{1}{2}B_0M \tan^2 \varphi = 0 \quad \text{or} \quad \tan^2 \varphi = -2GM/B_0M$$

With both GM and B_0M positive there is no solution to this meaning that the upright position is the only one of equilibrium. This also applies to the case of zero GM; it being noted that in the upright position the ship has stable, not neutral, equilibrium due to the term in B_0M.

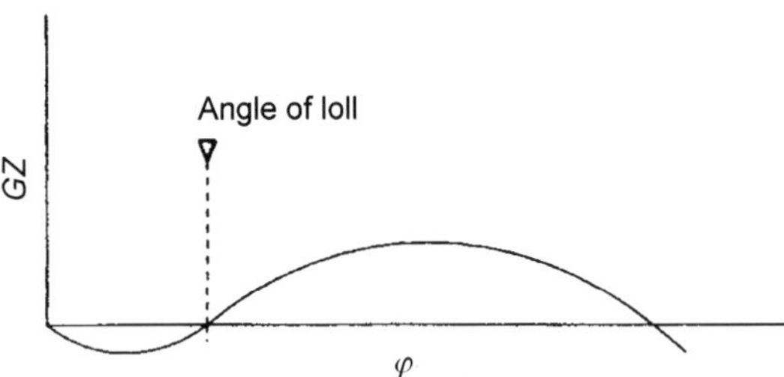

FIGURE 5.6 Angle of loll.

When, however, the ship has a negative GM there are two possible solutions for φ in addition to that of zero, which in this case would be a position of unstable equilibrium. These other solutions are at φ either side of the upright φ being given by

$$\tan \varphi = \left(\frac{2GM}{B_0M}\right)^{0.5}$$

The ship would show no preference for one side or the other. Such an angle is known as an *angle of loll*. The ship does not necessarily capsize although if φ is large enough the vessel may take water on board through side openings. The GZ curve for a ship lolling is shown in Figure 5.6.

If the ship has a negative GM of 0.08 m, associated with a B_0M of 5 m, φ, which can be positive or negative, is

$$\varphi = \tan^{-1}\left(\frac{2 \times 0.08}{5}\right)^{0.5} = \tan^{-1} 0.179 = 10.1^\circ$$

This shows that small negative GM can lead to significant loll angles. A ship with a negative GM will loll first to one side and then the other in response to wave action. When this happens the master should investigate the reasons, although the ship may still be safe in this condition.

Metacentric Height in the Lolled Condition

Continuing with the wall-side assumption, if φ_1 is the angle of loll, the value of GM for small inclinations about the loll position will be given by the slope of the GZ curve at that point. Now:

$$GZ = \sin \varphi(GM + \tfrac{1}{2}B_0M \tan^2 \varphi)$$

$$\frac{dGZ}{d\varphi} = \cos \varphi(GM + B_0M \tan^2 \varphi) + \sin \varphi \, B_0M \tan \varphi \sec^2 \varphi$$

substituting φ_1 for φ gives $dGZ/d\varphi = 0 + B_0M \tan^2 \varphi_1/\cos \varphi_1 = -2GM/\cos \varphi_1$.

Unless φ_1 is large, the metacentric height in the lolled position will be effectively numerically twice that in the upright position although of opposite sign.

STATICAL STABILITY (*GZ*) CURVES

By evaluating v and $h_e h_i$ for a range of angles of inclination it is possible to plot a curve of GZ against φ. A typical example is shown in Figure 5.7.

GZ increases from zero when upright to reach a maximum and then decreases becoming zero again. The curve of GZ against φ is termed the GZ *curve* or *curve of statical stability*. The main features of the GZ curve are discussed later.

Because ships are not wall-sided, it is not easy to determine the position of S and so find the volume and centroid positions of the emerged and immersed wedges. One method of dealing with this problem is illustrated in Figure 5.8. The ship is first inclined about a fore and aft axis through O on the centreline. This leads to unequal volumes of emerged and immersed wedges which must be compensated for by a bodily rise or sinkage. In the case illustrated the ship rises.

Using subscripts e and i for the emerged and immersed wedges, respectively, the geometry of Figure 5.8 gives

$$B_0R = \frac{v_e(h_eO) + v_i(h_iO) - \lambda(v_i - v_e)}{\nabla}$$

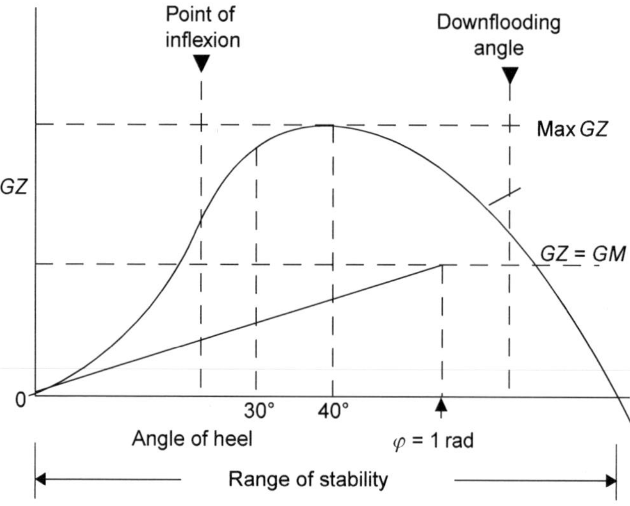

FIGURE 5.7 Curve of statical stability.

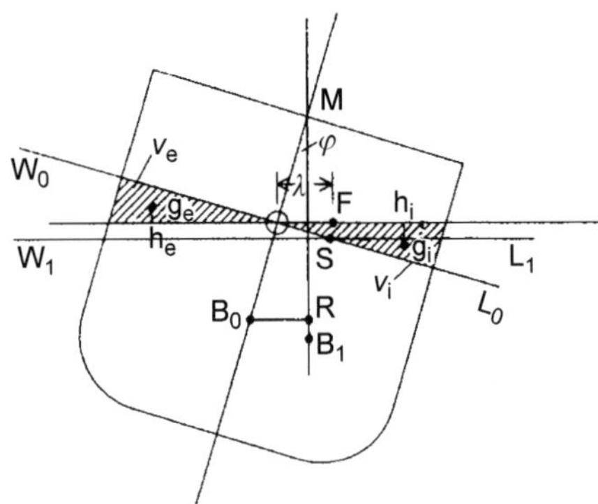

FIGURE 5.8 Determining GZ.

and

$$GZ = B_0R - B_0G \sin \varphi$$

$$= \frac{v_e(h_e\mathrm{O}) + v_i(h_iO) - \lambda(v_i - v_e)}{\nabla} - B_0G \sin \varphi$$

For very small angles GZ still equates to $GM\varphi$, so the slope of the GZ curve at the origin equals the metacentric height. That is, $GM = \mathrm{d}GZ/\mathrm{d}\varphi$ at $\varphi = 0$. It is useful in drawing a GZ curve to erect an ordinate at $\varphi = 1$ rad, equal to the metacentric height and joining the top of this ordinate to the origin to give the slope of the GZ curve at the origin.

CROSS CURVES OF STABILITY

Cross curves of stability are drawn to overcome the difficulty in defining waterlines of equal displacement at various angles of heel, particularly when deck edges are immersed and the bilges exposed.

Figure 5.9 shows a ship inclined to some angle φ about an arbitrary point, S, on the ship's centreline. S is sometimes taken at the keel (note that this S is not the same as that in Figure 5.8). By calculating, for a range of waterlines, the displacement and perpendicular distances, SZ, of the centroids of these volumes of displacement from the line YY through S, curves such as those shown in Figure 5.10 can be drawn. These are known as *cross curves of stability* and depend only upon the geometry of the ship, not upon its loading. For any given waterline they are also independent of the density of

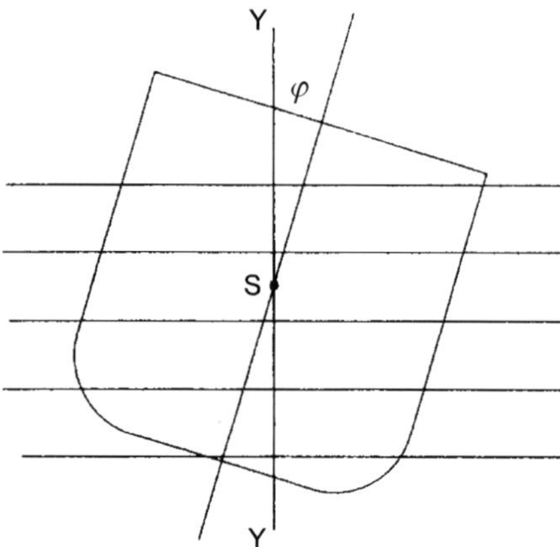

FIGURE 5.9 Deriving cross curves of stability.

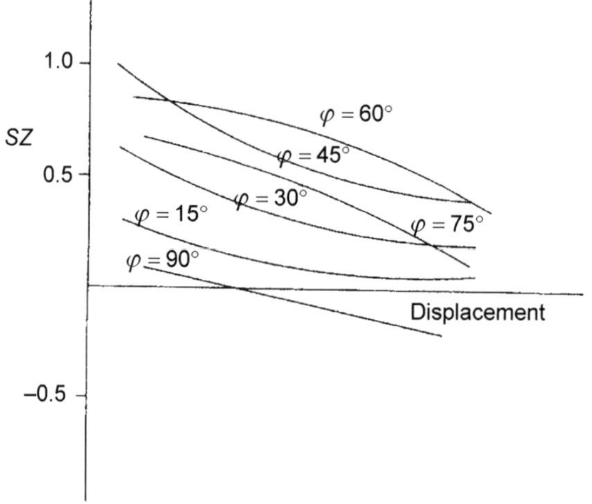

FIGURE 5.10 Cross curves of stability.

the water in which the ship is floating. They therefore apply to all conditions in which the ship may operate.

Deriving Curves of Statical Stability from the Cross Curves

For any desired displacement of the ship the values of SZ can be read from the cross curves. Knowing the position of G for that particular loading enables SZ to be corrected to GZ by adding or subtracting $SG \sin \varphi$, when G is below or above S, respectively.

Features of the Statical Stability Curve

There are a number of features of the *GZ* curve (Figure 5.7) which are useful in describing a ship's stability:

- The slope of the *GZ* curve at the origin is a measure of the initial stability *GM*.
- The maximum ordinate of the curve multiplied by the displacement gives the largest steady heeling moment the ship can withstand without capsizing.
- The angle at which *GZ* becomes zero is the largest angle from which a ship will return once any disturbing moment is removed. It is called the *angle of vanishing stability*.
- The range of angle over which *GZ* is positive is termed the *range of stability*. Important factors in determining the range of stability are freeboard and reserve of buoyancy.
- The angles of deck edge immersion and bilge emersion vary along the length of the ship. However, often they occur over a reasonable length within a small angle band. In such cases the *GZ* curve will exhibit a point of inflexion in that angle band.
- The *GZ* values at 30° and 40° help to define the curve and are involved in a number of stability criteria as will be seen later.
- There will be an angle, usually less than the angle of vanishing stability, at which significant amounts of water can enter the ship through openings in the superstructure. This is known as the *downflooding angle*.

Because of the problems in ensuring superstructure is fully watertight and because it only becomes submerged at large angles, some authorities ignore its influence on stability.

Worked Example 5.1

The angles of inclination and corresponding righting lever for a ship at an assumed *KS* of 6.5 m are:

Inclination (°)	0	15	30	45	60	75	90
Righting lever (m)	0	0.11	0.36	0.58	0.38	−0.05	−0.60

In a particular loaded condition the displacement mass is made up of:

Item	Mass (tonnes)	*KG* (m)
Lightship	4200	6.0
Cargo	9100	7.0
Fuel	1500	1.1
Stores	200	7.5

Plot the curve of statical stability for this loaded condition and determine the range of stability.

Table 5.1 Calculating GZ

Inclination (°)	sin φ	SG *sin* φ *(m)*	SZ *(m)*	GZ *(m)*
0	0	0	0	0
15	0.259	0.093	0.11	0.203
30	0.500	0.180	0.36	0.540
45	0.707	0.255	0.58	0.835
60	0.866	0.312	0.38	0.692
75	0.966	0.348	−0.05	0.298
90	1.000	0.360	−0.60	−0.240

Solution

The height of the centre of gravity is first found by taking moments about the keel:

$$(4200 + 9100 + 1500 + 200)KG = (4200 \times 6.0) + (9100 \times 7.0) + (1500 \times 1.1) + (200 \times 7.5)$$

$$KG = \frac{25200 + 63700 + 1650 + 1500}{15000} = 6.14 \text{ m}$$

Since G is below S the actual righting lever values are given by

$$GZ = SZ + SG \sin \varphi \quad \text{and} \quad SG = 6.5 - 6.14 = 0.36 \text{ m}$$

The GZ values for the various angles of inclination can be determined in tabular form as given in Table 5.1. By plotting GZ against inclination the range of stability is found to be 82°.

WEIGHT MOVEMENTS

Transverse Movement of Weight

If a weight, w, moves permanently across the ship moving horizontally through a distance h, there will be a corresponding horizontal shift of the ship's centre of gravity, $GG_1 = wh/W$, where W is the weight of the ship (Figure 5.11). The value of GZ is reduced by $GG_1 \cos \varphi$ and the modified righting arm $= GZ - (wh/W)\cos \varphi$. If the weight also moves vertically the effect of this must be included in the calculation.

The weight will not, in general, return to its original position when the ship rolls in the opposite direction. Assuming it does not, the righting lever, and righting moment, are reduced for inclinations to one side and increased for angles on the other side. If $GG_1 \cos \varphi$ is plotted on the stability curve,

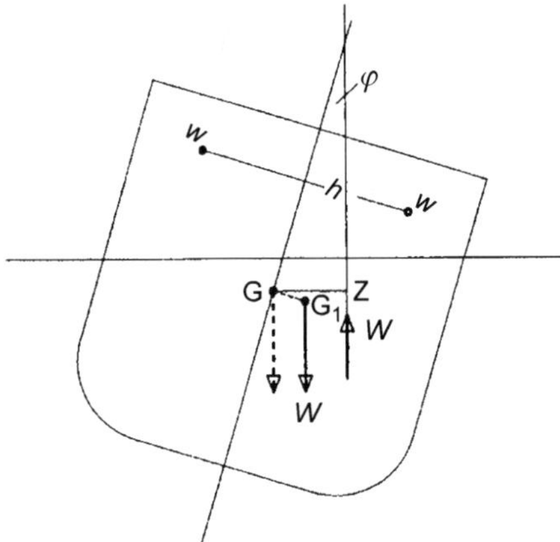

FIGURE 5.11 Transverse weight shift.

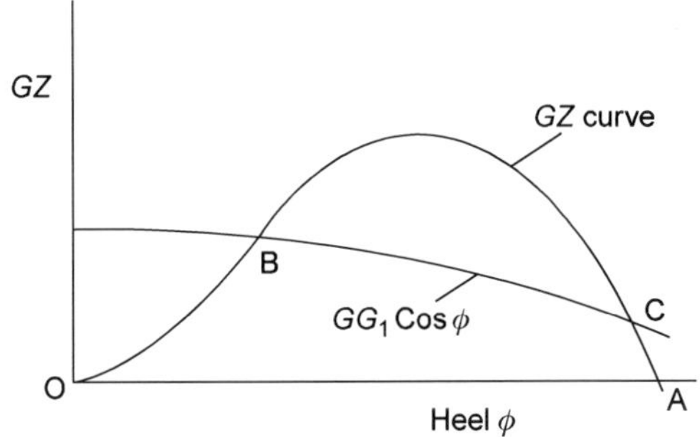

FIGURE 5.12 Modified *GZ* curve.

Figure 5.12, for the particular condition of loading of the ship, the two curves intersect at B and C. B gives the new equilibrium position of the ship in still water and C the new angle of vanishing stability. The range of stability and the maximum righting arm are greatly reduced on the side to which the ship lists. For heeling to the opposite side the values are increased but it is the worse case which is of greater concern.

Bulk Cargoes

A related situation can occur in the carriage of dry bulk cargoes such as grain, ore and coal. Bulk cargoes settle down when the ship goes to sea so that holds which were full initially have void spaces at the top. All materials of this type have an *angle of repose* – the angle the surface of the cargo can take up without sliding. If the ship rolls to a greater angle then this the cargo may move to one side and not move back later. Consequently there can be a

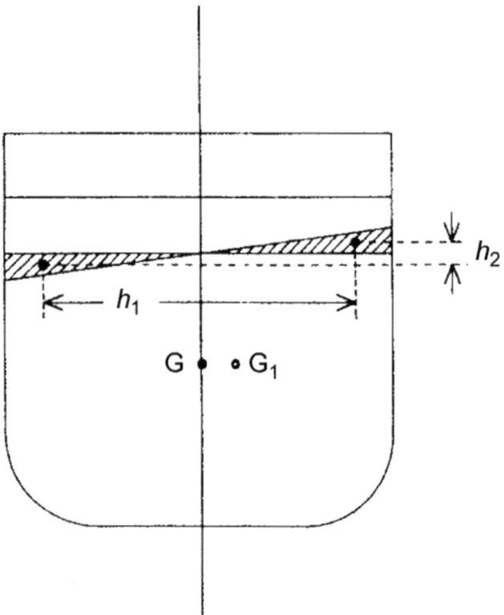

FIGURE 5.13 Cargo shift.

permanent transfer of weight to one side resulting in a permanent list with a reduction of stability on that side.

Figure 5.13 shows a section through the hold. When the cargo settles down at sea its centre of gravity is at G. If the ship rolls the cargo could take up a new position shown by the inclined line, causing some weight, w, to move horizontally by h_1 and vertically by h_2. The new surface will, in general, not be flat but is assumed to be so in this calculation. As a result the ship's G will move:

$$\frac{wh_1}{\Delta} \text{ to one side} \quad \text{and} \quad \frac{wh_2}{\Delta} \text{ higher}$$

The modified righting arm becomes

$$G_1Z_1 = GZ - \frac{w}{\Delta}[h_1 \cos\varphi + h_2 \sin\varphi]$$

where GZ is the righting arm before the cargo shifted.

Due to the settling of the cargo during the early stages of a voyage there will usually have been a slight improvement in stability compared with that on initial loading.

DYNAMICAL STABILITY

The work done in heeling a ship very slowly through an angle $\delta\varphi$ (Figure 5.14) will be given by the product of the displacement, GZ, at the instantaneous angle and $\delta\varphi$. Thus the area under the GZ curve, up to a given angle, is proportional to the energy needed to heel the ship to that angle. It is

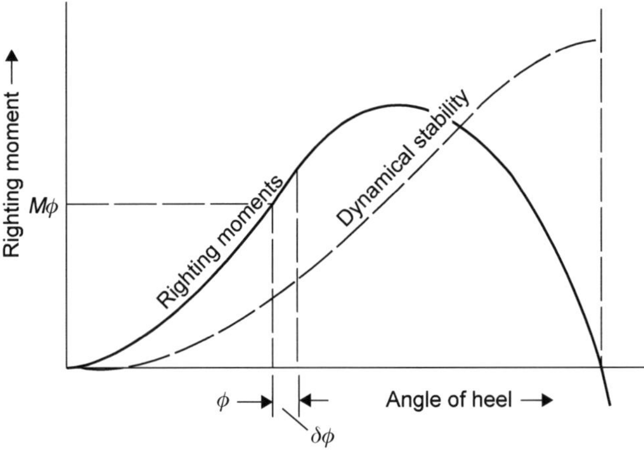

FIGURE 5.14 Dynamical stability curve.

a measure of the energy the ship can absorb from wind and waves. The plot of energy against angle of heel is known as the *curve of dynamical* stability. This energy is solely potential energy because the ship is assumed to be heeled slowly. In practice a moving ship will also have kinetic energy of roll.

Worked Example 5.2

Using the tabulated values of GZ from the previous example determine the dynamical stability of the vessel at 60° inclination.

Solution

The dynamical stability is given by

$$\int \Delta GZ \, d\varphi = \Delta \int GZ \, d\varphi$$

This integral can be evaluated, as is given in Table 5.2, using Simpson's 1,4,1 rule and the ordinate heights from Table 5.1 (associated with worked example 5.1)

Table 5.2 Calculation of Dynamical Stability

Inclination (°)	GZ *(m)*	Simpson's Multiplier	Area Product
0	0	1	0
15	0.203	4	0.812
30	0.540	2	1.080
45	0.835	4	3.340
60	0.692	1	0.692
			Summation = 5.924

$$\text{The area under the curve to } 60^\circ = \frac{15}{57.3} \times \frac{1}{3} \times 5.924$$

$$= 0.517 \text{ m rad}$$

$$\text{Dynamical stability} = 15000 \times 9.81 \times 0.517 = 76.08 \text{ M Nm}$$

EXTERNAL INFLUENCES

Wind

For a beam wind the force generated on the above water surfaces of the ship is resisted by the hydrodynamic force produced by the slow sideways movement of the ship through the water. The wind force may be taken to act through the centroid of the above water area, including the superstructure, and the hydrodynamic force as acting at half draught (Figure 5.15). For ships with high freeboard/superstructure the variation of wind speed with height above the sea surface can be allowed for. However, in view of the other approximations involved it is debatable whether this is worthwhile. For all practical purposes the two forces can be assumed equal.

Let the vertical distance between the lines of action of the two forces be h and the projected area of the above water form be A. To a first order as the ship heels, both h and A will be reduced in proportion to $\cos\varphi$.

The wind force will be proportional to the square of the wind velocity, V_w, and can be written as

$$\text{Wind force} = kAV_w^2 \cos\varphi$$

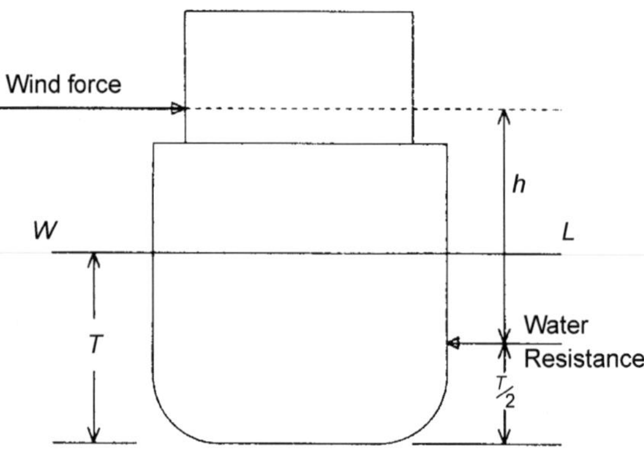

FIGURE 5.15 Heeling due to wind.

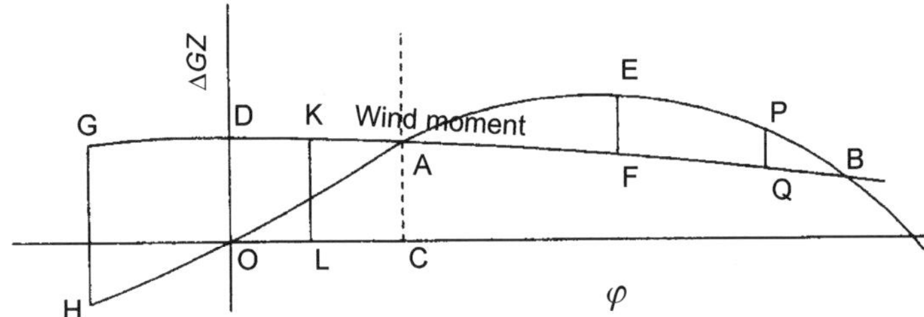

FIGURE 5.16 GZ curve with wind heeling moment.

where k is an empirical constant. The moment will be

$$M = kAhV_w^2 \cos^2 \varphi$$

The curve of wind moment can be plotted with the ΔGZ curve as shown in Figure 5.16. If the wind moment builds up or is applied slowly the ship will heel to an angle represented by A and in this condition the remaining range of stability will be from A to B. The problem would then be analogous to that of the shifted weight. On the other hand, if the moment is applied suddenly, say by a gust of wind, the amount of energy applied to the ship as it heeled to A would be represented by the area $DACO$. The ship would only absorb energy represented by area OAC and the remaining energy would carry it beyond A to some angle F such that area AEF = area DAO. Should F be beyond B the ship will capsize, assuming the wind is still acting.

A severe case for a rolling ship is if it is inclined to its maximum angle to windward and about to return to the vertical when the gust hits it. Suppose this position is represented by GH in Figure 5.16. The ship would already have sufficient energy to carry it to some angle past the upright, say KL in the figure. Due to damping this would be somewhat less than the initial windward angle. The energy put into the ship by the wind, up to angle L, is now represented by the area between the GZ and wind heeling curves between G and K. The ship will continue to heel until this energy is absorbed, perhaps reaching angle Q.

A Simple Wave Allowance

So far only the 'static' behaviour of the ship as it heels has been considered – any angular momentum is ignored. Continuing this approach, and to give some idea of the possible loss of stability in waves, it is possible to produce GZ values with the ship balanced on a wave with a length equal to the length of the ship on the waterline – first with the crest and then the trough amidships. This is still a 'quasi-static' analysis ignoring the periodicity of the GZ variation and wave forces. Both can be important as discussed later.

A *GZ* curve is produced from the mean of the two values and typically shows reductions in *GZ*.

Turning

When a ship is turning under the action of its rudder, the rudder holds the hull at an angle of attack relative to the direction of advance. The hydrodynamic force on the hull, due to this angle, acts towards the centre of the turning circle causing the ship to turn. Under the action of the rudder and hull forces the ship will heel to an angle that can be determined in a similar way to the above. Usually a ship will heel into the turn initially (i.e. to starboard for a starboard turn) and outwards once it is in a steady turn.

Icing

Increasingly ships are required to operate in areas where ice may form on exposed upperworks and rigging. This will represent a weight added high in the ship and therefore increase *KG* and lead to smaller *GZ* values. It must be allowed for in ships likely to operate in cold climates – particularly in polar regions.

STABILITY STANDARDS FOR THE INTACT SHIP

So far we have discussed stability for small and large angles; how that stability can be calculated and represented; the influence of free surfaces and icing together with some of the upsetting situations a ship is likely to experience.

Longitudinal stability is not a concern for normal ship operations as it is so much greater than the transverse. This may not be true for unconventional forms such as offshore platforms and catamarans. Such cases must be considered on their merits. The stability of planing craft, hydrofoils and surface effect craft also require special analysis because the forces supporting the weight of the craft, which will determine their stability, are at least partly dynamic in origin. In what follows attention is focused on transverse stability of intact conventional monohulls.

The stability of the ship is represented by its *GZ* curve, noting that this must include any necessary allowances for free surface effects, icing, possible reductions in a seaway and the downflooding angle.

Although only a simplified appreciation, the various features of the *GZ* curve can be compared with the design standards which are regulated for both merchant ships and warships. What, then, governs the standards to be adopted? Some stability is needed or else the ship will not float upright, but loll to one side or the other or even capsize. In theory a very small positive metacentric height would be enough to avoid this. In practice more stability is needed to allow for differing loading conditions, growth in displacement during the service life, bad weather, icing, and so on.

The designer must decide what situations to allow for in designing the ship and the level of stability needed to cope with each. Typically modern ships are designed to cope with:

- the action of winds, up to the order of 100 knots, noting they may be steady or gusting;
- the action of waves in rolling a ship;
- the heel generated in a high-speed turn;
- the possibility of ice having formed on exposed surfaces;
- the heel generated by lifting heavy weights over the side;
- the crowding of passengers to one side;
- the possible combinations of the above.

Stability standards are updated as more seagoing experience is obtained. They are quite complex with variations for different ship size and type and with a number of special circumstances to be taken into account. For these reasons reference should be made to the latest version of the regulations. Some national bodies or classification societies may have additional requirements.

An early attempt to formalise the stability standards for intact ships was won for USN warships by Sarchin and Goldberg (1962). IMO sets mandatory and recommended requirements for intact stability recognising that there is a wide variety of ships and that many operating and environmental conditions related to stability are not yet fully understood. In particular, the safety of a ship in a seaway involves complex hydrodynamic phenomena and the ship's responses must be treated as those of an elastic body excited by waves and wind. Stability criteria development poses complex problems requiring further research.

It should be noted that:

- ships which experience large righting lever variations when moving between wave trough and wave crest may experience parametric roll or a pure loss of stability;
- ships without propulsion or which lose the ability to steer may suffer resonant roll while drifting;
- in following and quartering seas ships may broach-to;
- ships should have a safe margin of stability throughout the voyage, taking account of consumption of fuel (with free surface effects) and stores and the possibility of natural phenomena such as ice build-up on rigging.

The various criteria laid down can best be understood by reference to Figure 5.17.

In Figure 5.17 the basic curve may allow for free surfaces, icing and for balancing on a standard wave (Figure 5.18).

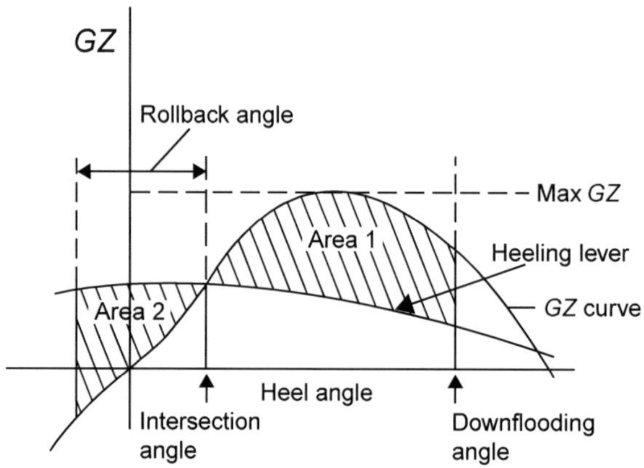

FIGURE 5.17 *GZ* and heeling lever curves.

IMO CRITERIA

The mandatory minimum criteria (maximum values are not specified) laid down by IMO for the majority of merchant ships are set out in terms of features of the basic *GZ* curve:

- The area under the *GZ* curve shall not be less than 0.055 m radians up to 30° heel; not less than 0.09 m radians up to 40° or up to the downflooding angle (if less than 40°) and not less than 0.03 m radians between these two angles.
- *GZ* must be at least 0.20 m at an angle of heel of 30° or greater.
- The maximum *GZ* must occur at an angle not less than 25°.
- The metacentric height shall not be less than 0.15 m.

IMO also lays down requirements in the form of a severe weather and rolling criterion based on the *GZ* curve. It is assumed that:

- The ship is subject to a steady wind (pressure 504 Pa) heeling lever which in this case is assumed constant for all angles of heel, rather than the cosine curve shown in Figure 5.17. This wind causes a steady heel angle which must not exceed 16° or 80% of the angle of deck edge immersion whichever is less.
- The ship suffers a roll back angle due to wave action and is then subjected to a gusting wind lever which is 50% greater than that due to the steady wind. Relative to this gusting wind lever the requirement is that area 1 should be equal to, or greater than, area 2.
- The downflooding angle is taken to be the actual downflooding angle or 50° or the intercept of the wind heeling moment with the *GZ* curve, whichever is least.

The regulations define how the wind heeling levers and the roll back angle are to be calculated from a ship's characteristics including its natural period of roll.

In addition to the above, in passenger ships the angle of heel due to:

- passengers crowding to one side must not exceed 10°. The weights and cg positions to be assumed for passengers are defined in the regulations;
- the ship's heel on turning must not exceed 10° using a defined heeling moment.

Other special requirements are laid down for oil tankers and ships carrying grain in bulk. Alternative stability criteria are quoted for ships loaded with timber deck cargoes and for high-speed craft.

Besides the mandatory requirements IMO give recommendations for certain ship types including offshore supply vessels and fishing vessels, taking into account the activities carried out.

Some Special Cases

The standards must not be applied without considering whether a particular ship may meet potentially dangerous situations in its normal activities. Three such cases are:

- *Icebreakers*: These often ride up onto the ice in order to use the ship's weight to help break the ice. In this case there will be a force acting upwards on the hull close to the bow which will be destabilising.
- *Fishing vessels*: If the vessel retrieves its catch over the side using a derrick, the weight of the catch will cause the vessel to heel as the net leaves the water. The full weight will act as though it is at the derrick head and will cause a rise in G as well as a heeling moment. When the derrick head is lifted to swing the catch inboard the destabilising effect will increase.
- Tugs. If a tug swings beam on to the towing wire it may be heeled to a large angle by the force in the wire. To reduce the effect some tugs are fitted with special towing arrangements – a radial arm or radial track – to move the load point towards the tug's low side. A heeling lever curve can be calculated in terms of the bollard pull and height of the towing hook. The IMO lays down minimum requirements in terms of areas under/between the two curves.

WARSHIP INTACT STABILITY

The requirements for warships are somewhat more severe as warships often cannot take avoiding action to avoid, or mitigate, the effects of bad weather. They must also be able to survive damage inflicted by an enemy. Briefly, the standards set out in UK MOD Defence Specifications for conventional monohull forms are:

- The area under the *GZ* curve shall not be less than 0.080 m radians up to 30° heel; not less than 0.133 m radians up to 40° or up to the downflooding angle (if less than 40°) and not less than 0.048 m radians between these two angles.
- The downflooding angle must not be less than 70°.
- The maximum *GZ* must be at least 0.30 m at an angle of heel of 30° or greater.
- Where the *GZ* curve has two maximums or the downflooding angle is excessive special conditions apply.
- The fluid metacentric height shall not be less than 0.30 m.
- The angle of vanishing stability to be as large as possible.

Allowance for Icing

For all warships icing is assumed to be with ice, density 950 kg/m^3, 150 mm thick on all exposed horizontal surfaces, decks and platforms. The resulting *GZ* curve must meet the following criteria:

- The area under the *GZ* curve shall not be less than 0.051 m radians up to 30° heel; not less than 0.085 m radians up to 40° and not less than 0.03 m radians between these two angles.
- The maximum *GZ* must be at least 0.24 m at an angle of heel of 30° or greater.
- The fluid metacentric height shall not be less than 0.15 m.

Stability in a Beam Wind

The angle of heel under the action of a steady 90 knot beam wind must not exceed 30° and the *GZ* at this angle must not be greater than 60% of the maximum *GZ*. A roll back angle of 25° is assumed. Area 1 must not be less than 1.4 times area 2 (see Figure 5.17).

Stability with Icing and Wind

Simultaneously with icing having occurred, a beam wind of 63 knots is assumed. Then the *GZ*/wind heeling plot must meet similar conditions to those quoted above for the ship without icing.

Stability in a High-Speed Turn

The heeling lever generated during a tight high-speed turn is plotted on the *GZ* curve and the following criteria must be met:

- Steady angle of heel to be less than 20°.
- *GZ* at the steady heel angle must be less than 60% of the maximum *GZ*.

- The area above the heeling lever curve, up to the downflooding angle, must be more than 40% of the total area under the *GZ* curve up to that angle.

The method of calculating the heeling lever is defined.

Loading Conditions

Possible loading conditions of a ship are calculated and information is supplied to the master. It is usually in the form of a profile of the ship indicating the positions of all loads on board, a statement of the end draughts, the trim of the ship and the metacentric height. Stability information in the form of curves of statical stability is supplied. The usual loading conditions covered are:

- the lightship,
- fully loaded departure condition with homogeneous cargo,
- fully loaded arrival condition with homogeneous cargo,
- ballast condition,
- other likely service conditions.

A trim and stability booklet is prepared for the ship showing all these conditions of loading. Nowadays the supply of much of this data is compulsory and, indeed, is one of the conditions for the assignment of a freeboard. Other data supplied include hydrostatics, cross curves of stability and plans showing the position, capacity and position of centroids for all spaces on board. These are to help the master deal with non-standard conditions.

COMMENT ON STABILITY STANDARDS

As discussed above, the current standards are based on experience with many different ships in service. However, although related to operational scenarios, they are not a direct measure of a ship's ability to survive in rough seas. Although the standards seem to be satisfactory in practice there are reasons to question them as Perrault et al. (2010) did for naval vessels. In reporting on the work of a Naval Stability Standards Working Group (NSSWG), they argued:

- The level of assurance against loss the standards provide is not known.
- They deal with a dynamic situation using essentially static ship characteristics.
- Even an intact ship may suffer a loss of stability in a seaway.
- Modern hull forms are often different from those for which the seagoing experience is available.

They considered possible capsize situations.

Static capsize can occur if an applied heeling moment exceeds the stability available, allowing for free surface effects. In an intact ship this may be due to improper loading, shifting loads or lifting heavy loads from the jetty; topside icing; towing forces or wind.

Dynamic capsize is most likely to be due to the actions of wind and waves and will occur when the forcing function exceeds the available restoring force. Possible scenarios (see Chapter 10) include sympathetic rolling, resonant excitation in a beam seas, parametric rolling in following seas and broaching in following and quartering overtaking seas. The need then is to establish criteria which take such situations into account or determine whether the features of the *GZ* curve are strongly related to whether a ship is likely to be lost in dynamic conditions.

In the NSSWG study the approach adopted was to take a group of frigates representative of past and present design. They determined the probability of each design exceeding a critical roll angle (assumed to be the same as the probability of capsize), taking account of the full range of wind and wave conditions likely to be met during the ship's life. Oceanographic data were used to give the probability of occurrence of different wave conditions. It was assumed that winds were co-directional with waves and linearly related to the significant wave height. Any voluntary reductions in speed, or change of heading, the command might take were ignored. They went on to see whether the probabilities of capsize correlated well with a number of hull form coefficients and with the 'quasi-static' features of:

- The straightforward *GZ* curve.
- The *GZ* curve with wind heeling lever superimposed leading to areas (energy measures) between the two curves up to the downflooding angle.
- a *GZ* curve using the 'seaway' righting arm. For this the ship was balanced on a wave of length, λ, equal to the ship waterline length and height $\lambda/(10 + 0.05\lambda)$. The *GZ*s for the trough and then the crest amidships were calculated and the mean of the two termed the seaway *GZ*.
- The seaway *GZ* combined with the wind heeling levers.

The study was not conclusive and work continues, but it was found that:

- The correlation with hull form parameters was not good, probably because they do not allow for the ship's inertial properties.
- There was some correlation with features of the *GZ* curve.
- Correlation was better using the 'seaway' *GZ* curve.

Model Testing of Ship Stability

ITTC 7.5-02-07-04.1 sets out a procedure for model experiments on an intact ship in waves to determine its behaviour in extreme conditions and to establish thresholds for extreme motions including capsizing. The capsizing modes of an intact ship include loss of static and dynamical stability and broaching, where loss of dynamic stability may be associated with dynamic rolling, parametric excitation, resonant excitation or impact excitation. Procedure 7.5-02-07-04.2 deals with the procedures for tests of ship stability in the damaged state.

STABILITY OF SMALL CRAFT

Because of the many types of boat, the different environments in which they are used and inconsistencies in national standards, it was not until 2002 that an international standard for stability and buoyancy of small craft (boats under 24 m in length) was established (ISO 12217). The ISO is in three parts: Part 1 for non-sailing boats of hull length ≥ 6 m; Part 2 for sailing boats of hull length ≥ 6 m and Part 3 for boats of hull length <6 m.

The need for a standard arose from the increasingly international nature of the small craft market. Blyth (2005) describes work done under an EC Directive, including tests carried out and a critique of the standards. Briefly:

- Sailing boats were defined as having a sail area/(displacement)$^{2/3} \geq 0.07$ with sail area in square metre and displacement in kilogram.
- Four categories of environmental conditions (A to D, with A the most severe) were set to which a boat can be designed.
- A Category A or B boat's ability to withstand waves and wind is based on a *GZ* curve with a wind heeling moment superimposed and a roll back angle that is a function of displacement and category.
- The required angles of vanishing stability in the *GZ* curve vary from $>100^\circ$ (Category A) to $\geq 75^\circ$ (Category D).
- A stability index (STIX) based on length multiplied by a function of seven factors (including the ability to recover from inversion unaided, ability to recover from a knockdown and the risk of downflooding) was evolved. The minimum STIX value ranged from 32 for Category A designs to 5 for Category D designs.

Blyth (2005) describes how various tests are carried out – the knockdown recovery, the wind stiffness and capsize recovery tests. He also covers multi-hull sailing boats and points out some of the shortcomings of the standards which are under review.

The UK MCA issues a Large Commercial Yacht Code – MSN 1792 (M) which includes damage stability.

FLOODING AND DAMAGED STABILITY

General

So far only the intact stability of a ship has been considered. Damage, from an internal explosion, say, may destroy some of the structure, but the overall effect on stability is unlikely to be severe. Such damage, or fire, is unlikely to sink a ship, although it may require it to be evacuated. A ship will sink only if enough water gets into the hull to destroy the reserve of buoyancy or reduce stability so that it overturns. Water may enter a ship following a collision, grounding on rocks or merely springing a leak. It is necessary to consider the amounts of transverse and longitudinal stability remaining since both may be reduced significantly. There are three scenarios:

- The ship's remaining buoyancy is inadequate to support its weight and it settles in the water; it *founders*.
- The transverse stability is inadequate and it turns over; it *capsizes*.
- The longitudinal stability is inadequate and the ship goes down by the bow or the stern; it *plunges*.

If, due to a ship's subdivision, flooding is confined to a small part of the ship the effects may not be too great. The degree of subdivision varies with the size of ship and the service. The greatest subdivision is applied to ships carrying passengers. Penetrations of decks and bulkheads between watertight zones low down in the ship are kept to a minimum and the small number of watertight doors and hatches so fitted would normally be closed at sea. Usually doors can be closed from remote positions or locally. If stability were the only consideration, zones would be numerous and small but this would make the general running of the ship difficult and a compromise is needed. Warships have more extensive subdivision so that they can better withstand enemy action.

Floodable Length

Early deterministic assessments of damaged stability (now being replaced for some ship types by probabilistic methods) illustrate the basic principles involved. The deterministic methods consider what length of ship can be flooded without loss of the ship. Loss is assumed when the damaged waterline is tangent to the *bulkhead deck* (the uppermost weathertight deck to which transverse watertight bulkheads are carried) line at side. To provide a margin, the limit is taken with the waterline tangent to a line 76 mm below the bulkhead deck at side. This line is called the *margin line*. The *floodable length* at any point along the length of the ship is the length, with that point as centre, which can be flooded without immersing any part of the margin line when the ship has no list.

Using Figure 5.18 and subscripts 0 and 1 to denote the ship data for the intact and damaged waterlines. Loss of buoyancy = $V_1 - V_0$ and this must be at such a position that B_1 moves back to B_0 so that B is again below G. Hence:

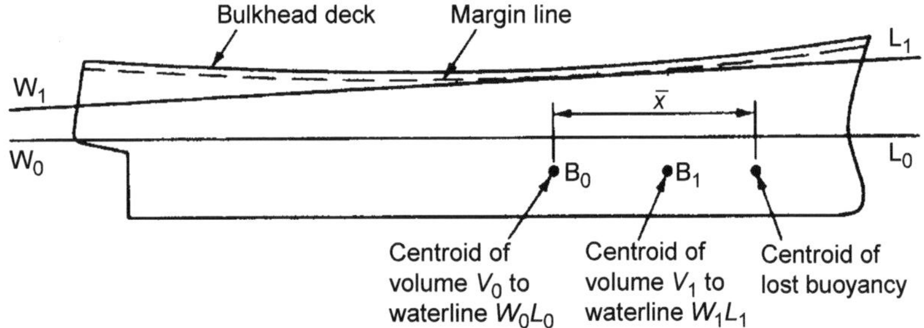

FIGURE 5.18 The margin line.

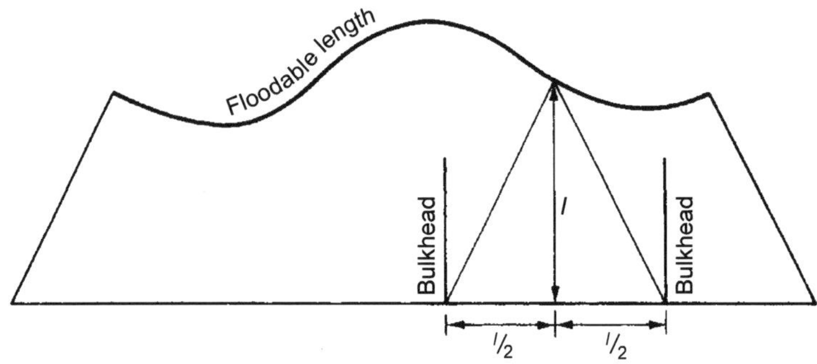

FIGURE 5.19 Floodable length.

$$\bar{x} = \frac{V_1 \times B_0B_1}{V_1 - V_0}$$

This gives the centroid of the lost buoyancy and, knowing $(V_1 - V_0)$ it is possible to convert this into a length of ship that can be flooded without immersing the margin line. The calculation is one of reiteration until reasonable figures are obtained.

The calculations can be repeated for a series of waterlines tangent to the margin line at different positions along the length. This leads to a curve of floodable length as shown in Figure 5.19. The ordinate at any point represents the length which can be flooded with its centre at the point concerned. Thus if l is the floodable length at some point the positions of bulkheads giving the required compartment length are given by setting off distances $l/2$ either side of the point. The lines at the ends of the curves, called the *forward and after terminals*, will be at an angle $\tan^{-1} 2$ to the base if the base and ordinate scales are the same. One advantage of the use of floodable length is that it provides guidance on the desirable spacing of transverse watertight bulkheads. The new probabilistic method does not provide this – it analyses the situation with a proposed bulkhead arrangement.

Permeability

Most compartments in the ship contain equipment and fittings and tanks may contain liquids. Even a void space will contain stiffeners and brackets. Thus when opened to the sea the volume that the incoming water can occupy is less than 100% of the nominal volume. The ratio of volume of water that can enter to the nominal volume of the compartment is the permeability. Typical values are given in Table 5.3. Formulae for calculating permeabilities are laid down in some regulations. Although not strictly accurate, in the absence of better data, the same values of permeability are usually applied as factors when assessing the area and inertias of the waterplane in way of damage.

The Merchant Shipping Regulations set out formulae for calculating permeabilities and a *factor of subdivision* to be applied to the floodable length curves giving *permissible length*. The permissible length is the product of the floodable length and the factor of subdivision. The factor of subdivision depends upon the length of the ship and a *criterion of service numeral* or more simply *criterion numeral*. This numeral represents the criterion of service of the ship and takes account of the number of passengers, the volumes of the machinery and accommodation spaces and the total ship volume. Broadly the factor of subdivision ensures that one, two or three adjacent compartments can be flooded before the margin line is immersed leading to what are called *one-, two- or three-compartment ships*. That is, compartment standard is the inverse of the factor of subdivision. In general terms the factor of subdivision decreases with length of ship and is lower for passenger ships than cargo ships.

For purposes of calculating stability after damage, tanks normally containing liquid are assumed to be full or empty depending upon which assumption

TABLE 5.3 Permeabilities

Space	Permeability (%)
Void watertight compartment	97 (warship) 95 (merchant ship)
Accommodation spaces	95 (passengers or crew)
Machinery compartments	85
Spaces for liquids	0 of 0.95 whichever is worse for stability
Stores	60
Dry cargo and container spaces	0,70, 0.80, 0.95
Ro–Ro spaces	0.90, 0.90, 0.95
Cargo liquids	0.70, 0.80, 0.95

Note: The three figures quoted for cargo spaces are for the ship at light service draught, the partial subdivision draught and deep subdivision draught, respectively, as defined in IMO regulations.

is more severe as there is no way of knowing, in advance, the state of the tanks at the time of the flooding.

If the hull is opened up over a sufficient length several compartments can be flooded. This was the case with *Titanic* which lost enough longitudinal stability to cause it to plunge by the bow. Any flooding can cause a reduction in stability and if this reduction becomes great enough the ship will capsize or plunge. Even if the reduction does not cause loss it may lead to an angle of heel at which it is difficult, or impossible, to launch lifeboats. The losses of buoyancy and stability due to flooding are considered in the following sections.

A major consideration for any ship, but particularly important for one carrying large numbers of people, is the ability to get everyone off safely in the event that the master has to order 'abandon ship'. This means not only having boats and rafts capable of taking all on board but the ability to launch them and get people into them. This may require limiting the angles the ship may take up when damaged, marking escape routes clearly and the provision of chutes to get people from the ship. Time is a major consideration and allowance must be made for the time from initial damage to the acceptance that the ship has to be abandoned – by the passengers as well as the master. Passengers may feel disorientated if woken from deep sleep and some may panic or act irrationally. Some may not be physically fit. The current thinking is that the ship itself is the best lifeboat.

Sinkage and Trim When a Compartment Is Open to the Sea

Suppose a forward compartment is open to the sea (Figure 5.20). The buoyancy of the ship between the containing bulkheads is lost and the ship settles in the water until it picks up enough buoyancy from the rest of the ship to restore equilibrium. The position of the *LCB* moves and the ship must trim until G and B are again in a vertical line. The ship which was originally floating at waterline W_0L_0 now floats at W_1L_1. Should W_1L_1 be higher at any point then the deck at which the bulkheads stop (the *bulkhead deck*) it is usually assumed that the ship would be lost as a result of the water pressure in the damaged compartment forcing off the hatches and leading to unrestricted flooding fore and aft. In practice the ship might still remain afloat for a considerable time.

Allowance must be made for the permeability of compartments.

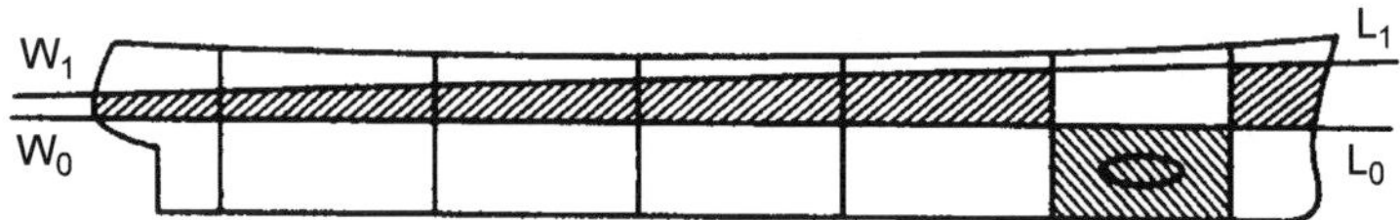

FIGURE 5.20 Compartment open to the sea.

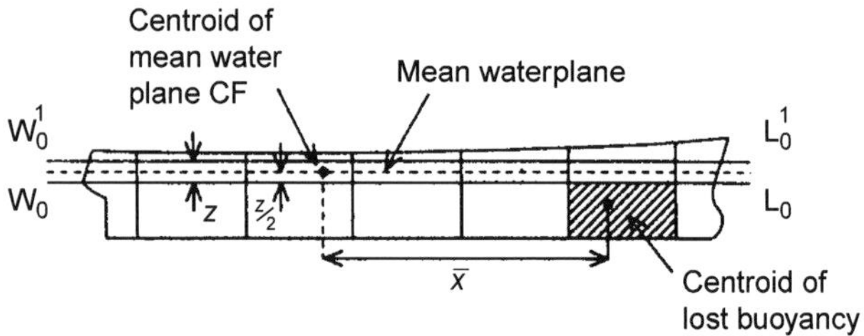

FIGURE 5.21 Lost buoyancy method.

To calculate the damaged waterline successive approximation are needed as the assumptions of small changes will not apply. There are two approaches: the *lost buoyancy* and *added weight methods*. These will give different *GM* values but the same righting moment. The final draughts must be the same, whichever of the two methods is used.

Lost Buoyancy Method

First the volume of the damaged compartment (Figure 5.21) up to the original waterplane and the area of waterplane lost are calculated making allowance for the permeability. Suppose the area of original waterplane is A and the area lost is μa, where μ is the permeability. Let the lost volume of buoyancy be μv. A first approximation to the parallel sinkage suffered is given by

$$z = \frac{\mu v}{A - \mu a}$$

A second approximation will usually be needed because of the variations in waterplane area with draught. This can be made by taking the characteristics of a waterplane at sinkage $z/2$. The longitudinal centre of flotation and the moment to change trim (MCT) can be calculated for this intermediate waterplane, again allowing for the permeability. Using subscript m to denote the values for the intermediate waterplane:

$$\text{Sinkage} = \frac{\mu v}{A_m} \quad \text{and} \quad \text{trim} = \frac{\mu v \bar{x}}{MCT_m}$$

where $\bar{x}$ is the centroid of the lost volume from the *CF*.

The new draughts can be calculated from the sinkage and trim. A further approximation can be made if either of these is very large, or the results can be checked from first principles using the Bonjean curves allowing for the flooding and permeability.

In the lost buoyancy method the position of G remains unaltered unless the damage has been so severe as to remove structure or equipment from the ship. It is the method usually used in damage stability calculations.

Added Weight Method

In this method the water entering the damaged compartment is regarded as an added weight. Permeability would have to be allowed for in assessing this weight, and allowance must be made for the free surface of the water that has entered, but all the hydrostatic data used are those for the intact ship. Initially the calculation can proceed as for any added weight, but when the new waterline is established allowance must be made for the extra water that would enter the ship up to that waterplane. Again further iterations may be needed.

In the description of both methods it has been assumed that the compartment that has been breached extends above the original and the final waterlines. If it does not then the actual floodable volumes must be used, and the assumed waterplane characteristics amended accordingly. Clearly it is highly desirable that the ship have a reasonable amount of potential buoyancy above the intact waterplane as a 'reserve' – the *reserve of buoyancy*.

Worked Example 5.3

A vessel of constant rectangular cross section is 60 m long and 10 m wide. It floats at a level keel draught of 3 m and has a centre of gravity 2.5 m above the keel. Determine the fore and aft draughts if an empty, full width, fore-end compartment 8 m long is opened to the sea. For simplicity a permeability of 100% is assumed.

Solution

Lost Buoyancy Method

Area of intact waterplane, $A = 52 \times 10 = 520\ \text{m}^2$.

Volume of lost buoyancy, $v = 8 \times 10 \times 3 = 240\ \text{m}^3$.

Parallel sinkage, $z = 240/520 = 0.46$ m.

The vessel now trims about the new centre of flotation, F_1 from amidships. Taking moments about amidships, and using subscript 1 to denote damaged values:

$$(60 \times 10 \times 0) - (8 \times 10 \times 26) = [(60 \times 10) - (8 \times 10)]F_1, \ \text{giving}\ F_1 = -4\ \text{m}$$

That is, the centre of flotation is 4 m aft of amidships or 30 m aft of the centroid of the damaged compartment.

$$KB_1 = \frac{T_1}{2} = \frac{3 + 0.46}{2} = 1.73\ \text{m}$$

$$B_1M_L = \frac{I_L}{\nabla} = \frac{1}{12} \times \frac{52^3 \times 10}{60 \times 10 \times 3} = 65.10\ \text{m}$$

$KG = 2.50$ m (constant).

$GM_L = 1.73 + 65.10 - 2.5 = 64.33$ m.

$$\text{Hence MCT 1 m} = W \times GM_L/L$$

$$= \frac{60 \times 10 \times 3 \times 1.025 \times 9.81 \times 64.33}{60}$$

$$= 19406 \text{ kNm}$$

$$\text{Trim} = \frac{\rho g v \bar{x}}{MCT1\text{ m}} = \frac{1.025 \times 9.81 \times 240 \times 30}{19406} = 3.73\text{ m}$$

$$\text{Thus draught aft} = 3 + 0.46 - \frac{26 \times 3.73}{60} = 1.84\text{ m}$$

$$\text{Draught forward} = 3 + 0.46 + \frac{34 \times 3.73}{60} = 5.57\text{ m}$$

Added Mass Method

$$\text{Mass added at 3 m draught} = 8 \times 10 \times 3 \times 1.025$$

$$= 246\text{ tonnes [364.9]}$$

$$\text{Parallel sinkge} = \frac{246}{1.025 \times 60 \times 10} = 0.4\text{ m [0.593]}$$

$$\text{New displacement mass} = 60 \times 10 \times 3.4 \times 1.025$$

$$= 2091\text{ tonnes [2210]}$$

$$KB_1 = \frac{3.4}{2} = 1.7\text{ m [1.797]}$$

$$BM_1 = \frac{I_L}{\nabla} = \frac{1}{12} \times \frac{(60^3 - 8^3) \times 10}{60 \times 10 \times 3.4} = 88.0\text{ m [83.3]}$$

$$KG_1 = \frac{(60 \times 10 \times 3 \times 1.025 \times 2.5) + (246 \times 1.5)}{2091}$$

$$= 2.38\text{ m [2.45]}$$

$$\text{MCT 1 m} = \frac{2091 \times 9.81 \times (1.7 + 88.0 - 2.38)}{60}$$

$$= 29850\text{ kNm [29860]}$$

$$\text{Trim} = \frac{246 \times 9.81 \times 26}{29850} = 2.10\text{ m [3.12]}$$

$$\text{Thus draught aft} = 3 + 0.4 - \frac{2.10}{2} = 2.35\text{ m}$$

and

$$\text{Draught forward} = 3 + 0.4 + \frac{2.10}{2} = 4.45\text{ m}$$

A second calculation considering the mass of water entering at 4.45 m draught forward will give a trim of 3.12 m and draughts of 2.03 m aft and 5.15 m forward. Results of the intermediate steps in the calculation are given in square brackets above. A third calculation yields draughts of 1.88 m aft and 5.49 m forward.

In this case since a rectangular body is involved the draughts can be deduced directly by simple calculation using the lost buoyancy approach and treating the underwater fore and aft sections as trapezia. The body effectively becomes a rectangular vessel 60 m long (but with buoyancy only over the aftermost 52 m) by 10 m wide with an *LCG* 30 m from one end and the *LCB* 26 m from aft. It will trim by the bow until the *LCB* is 30 m from aft. It will be found that the draught aft = 1.863 m and the draught 52 m forward of the after end = 5.059 m. The draught right forward will be

$$1.863 + (5.059 - 1.863) \times \frac{60}{52} = 5.551 \text{ m}$$

STABILITY IN THE DAMAGED STATE

Consider first the lost buoyancy method and the metacentric height. The effect of the loss of buoyancy in the damaged compartment is to remove buoyancy (volume v) from a position below the original waterline to some position above this waterline so that the centre of buoyancy will rise. If the vertical distance between the centroids of the lost and gained buoyancy is bb_1 the rise in centre of buoyancy is given by

$$\text{Rise} = \mu v bb_1/\nabla$$

BM will decrease because of the loss of waterplane inertia in way of the damage. If the damaged inertia is I_d:

$$BM_d = I_d/\nabla$$

The value of *KG* remains unchanged so that the damaged *GM*, which may be more, but is generally less, than the intact *GM* is

$$\text{Damaged } GM = GM \text{ (intact)} + \frac{\mu v bb_1}{\nabla} - \text{BM (intact)} + \frac{I_d}{\nabla}$$

If the added weight method is used then the value of *KG* will change and the height of M can be found from the hydrostatics for the intact ship at the increased draught. The free surface of the water in the damaged compartment must be allowed for.

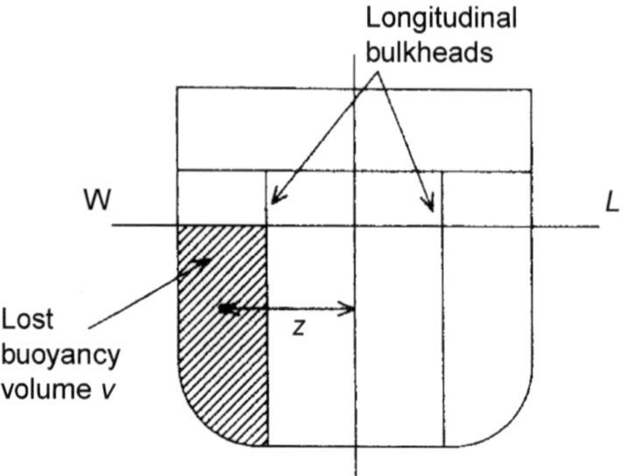

FIGURE 5.22 Asymmetrical flooding.

A GZ curve can be produced for the damaged ship but the calculations are more difficult than those for the intact ship.

Asymmetrical Flooding

When there are watertight longitudinal bulkheads in the ship the flooding will not extend right across the ship causing the ship to heel. In deciding whether a longitudinal bulkhead will be breached it has been usual to assume that damage does not penetrate more than 20% of the breadth of the ship. In more recent approaches statistical data is used instead of a fixed percentage. Taking the case illustrated in Figure 5.22 and using the added weight approach, the ship will heel until:

$$\rho g\, \nabla GM \sin\varphi = \mu\rho g v z \quad \text{or} \quad \sin\varphi = \frac{\mu v z}{\nabla GM}$$

As with the calculation for trim, this first angle will need to be corrected for the additional weight of water at the new waterline, and the process repeated if necessary.

Large heels should be avoided and the crew will take corrective action to counter it perhaps by flooding a compartment on the opposite side of the ship. This is termed *counterflooding*. The ship will sink deeper in the water, but this is usually a lesser danger than that posed by the heel.

THE PROBABILISTIC APPROACH

The earlier deterministic methods for assessing damage stability assumed that a ship was equally likely to be holed anywhere along its length. Standard penetration widths and depths were assumed in calculating the flooding that would be likely to occur between transverse bulkheads. Analysis of historical data from past incidents has enabled designers to assess the probability of the ship being holed at different points along their length and the probability distributions of the depth and width of the resulting penetration.

Such analyses mean the designer can establish the most likely damage scenarios and calculate the damage stability for these. When calculations were carried out manually there was time for only a few (possibly even only one) damage cases to be investigated. The cases taken would be those judged the most dangerous, taking account of their likelihood and consequences. With the increasing power of computers many more cases (some thousands in the case of a passenger ship) can be examined, and guidance given to the master covering a number of possible scenarios. The greatly increased computer power has also enabled consideration to be given to the progress of flooding with time (through ventilation trunks, say, or through non-watertight structure) rather than just consider the final equilibrium position. In other words the probabilistic approach, now adopted by IMO, became possible.

Probabilistic methods of damage stability assessment were introduced in 1992 for dry cargo ships of length greater than 100 m. Then in 2009 the latest IMO (SOLAS 2009) regulations were brought in with two versions, one for dry cargo ships of 80 m and longer and the other for passenger ships. Some special types (e.g. offshore supply vessels and drilling rigs) are still excluded and the deterministic method is used for these. The latest version of the regulations should be consulted as, apart from periodic changes, they are quite complex and only a brief outline can be given here.

The new regulations are based on a probabilistic concept – see Dodman (2010) – which uses the probability of survival after collision as a measure of a ship's safety in a damaged condition and leads to an *attained subdivision index*, A. The philosophy is that two different ships with the same attained index are of equal safety and that there is no need for special treatment of specific parts of the ship, even if they are able to survive different damages. The only areas given special attention in the regulations are the forward and bottom regions, which are dealt with by special subdivision rules provided for cases of ramming and grounding. The regulations for passenger ships include some deterministic elements to ensure it is not unduly vulnerable to minor damage at some point along its length.

The margin line is no longer used as a criterion for acceptable flooding. These new regulations use horizontal escape routes, openings (vents, doors, and so on) and emergency control spaces to determine the maximum acceptable waterplane levels both after and during flooding. This is a more realistic representation of the real-life situation. It should be appreciated that the experience of the maritime industry with these regulations is limited. There are still concerns and uncertainties regarding the interpretation of the new regulations and the database on which they rely. These are under constant review by IMO and a significant research programme is under way to throw more light on contentious issues. IMO has no funding for research directly, but it encourages member states to carry it out.

The Attained Subdivision Index, *A*

There are many factors that will affect the final consequences of hull damage to a ship, including the actions of the crew. These factors are random and their influence differs from ship to ship. The mass and velocity of the ramming ship is such a random variable. Other factors include:

- which particular space or group of adjacent spaces is flooded;
- the draught, trim and intact stability at the time;
- the permeability of flooded spaces which may vary during a voyage;
- the sea state at the time and for a time after the accident;
- possible heeling moments acting;
- the effect of hull strength on penetration in the case of ramming or grounding, and the consequent damage to that structure.

Although damage is random, the probability of flooding a given space can be determined if the probability of occurrence of certain damages is known from experience, i.e. damage statistics. Such statistics are not as extensive as desirable to give good probabilities of some factors and those available must be analysed to reflect ship type and size. Older statistics would not be relevant to modern ships because of design changes over the years.

The probability of flooding a space is given by the probability of occurrence of all damages which open that particular space to the sea. Because of the mathematical complexity and insufficiency of data, it is not practicable to make an exact or direct assessment of their effect on the probability that a particular ship will survive a random damage if it occurs. However, with some qualitative judgements, a logical treatment may be achieved using the probability approach as the basis for a comparative method for the assessment of ship safety. Using probability theory the probability of ship survival can be calculated as the sum of probabilities of its survival after flooding each single compartment, each group of two, three or more, adjacent compartments multiplied, respectively, by the probabilities of occurrence of such damages. This leads to an attained subdivision index, A, as a measure for the ship's ability to sustain collision damage.

The probability that a ship will remain afloat without sinking or capsizing as a result of an arbitrary collision in a given longitudinal position breaks down to the probability that:

- the longitudinal centre of damage occurs in just the region of the ship under consideration;
- this damage has a longitudinal extent that only includes spaces between the transverse watertight bulkheads found in this region;

- the damage has a vertical extent that will flood only the spaces below a given horizontal boundary, such as a watertight deck;
- the damage has a transverse penetration not greater than the distance to a given longitudinal boundary;
- the watertight integrity and the stability throughout the flooding sequence is sufficient to avoid capsizing or sinking.

By grouping these probabilities, calculations of the probability of survival, or attained index A, have been formulated to include the following probabilities:

- of flooding each single compartment and each possible group of two or more adjacent compartments;
- that the stability after flooding will be sufficient to prevent capsizing or dangerous heeling due to loss of stability or to heeling moments in the intermediate or final stages of flooding.

Producing the index requires the calculation of various damage scenarios defined by the extent of damage and the initial loading conditions of the ship before damage.

Three loading conditions are specified, with different draughts and trim. To produce the overall index the results for these three conditions are weighted as follows:

$A = 0.4A_s + 0.4A_p + 0.2A_l$, where subscripts s, p and l denote the three loading conditions, and:

A_s = the attained index at deep subdivision draught
A_p = the attained index at partial subdivision draught
A_l = the attained index at light service draught.

The index A for each loading condition is expressed by

$$A = \sum p_i[v_i s_i]$$

where

- p_i is the probability of a compartment, or group of compartments (forming zone 'i') being flooded, disregarding any horizontal subdivision, but taking transverse subdivision into account. p depends only on the geometry of the ship and its subdivision;
- s_i is the probability of survival after flooding zone 'i';
- v_i depends on the geometry of the horizontal watertight arrangement (decks) of the ship and the draught in the initial loading condition. It represents the probability that the spaces above the horizontal subdivision will not be flooded;
- the summation for A is carried out over the subdivision length of the ship – basically this is the length embracing the buoyant hull and the reserve of buoyancy.

The regulations give detailed guidance on the calculation of these factors. The summation is carried out over the full range of damage scenarios. Ideally this would be all possible scenarios but in practice the number of cases considered is less as some make insignificant contributions to A and others exceed the maximum possible damage length. Longitudinal subdivision within the zone will result in additional flooding scenarios, each with its own probability of occurrence.

A few deterministic elements, which were necessary to make the concept practicable, have been included in the regulations.

STABILITY STANDARDS FOR THE DAMAGED SHIP

The probabilistic methods of calculating the stability of a ship after damage have been outlined as well as the earlier approaches. It remains to consider the standards of that stability that should be aimed at in the damaged state.

Stability Standards Based on SOLAS (1960)

As an illustration of standards applying to passenger ships (many existing ships will have been designed using these standards) under the earlier IMO regulations, SOLAS 1960 required of the ship, in its final condition following flooding, that:

- in the case of symmetrical flooding, the ship should have a positive residual metacentric height of at least 0.05 m, as calculated by the constant displacement method;
- for asymmetrical flooding the heel angle should not exceed 7°, except that, in exceptional circumstances, the administration might allow up to 15°;
- in no case should the margin line be submerged in the final stage of flooding. The possibility of the margin line being submerged at an intermediate stage was to be considered.

The Required Subdivision Index, *R*

With the new probabilistic approach the stability standard demanded of the damaged ship is represented by a *required subdivision index, R*. R is a comparative index used to define a certain level of safety. It is based on calculations of A values for existing ships which met the previous minimum standards of damage stability. The attained index, A, is difficult to interpret in real terms. Generally the larger the index (i.e. the closer it is to unity) the safer the ship. Since, however, A is based on a ship's ability to survive a whole range of possible damage scenarios it is in effect an 'average' value. For a particular case of damage a ship with a given value of A may not be safer than a ship with a lower A value. The design of the latter may be such as to make it less vulnerable to the particular damage case being considered. This is one reason why the regulations include a deterministic minor damage requirement on top of the probabilistic regulations for passenger ships to avoid unacceptably vulnerable spots in some part of their length. In such cases the currently accepted minimum levels of heel and residual stability must be met.

As a stability standard the regulations require that the attained index must be above a required subdivision index, R. R is a relatively simple formulation based on ship length and the number of people carried.

Other factors that might be important for certain ship types can be included if deemed necessary. A value of 1.0 cannot be required since no ship can be completely safe. This was dramatically shown by the loss of the *Titanic* in 1912 and reinforced in 2012 by the *Costa Concordia* incident. It has been suggested that values in excess of 0.9 should be achievable and it is likely that the required R values will increase as more experience is gained and better designs emerge.

One complication in applying the probabilistic methods is that the damage stability calculations have to be carried out at a relatively early design stage. For instance, the maximum head of water after damage is required as an input to strength calculations for bulkheads. In the deterministic approach it would have been obvious to the designer if later changes in design were likely to affect the results obtained from earlier calculations – a change in watertight bulkhead spacing, say. The designer needs to be more vigilant now as so many other design features (trunking, valves, escape routes and so on) are involved.

Application to Cargo Ships

Gullaksen (2011) describes an investigation into cargo ships more than 100 m in length. For these R is given by

$$R = 1 - 128/(L_s + 152)$$

where L_s is the subdivision length.

The attained index, A, is based on the three loading conditions as described above. The subdivision length is divided into a number of discrete damage zones which determine the specific damages to be calculated. Although there are no rules for subdividing it must be done carefully to achieve a good result. All zones and combinations of zones contribute to A. The p factors are modified by an r factor (defined in the regulations) if there is longitudinal subdivision and the distance of that division from the shell. The s factors for the three draughts are calculated using the damaged righting arm curve characteristics. The formula used for calculating s is

$$S_{\text{final,i}} = K[(GZ_{\max}/0.12)(\text{range}/16)]^{0.25}$$

where

$GZ_{\max}$ is not to be taken as more than 0.12 m.
Range is not to be taken as more than 16°.

The value of K is taken as

$$1 \text{ if } \theta_e \leq \theta_{\min};$$
$$0 \text{ if } \theta_e \leq \theta_{\max}$$

or otherwise as

$$K = [(\theta_{\max} - \theta_e)/(\theta_{\max} - \theta_{\min})]^{0.5}$$

where

θ_e is the equilibrium heel angle at any stage of flooding, degrees.
For cargo ships $\theta_{\min}$ is 25° and $\theta_{\max}$ is 30°.

Gullaksen goes on to describe how the ν factors are calculated, ν being a reduction factor depending on the horizontal subdivision (decks) and the initial draught. It will be appreciated from this brief summary that the calculations involve many factors and the full IMO regulations should be consulted to ensure that the assessments are carried out properly. Software programmes are available to help.

On Board Data Concerning Damage Stability

Documents are provided to ships' masters so that, in the event of damage, flooding can be limited and steps taken to restore some of the lost stability. Included are damage control plans and booklets. The plans include information on watertight subdivision, external and internal hull openings and equipment provided to maintain compartment boundaries. They show penetration lines and data on limiting progressive flooding. The booklets include stability limiting values, cross flooding and downflooding arrangements and summaries of damage stability calculations. Software and computers for carrying out calculations may be carried.

DAMAGE STABILITY STANDARDS FOR WARSHIPS

Warships are not subject to IMO regulations. Generally their damage stability will be studied as part of a much wider vulnerability assessment taking into account the likely forms of attack. As well as sustaining damage a warship must be able to operate in an environment in which nuclear, biological and chemical contamination is a possibility. To protect it various defensive measures are provided. These include marking closures to indicate which should be closed when the threat reaches certain levels. As far as stability following damage is concerned the standards adopted by the UK Royal Navy are set out in MOD Defence Standards. In essence:

- the initial loading condition of the ship and the transverse and vertical extent of damage are assumed to be those which result in the least stability after damage;
- the extent of damage (damage length and number of compartments flooded) assumed is specified and is dependent upon the ship length;
- permeabilities are specified.

A *GZ* curve is prepared for the damaged ship with a wind heeling moment superimposed upon it; the wind speed specified varying with displacement. The curves must meet certain criteria, including:

- any angle of loll or list must be less than 20°;
- the *GZ* at the point where the wind heeling moment intercepts the *GZ* curve must be less than 60% of the maximum *GZ*;
- the minimum area under the *GZ* curve and above the wind heeling lever, from the point they cross up to the downflooding angle is specified;
- longitudinal *GM* must be positive and any trim must not lead to downflooding.

Associated with warship damage stability the watertight subdivision is of a high standard. Points relating to the design include:

- a *damage control deck* which is the lowest deck on which continuous fore and aft access is provided. Access is by watertight doors. Below this deck no openings are permitted in the main watertight bulkheads;
- a *red risk zone* which is that part of the ship at immediate risk of flooding after damage;
- a set of *V-lines* defining that part of the ship at some risk, not immediate, of flooding;
- freeing ports are to be provided in bulwarks to prevent water being retained on exposed decks.

The Red Risk Zone

This zone (Figure 5.23) is defined by the highest level of flooding following damage. The deepest centreline immersion is found for each bulkhead when it bounds a flooded volume determined from applying the stability standards using the deep displacement (allowing for growth during ship life) and the maximum extents of flooding. These points of deepest immersion are plotted on the ship's profile leading to an envelope of damaged waterlines. Using this envelope, the red risk zone is defined by *red risk lines* drawn from the centreline point at an angle to the horizontal equal to the static heel due to damage

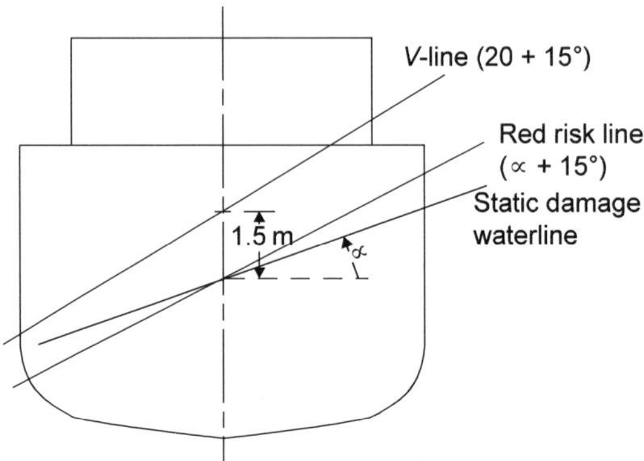

FIGURE 5.23 *V* and red risk lines.

plus 15° as a dynamic allowance for transient heel angles and rolling in waves.

In the red risk zone all watertight closures need to be closed rapidly in the event of damage. Some closures will already be closed depending upon the NBCD state at the time.

The V-Lines

These lines are drawn through a point on the centreline 1.5 m above the point at which the red risk line intersects it. This distance is to allow for the motion of the vessel in waves. The lines are drawn at an angle of 35° to the horizontal. The 35° is the sum of a heel angle of 20° following damage and a 15° dynamic allowance. In the associated damage stability calculations the *GZ* curve is terminated at 45° or the downflooding angle whichever is less.

The ship's watertight structure should extend up to the V-lines, but watertight access and system penetrations are allowed. This helps in deciding where fore and aft access and systems are provided.

Freeing Ports

If bulwarks are fitted, their area is specified in terms of the length of bulwark and its height.

Marshall (2011) deals with the damage stability of naval ships and includes application of IMO standards to these ships. He points out that warships are designed to absorb substantial damage before becoming non-operational. This represents an additional capability over and above ship safety in normal operations – the accidental damage suffered by merchant ships as well. Figure 5.24 shows the different subdivision resulting from applying SOLAS 2009 and from applying the Naval Ship Code.

It has generally been assumed that designing to the naval code is adequate to cover accidents such as collision or grounding. The grounding

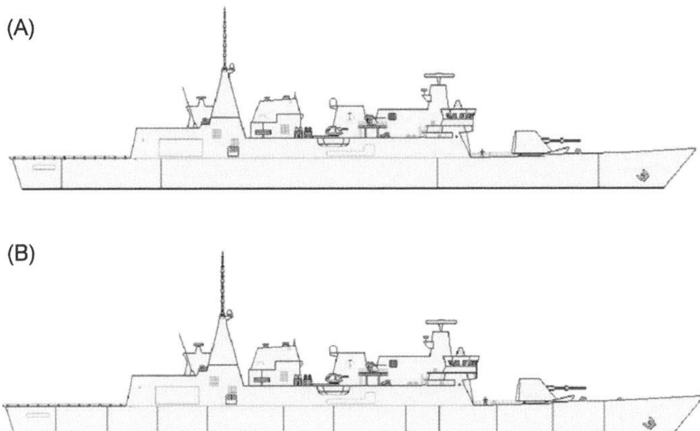

FIGURE 5.24 Comparison of standards: (A) subdivision to SOLAS standards and (B) subdivision to warship standards. *Courtesy of RINA.*

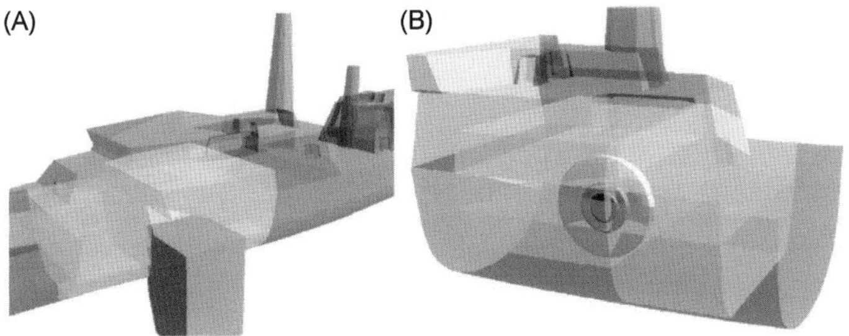

FIGURE 5.25 Accidental and hostile damage templates: (A) collision and (B) above water attack. *Courtesy of RINA.*

of *HMS Nottingham* (2002), when nine watertight zones were damaged, showed that this is not necessarily true.

Marshall goes on to consider a range of 'loss templates' reflecting a ship's ability to survive, both in terms of stability and strength. These templates (Figure 5.25) are applied within the Paramarine CAD system (see Chapter 14) and are of two types:

- Accidental damage templates – collision and grounding.
- Hostile damage templates – for above and below water attacks.

CONTINUING WORK

The IMO regulations give guidance on the process of damage stability analysis which aids the approval process. A damage stability analysis provides proof of the damage stability standard required for the respective ship type.

The guidance provided includes such things as calculating p and s values in the equation for A; the treatment of longitudinal subdivision; penetration of wing compartments; treatment of pipes and valves and progressive flooding. Following this guidance leads to a useful comparative measure of the merits of the longitudinal, transverse and horizontal subdivision of a ship. However, at present, to make the probability concept practicable, many simplifications are necessary and it is not possible to calculate the exact probability of survival.

A fuller description of the damage is restricted by the presently available damage statistics and our knowledge of progressive flooding in assessing the probability that a ship will not capsize or sink during intermediate stages of flooding. As more historical data become available, and more research is done, this situation will improve.

IMO is putting considerable effort into improving the regulations and reducing, or removing, some of the present limitations. Such work includes developing guidance on the impact of open watertight doors. IMO recognise that future large passenger ships should be designed on the basis that they are their own best lifeboat and are investigating their design such that when damaged in a seaway they can return to port or survive for 3 h to allow a safe evacuation.

Gullaksen (2011) and Hutchinson et al. (2011) are interesting commentaries on the application of the IMO probabilistic method. The latter deals with cargo vessels over 100 m in length.

Roll on/Roll off (Ro–Ro) Vessels

Ro-Pax ships (Ro–Ro passengers ships) often have long holds low in the ship which can be susceptible to flooding. This flooding can create a large free surface effect and generate an adverse trim. The ability of such designs to survive damage depends greatly upon the reserve of buoyancy and its longitudinal distribution.

The special problems of Ro–Ro ships, with their large open decks for vehicle stowage, were highlighted by the loss of the *Herald of Free Enterprise* (1987) and the *Estonia* (1994). This led to what became known as the Stockholm Agreement which required damage stability assessments for such ships to:

- specifically allow for water on the Ro–Ro deck;
- link damage survivability to operational sea states;
- encouraged first principle approaches to ship safety.

This provided a satisfactory answer in conjunction with SOLAS 90 deterministic standards. In principle the 2009 probabilistic methods should cover these ships as their special features are covered by the survivability function, s. Opinions on that are not unanimous, however, and more work is being carried out on this topic. Features of SOLAS 2009 to be considered are the change in allowed inboard position of longitudinal bulkheads (from B/5 to B/10) and the way the required safety index, *R*, varies with the number of passengers. Ships with fewer passengers have a lower *R* value which may be less demanding than the earlier requirements. The European Union currently requires new Ro-Pax ships to meet the Stockholm Agreement in addition to SOLAS 2009 (EU Directive 2003/25/EC, as amended). The European Maritime Safety Agency (EMSA www.emsa.europa.eu) (2009) published a report on the 'Damage Stability of RoPax Vessels'.

Tankers

Tankers also present a special problem, as discussed by Hutchinson et al. (2011). It arises from the so-called fluid deadweight dropout. Tanks may initially be filled to different levels with cargoes of different density. If the oil is denser than seawater the ship can lose weight on the damaged side and heel away from the damage. Combined with the number of tanks, their state at time of the incident and possible damage to side and bottom of the ship, it is difficult to show compliance with the damage stability criteria. For some ships design assessments can be made against predictable loading patterns. If the ship is loaded in a different manner then strictly speaking a new calculation should be made and approved before the ship sails.

SUMMARY

A ship must have adequate stability to carry out its normal operations and to survive a reasonable amount of flooding following damage. There are methods for calculating a number of criteria related to a ship's stability for small and large inclinations in the intact state. These criteria are compared with accepted standards based on those criteria. These standards, however, do not directly show the degree of risk of losing a ship. Probabilistic methods are now used to assess a ship's damage stability. Much has been done to improve our understanding of the dynamic behaviour of ships at sea and the effects of progressive flooding following damage. More is needed. A number of ship types, including Ro–Ro passenger ships still need special consideration.

An inclining experiment is conducted to determine the ship's displacement and *KG* as built.

This page intentionally left blank

Chapter 6

Launching, Docking and Grounding

INTRODUCTION

The 'natural' condition for a ship is floating freely in water. The water surface may be rough and this will cause unpleasant motions and may apply high loads to the structure. However, these are the conditions which the designer will have had constantly in mind as the design progressed and they provide the norm for the mariner. On occasion, the ship is subject to a different environment. At some stage, it will be transferred from dry land where it was built to the water where it will be launched. Periodically during its life it will be taken out of the water for repair and maintenance. It will be docked or lifted out of the water. Launching and docking a ship are potentially hazardous operations as the ship passes between the dry and waterborne conditions. It is important they be studied in some depth.

Docking is now less frequent because hull coatings, to reduce corrosion and fouling, remain effective for longer. Also more can be achieved in the way of repairs with a vessel still afloat.

A ship may run aground either due to human errors in navigation, due to obstacles not recorded on charts, or due to the failure of the ship's control systems.

The designer must legislate for all these eventualities. Launching and docking are operations carried out under predictable conditions and are carefully planned. The naval architect produces the ship data needed for the operations. Those responsible for the operation use this data together with details of the actual loading condition of the ship at the time. Grounding is unpredictable. It involves a more variable set of circumstances including the point of grounding (along the length and transversely), the nature of the seabed, the prevailing weather and tide conditions and the actions of the crew. All influence what happens to the ship in terms of structural damage and flooding. In design the study of what might happen to a ship if it does run aground must be based upon a statistical analysis of past occurrences. From this it is possible to develop a number of situations to be dealt with, or a worst case scenario can be selected. A designer must make assumptions

Introduction to Naval Architecture. DOI: http://dx.doi.org/10.1016/B978-0-08-098237-3.00006-0

about the state of the ship (loading and structural integrity) at the time of the incident, the loads upon the ship and the associated safety implications.

LAUNCHING

Large ships may be built in docks, and in this case the 'launch' is like an undocking except that the ship is only partially complete and the weights built in must be carefully assessed to establish the displacement and centre of gravity position. Some weight adjustments may be necessary to ensure that the ship does not have excessive trim or heel when it floats off. The ship will not be structurally complete and some temporary strengthening may be necessary. In general, the ship is launched down inclined ways and one end, usually the stern, enters the water first. This is the situation considered here.

The Building Slip

Typically the floor of a building slip has a slope of about 1 in 20 and is of masonry with inserts of wood to facilitate the securing of blocks, shores, and so on. There is usually a line of transverse wooden blocks, the *building blocks* running down the centre of the slipway. These support the keel and most of the ship's weight during build. Shores are used to provide support at the bilges and for overhangs. For large ships, additional fore and aft lines of blocks may be provided on each side. Either side of the building blocks are the *groundways* which run parallel to the building blocks and are set about a third of the ship's beam apart. They provide the surface on which the vessel slides during the actual launch.

The Building Blocks

These run along the intended line of keel. They are vertical, rather than normal to the slip floor, to reduce the risk of tripping. The design office will decide the number of blocks and their distribution to reflect the spread of loading likely to be exerted on them by the ship. The height of the blocks, typically 1.5 m, must be adequate to provide space under the ship to enable the outer bottom and its fittings to be worked on, to facilitate the insertion of the launching cradle and to ensure that the forefoot does not strike the slip floor on launch. Individual sets of blocks can be removed to enable areas of the hull to be worked on. Wedges are incorporated to provide adjustment to height and facilitate the removal of the upper part of the blocks.

Just before launch, the ship's weight is transferred to cradles which are built to support the ship on its way down the slipway. The cradles rest on *sliding ways* which slide over the greased *groundways*.

The Groundways

The slope of the groundways must be such that the component of the launching weight down the ways is enough to overcome the initial sliding resistance of the grease which is applied to the ways just before launch. In the later stages of the launch, the slope must be adequate to overcome the resistance of the grease and water resistance. On the other hand, the ship should not enter the water at too great a velocity. The angle of the lower stretch of the ways is important to the rate at which buoyancy builds up and its moment about the *fore poppet*. It is often found that cambering the ways fore and aft is a good way of establishing the best compromise between the competing requirements. The grease used is tested to establish its properties at the temperatures and pressures likely at launch.

Sliding Ways and Cradles

The sliding ways must have enough area that the pressure exerted on the grease (up to about 20 tonnes/m^2) does not squeeze the grease out from between the sliding- and groundways. The exact spacing will depend upon the ship's beam and arrangement of internal structure. The naval architect must ensure that the loads are safely transmitted between the ways and the hull through the *launch cradles*. Because of the ship's form, those parts of the cradle nearest the bow and stern may be quite high. These two parts are known as the *fore* and *after poppets*. The fore poppets are particularly important as it about them that the ship pivots as it approaches the end of the ways. The load they then carry must be carefully calculated and it may be about 20% of the total weight. The cradles are secured to the ship before launch so that the grease can be inserted. These securings also hold the cradle, as the ship travels down the slip, so preventing it falling and damaging the bottom of the ship.

The Launch

Before launch the building blocks are removed, the ship's weight being transferred to the sliding ways and hence to the grease and groundways. The number of shores is reduced to a minimum and a trigger prevents the ship sliding down the ways prematurely. At launch the remaining shores are removed and the trigger released. Hydraulic rams are provided to push the ship if its component of weight down the slip is inadequate to set it in motion.

The ship follows the curve of the ways, the stern enters the water and the increasing buoyancy creates a moment tending to lift the stern. When the moment of buoyancy about the fore poppet exceeds that of the weight the stern lifts. If the slipway is long enough, the vessel finally floats off. If the ways are not long enough for this, the bow will drop off the end. The depth of water at the ends of the ways must be enough to allow this to happen without the bow striking the bottom, bearing in mind that due to dynamic effects, the actual drop will be greater than that associated with the final draught forward.

The ship will have built up significant momentum by the time it is waterborne. Whilst water resistance will slow the ship down, additional measures are usually necessary to take the way-off the vessel. The usual methods are to fit 'water brakes', often in the form of wooden barriers built over the propellers, and drag chains which are set in motion progressively as the ship leaves the slipway. Finally tugs take over and manoeuvre the ship to its fitting out berth.

The slipway chosen for the build must project on to a sufficient depth and width of water. Although advantage may be taken of a high tide, it is desirable to have as big a window of opportunity as possible for launch. The ground must be firm enough to take the weight of the ship during build. Dredging and piling can be used to improve existing conditions to enable old slipways to be used for a larger than usual ship. The arrangement and height of blocks must reflect the dynamics of the launch to ensure that the ship will enter the water smoothly and safely. All these considerations must be dealt with before the build begins (Figure 6.1).

The calculations the naval architect must carry out follow from the physical description of the launch process. An assessment must be made of the weight and centre of gravity position at the time of launch. To facilitate this, a detailed record of all materials built into the ship is kept, often backed up by actual weighing. In manual calculations, the procedure adopted is to treat the launch as a quasi-static operation. That is, it assumed that all forces and moments are in balance at every moment. A profile of the ship is moved progressively down a profile of the launch ways, taking account of the launching cradle. The moments of weight about the fore poppet and the after end of the ways are calculated at a number of positions. As the ship moves the waterline at various distances down the ways can be noted on the profile. From the Bonjean curves, the immersed

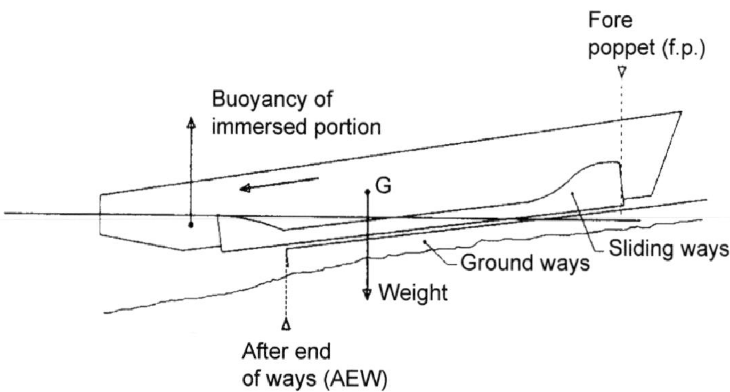

FIGURE 6.1 Launching.

sectional areas can be read off and the buoyancy and its longitudinal centre computed. The ship will continue in this fashion until the moment of buoyancy about the fore poppet equals that of the moment of weight about the same position. At that point, the ship pivots about the poppet and the force on it can be large and stability can be critical. The maximum force on the fore poppet will be the difference between the weight and the buoyancy at the moment the ship pivots. By continuing to plot the way the buoyancy increases with travel, it can be determined whether the ship will float off or the bow will drop. The ship becomes fully waterborne when the buoyancy equals the weight. To ensure the ship does not tip about the after end of the ways, the moment of buoyancy about that point must always be greater than the moment of weight about it. The analysis will be more complicated if the launching ways are curved in the longitudinal direction to increase the rate at which buoyancy builds up in the later stages.

The data are usually presented as a series of curves, the *launching curves* (Figure 6.2).

The curves plotted are the weight which will be constant, the buoyancy which increases as the ship travels down the ways, the moment of weight about the fore poppet which is also effectively constant, the moment of buoyancy about the fore poppet, the moment of weight about the after end of the ways and the moment of buoyancy about the after end of the ways.

The stability at the point of pivoting can be calculated in a similar way to that adopted for docking, as described later. There will be a large bending moment acting on the hull girder which must be assessed. The maximum force on the fore poppet will be the difference between the weight and the buoyancy at the moment the ship pivots about the fore poppet. The forces acting are also needed to ensure that the launching structures are adequately

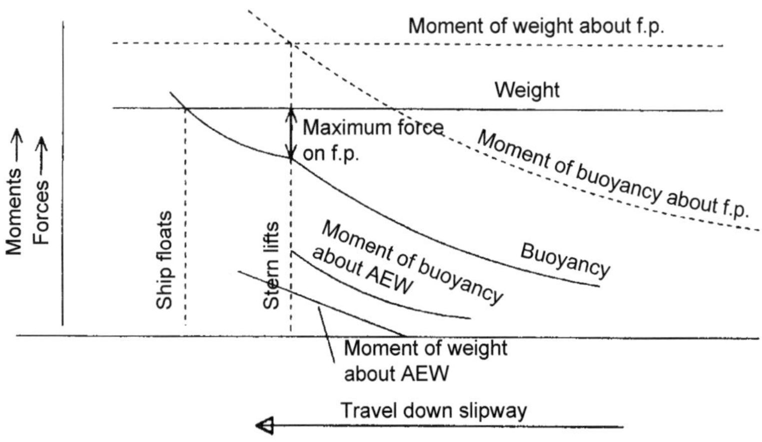

FIGURE 6.2 Launching curves.

strong, bearing in mind that at that stage of build, some elements of structure may be incomplete and some temporary stiffening may be needed.

The calculations associated with launching have remained substantially the same in principle over the years. However, these days, more detailed studies of the strength of the ship, both overall and locally in way of the poppets, can be carried out using finite element analyses. A simulation can be prepared, in effect automating the process of moving the ship progressively down the slip and ensuring forces and moments are in balance at each point or allowing for the full dynamic nature of the launch.

Sideways Launching

Some shipyards, particularly those building small ships, lie on relatively narrow rivers into which a conventional stern first launch is not practicable. Then a sideways launch is adopted. The ship is built parallel to the river bank and the launch ways are normal to the line of keel and set 3–5 m apart, the supporting cradle being adjusted accordingly. One advantage is that the ship can be built on a level keel.

Typically the ship slides down the ways which at a certain point tilt and the ship 'drops' into the water creating a sizeable wave. Due to the resistance of the water to sideways movement, the ship does not travel far from the bank but it will roll fairly violently. Openings in the hull that might become immersed must be made watertight. In order that the ship travels far enough from the river bank to prevent it rolling back on to the launch ways, relatively high launch velocities are used.

DOCKING

Most large ports and shipyards have fixed dock installations. Wet docks are used to accommodate ships while they are fitting out, loading or unloading. If tidal there must be sufficient depth at low tide to enable the ship to remain afloat. Otherwise it must be checked that the ship can be allowed to ground on the dock floor without damage. *Dry*, or *graving, docks* are used to enable the ship's bottom and underwater fittings to be inspected and worked on. They are essentially large holes in the ground, lined with masonry, and provided with a means to close off the entrance once the ship is in the dock so that the water can be pumped out. Hinged gates or floating *caissons* are the usual means of closing the entrance. The caissons are either floated into position and then ballasted down or they slide across the entrance. In all cases the closure must be designed so that the water pressure from outside the dock makes it watertight. The dock has a line of blocks on which the ship sits when the water has receded. In addition the ship is supported by *breast shores* which are set up between the ship's side and the wall of the dock. Other shores support the turn of bilge and stern overhangs.

Docking in a Graving Dock

For each ship a *docking plan* will have been prepared, showing the following:

- A profile indicating points to which shore supplies of electricity, hydraulic power, cooling water and so on can be run. Ideally these will be aligned with the corresponding shore supply positions.
- Deck plans.
- Sections of the ship at which breast shores can be set up, usually at transverse bulkheads where the hull will be better able to take the forces exerted by the shores.
- Details of projections that might foul the dock entrance or the blocks. For instance, the propellers may project below the line of the keel and bilge keels must be allowed for.

The docking plan is used in conjunction with the plans of the dock to establish the acceptable combinations of tide, draught and trim for which docking is feasible. The dock entrance is smaller in section than the dock itself but the top of the dock blocks is usually higher than the sill of the entrance. The ship is docked on the centreline of the dock unless more than one ship is being docked at the same time. The position of the ship along the length of the dock is dictated by the positions of the shore supplies and access positions. Some dock blocks may have to be removed in way of projections below the line of keel. The alignment of the block tops is carefully checked before the dock is flooded up ready to receive the ship. Comparing the ship and dock sections allows the length of breast shores to be established, making allowance for the wedges which are driven home finally to hold the ship once it is sitting firmly on the blocks. The positions of the two ends of the breast shores are marked on the ship and at the side of the dock. The lengths of bilge shores and those under the stern overhang can be determined from the plans.

The dock is flooded, the dock closure device opened and the ship is drawn into the dock by winches. It is aligned with the fore and aft marks on the side of the dock and with ropes across the dock marked to show the centreline. The dock is closed and the water is pumped out. As the water level drops, the ship's keel approaches the top of the blocks. The ship's trim will have been adjusted so that it is not much different from the slope of the block tops and so that the after cut-up will touch the blocks first. The breast shores are held loosely in position on ropes. After the after cut-up touches the blocks, a force begins to build up at the cut-up as the water level drops further. This force causes the ship to trim by the bow. As the water recedes further, the ship trims until the keel touches the blocks along the length of the keel. Then the breast shores can be finally secured to hold the ship against tipping. Once the dock is dry, bilge shores and shores supporting overhangs can be positioned. Shore supplies are connected as required.

For vessels with a very rounded form, for instance submarines, a cradle is set up in the dock before flooding up. When the vessel enters the dock, it is manoeuvred above this cradle and the water is pumped out. Positioning is more critical in this case and divers may be sent down to ensure that the vessel is sitting properly in the cradle before the water level drops too far.

Floating Docks

There are many floating docks available worldwide ranging from small docks with a lift capacity of less than 500 tonnes to the ones capable of lifting ships of up to 100,000 tonnes. They have some of the following advantages:

- They can be taken to ports/harbours which have no graving dock facilities. In transit care must be taken to ensure the dock is seaworthy.
- They can be heeled and trimmed to match a damaged ship's condition and provide partial support while assessments are made of the damage.

A floating dock usually takes the form of a U-shaped box structure with side walls mounted on a base pontoon. A large part of the structure is devoted to ballast tanks which are free flooded to sink the dock so that the ship can be moved into the correct position within the dock. The dock, with the ship, is then raised by carefully controlled pumping out of the ballast tanks. The sequence of pumping is such as to limit the longitudinal deflection of the dock (and hence the ship in it) to avoid undue longitudinal bending moments.

The dock stability, transverse and longitudinal, is high when it is at its operating freeboard with the deck of the pontoon above the water level. A case of minimum transverse stability usually arises when the water level is between the pontoon deck and the top of the docking blocks. A metacentric height of 1.0 – 1.5 m is commonly accepted but some operators demand twice this.

Shiplifts

Shiplifts are devices providing a means of lifting ships vertically out of the water to a level where they can be worked on.

The main elements of a shiplift are as follows:

- An articulated steel platform, generally wood decked, arranged for end on or longitudinal transfer.
- Wire rope hoists along each side of the platform, operated by constant speed electric motors.
- A load-monitoring system to ensure a proper distribution of loads so as not to cause damage to the ship or the platform.
- A cradle configured to suit the ship's hull form.

The lifting capacity of a shiplift is expressed in terms of the maximum load per metre, the *maximum distributed load* that can be distributed along the centreline of the platform. The actual weight of ship that can be lifted depends upon the distribution of weight along the length. Although many shiplifts are for relatively small vessels of say 1000 tonnes, they can be designed to lift ships of 30,000 tonnes or more. One was designed for vessels of 80,000 tonnes deadweight.

A transfer system is usually provided on shore so that one lift can serve a number of positions at which the ship can be worked on, including refitting sheds. This increases the value, and the usage rate, of the lift considerably. Usually a rail-mounted system is used but other transfer methods can be adopted.

Economics

Docking a ship is an expensive business and over the years much effort has been devoted to increasing the intervals between dockings and the time need in dock. Measures taken include the following:

- Developing hull coatings which remain effective for longer, including so-called self-polishing paints. Coatings which might pollute the environment must be avoided.
- Using cathodic protection systems to protect the hull and its fittings against corrosion.
- Designing underwater features so that they can be removed and replaced with the ship still afloat.
- Developing means of repairing under water, such as underwater welding, using divers or providing watertight enclosures, called habitats, enabling underwater fittings to be worked on in the dry while the ship is afloat.
- Using mobile staging or mechanised platforms to enable the hull to be accessed in dock without using extensive scaffolding which is expensive of money and time.
- Using refit and repair by replacement.

Stability When Docking

When a ship is partially supported by the dock blocks, its stability will be different from that when floating freely (Figure 6.3) and it must be investigated.

A ship usually has a small trim by the stern as it enters dock, and as the water is pumped out it sits first on the blocks at the after end of the keel. As the water level drops, the ship trims until the keel touches the blocks over its entire length. It is then that the force on the sternframe or after

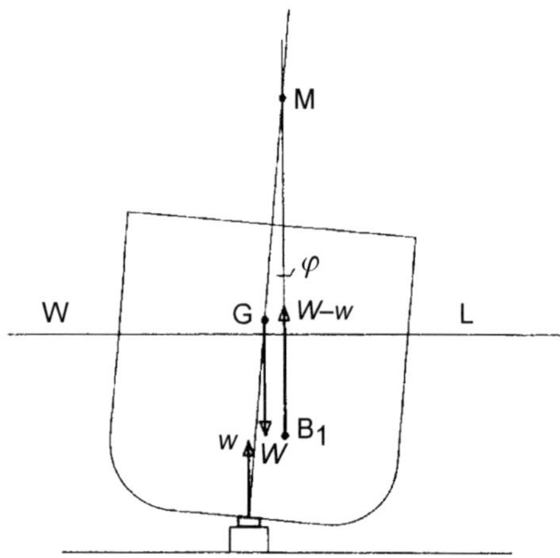

FIGURE 6.3 Docking.

cut-up will be greatest and the stability most critical. As the water drops further, the ship will be steadied by the breast shores.

Suppose the force at the time the keel touches along the whole length is w, and that it acts a distance x aft of the centre of flotation. Then, if t is the change of trim since entering dock:

$$wx = t(\mathrm{MCT})$$

The value of w can be found using the value of moment to change trim (MCT) read from the hydrostatics. The MCT value used should be that appropriate to the actual waterline at the instant concerned and the density of water in the dock. As the mean draught will itself be dependent upon w, an approximate value can be found using the mean draught on entering dock followed by a second calculation when this value of w has been used to calculate a new mean draught. Referring to the figure, the righting moment acting on the ship, assuming a very small heel, φ, is:

$$\begin{aligned}\text{Righting moment} &= (W - w)\,\mathrm{GM}\sin\varphi - wKG\sin\varphi \\ &= [WGM - w(GM - KG)]\sin\varphi \\ &= (WGM - wKM)\sin\varphi \\ &= \left(GM - \frac{w}{W}KM\right)W\sin\varphi\end{aligned}$$

Should the expression inside the brackets become negative, the ship will be unstable and may tip over. Whilst the breast shores will hold the ship to a degree, they are still held loosely and the ship may slip off the blocks.

Worked Example 6.1

Just before entering drydock, a ship of 5000 tonnes mass floats at draughts of 2.7 m forward and 4.2 m aft. The length between perpendiculars is 150 m and the water has a density of 1025 kg/m^3. Assuming the blocks are horizontal and the hydrostatic data given are constant over the variation in draught involved, find the force on the heel of the sternframe, which is at the after-perpendicular, when the ship is just about to settle on the dock blocks, and the metacentric height at that instant.

Hydrostatic data: $KG = 8.5$ m, $KM = 9.3$ m, MCT 1 m = 105 MN m, longitudinal centre of hydrostatic data: $KG = 8.5$ m, $KM = 9.3$ m, MCT 1 m = 105 MN m, longitudinal centre of flotation (LCF) = 2.7 m aft of amidships.

Solution

Trim lost when touching down = 4.2 − 2.7 = 1.5 m

$$\text{Distance from heel of sternframe to LCF} = \frac{150}{2} - 2.7 = 72.3 \text{ m}$$

Moment applied to ship when touching down is equal to $w \times 72.3$

Trimming moment lost by ship when touching down is equal to $1.5 \times 105 = 157.5$ MN m

$$\text{Hence, thrust on keel } (w) = \frac{157.5}{72.5} = 2.17 \text{ MN}$$

$$\text{Loss of } GM \text{ when touching down} = (w/W)KM = \frac{(2.17 \times 10^3 \times 9.3)}{(5000 \times 9.81)}$$

$$= 0.41 \text{ m}$$

$$\text{Metacentric height when touching down} = 9.3 - 8.5 - 0.41$$
$$= 0.39 \text{ m}$$

GROUNDING

With a few exceptions, ships are not intended to ground. When they do, the hull and underwater fittings may be damaged, the extent of damage depending upon a number of factors, including:

- the nature of the seabed;
- the speed and angle of impact;
- the sea state and tide at the time of grounding and up until the ship can be refloated and
- the area of ship's hull which impacts the seabed.

If the area of hull in contact with a smooth seabed is relatively small, the stability can be calculated in a manner similar to that described for docking. The same applies to an otherwise intact ship balanced on a rock. However, the force the ship experiences will not, in general, be on the centreline so that it will cause the ship to heel as well as trim. The value of the force can be calculated as that which will cause the ship to heel and trim so that, at the point of contact, the draught is equal to the depth of water at that point. The force will vary as the tide falls and rises and the tidal variations must be predicted to enable the maximum force to be assessed. The force will vary also due to any wave action. The master can determine the changes in ballast and load distribution to allow the ship to lift clear.

Stability of an Intact Ship when Partially Grounded in Mud

Grounding is usually as a result of an accident, although a small ship may, either deliberately or unwittingly, be allowed to settle on a muddy bottom at its unloading berth as the tide falls.

Although a somewhat artificial case, it is instructive to consider the principles involved in a rectangular barge partially supported by homogeneous mud. See Figure 6.4.

Assume:

- Length of barge $= L$
- Beam $= B$
- Draught is uniform and is equal to T initially when floating freely in water
- The centre of gravity is amidships and on the centreline

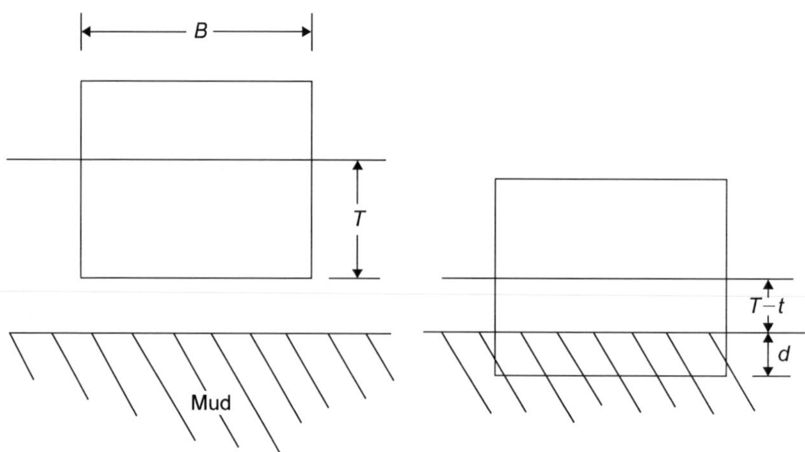

FIGURE 6.4 Stability when grounded in mud.

- Density of water $= \rho$
- Density of mud $= k\rho$

The weight of water displaced when floating freely is equal to $TBL\rho$.

This must equal the weight of water and mud displaced when partly supported by the mud.

If the tide falls t after the barge first touches the mud, then the depth of water becomes $T - t$. Assume that at that time, the depth of barge in mud is d, then:

$$TBL\rho = (T - t)\,BL\rho + dBLk\rho$$

Hence,

$$t = dk$$

To put numbers to this, if $k = 2$ and the tide falls by 2 m, $2 = 2d$, i.e. $d = 1$ m.

In this case, the vessel settles into the mud by an amount equal to half the fall in tide.

For the ship floating freely, the transverse stability can be calculated in the usual way.

That is,

$$KB = T/2;\; BM = I/V = (1/12)(B^3\,L/BLT) = B^2/12T$$
$$KM = KB + BM = T/2 + B^2/12T$$

When partly in the mud, if the barge is heeled through a small angle, it experiences righting moments due to the transfer of buoyancy between the wedges of water at the surface and due to a transfer of mud buoyancy, the latter being offset by a transfer the other way of the equivalent wedge of water. The net effect, felt by the barge, is due to the transfer of mud wedges which leads to a righting moment k times that experienced by the freely floating body.

Hence,

$$BM = kI/V$$

KB has changed and can be found by taking moments about the keel:

$$KB = \{(T - t)B[d + (T - t)/2] + kdBd/2\}/B(T - t + kd)$$
$$= \{kd^2 + 2d(T - t) + (T - t)^2\}/2(T - t + kd)$$

If the initial draught is 5 m and the fall of tide is 2 m, the ship sinks 1 m into the mud and

$$KB = \{2(1)^2 + 2 \times 1 \times 3 + 3^2\}/2(3 + 2 \times 1) = 1.7 \text{ m}$$

Knowing the length and beam of the barge will lead to values of BM and hence of KM. If KG is known GM values follow, noting that KG is constant.

Grounding on a Rock

A more serious situation arises if the seabed is rocky and the ship's outer bottom is punctured allowing water to enter and leading to further stability changes and structural damage. As far as the stability of the ship is concerned, the assessment needed is similar to the damaged stability case discussed in the chapter on stability but with a ground reaction superimposed. Also the residual strength of the damaged hull must be assessed.

Paik et al. (2012) proposed a method for assessing the structural safety of a ship following grounding on a rock. A sampling technique was used to select 50 credible damage scenarios from the total number of situations possible and calculations were carried out for four types of double-hull tankers – VLCC, Suezmax, Aframax and Panamax.

The authors argued that the main parameters involved are the following:

- grounding location and width of damage transversely;
- height of rock penetration and
- shape of rock.

They used International Maritime Organisation (IMO) data defining the probability density distributions of the first two sets of items and the breadth of the rock. For the damaged hull, they deduced a *grounding damage index* (GDI) defined in terms of areas of the outer bottom (OB) and inner bottom (IB) as:

$$\begin{aligned} \text{GDI} &= (\text{area of OB reduced by damage})/(\text{original area of the OB}) \\ &\quad + \alpha(\text{area of IB reduced by damage})/(\text{original area of the IB}) \end{aligned}$$

where α is a correction factor reflecting the IB's smaller contribution to overall longitudinal strength. The α values are as shown in Table 6.1.

The residual ultimate longitudinal strength, in hogging and sagging, was calculated for each damage scenario, ignoring the fact that the neutral axes of the damaged cross section will not be parallel to those of the undamaged section. A residual strength/damage index (R–D) diagram was then produced by plotting the ratio of ultimate longitudinal strengths for the damaged

TABLE 6.1 α Values

Condition	Tanker Size			
	VLCC	Suezmax	Aframax	Panamax
Hogging	0.5498	0.5604	0.5689	0.5975
Sagging	0.2847	0.2299	0.2044	0.2645

TABLE 6.2 Coefficients for R–D Diagram Ordinates

Condition	Tanker Size			
	VLCC	Suezmax	Aframax	Panamax
Hogging				
A	0.0511	0.0125	−0.0176	−0.0307
B	−0.3617	−0.3379	−0.2902	−0.2400
C	1.0	1.0	1.0	1.0
Sagging				
A	−0.2056	−0.2142	−0.2069	−0.1553
B	−0.1498	−0.1371	−0.1387	−0.1614
C	1.0	1.0	1.0	1.0

and intact ships against the GDI. It was found that the diagrams could be represented by a function of the GDI in the form:

For hogging or sagging: $M_D/M_U = A(GDI)^2 + B(GDI) + C$

where M_D is for the damaged ship and M_U is for the undamaged ship. The values they deduced for A, B and C for the different ship types for both hogging and sagging are given in Table 6.2.

It was concluded that providing the damage was not too severe a single formula would suffice for all four ship types. Viz.:

For hogging: $M_D/M_U = -0.0036(GDI)^2 - 0.3072(GDI) + 1.0$

For sagging: $M_D/M_U = -0.1941(GDI)^2 - 0.1476(GDI) + 1.0$

Assessments of residual hull strength are clearly of great use in guiding those responsible for salvaging a damaged vessel. Noting that IMO regulations require M_D/M_U to be not less than 0.9, it is possible to find the allowable GDI from the R–D curve or the formulae derived as above for the ship.

SUMMARY

Launching and docking are potentially hazardous activities for a ship as it moves between the waterborne and dry conditions. Successful launching and docking of ships can be achieved if the operations are well planned and supporting calculations are carried out. Grounding signifies an accident and is less predictable and a potentially more serious matter, particularly if the hull is punctured. The residual strength of a ship damaged by grounding can be calculated and a GDI assessed.

This page intentionally left blank

Chapter 7

Resistance

INTRODUCTION

Resistance and propulsion are dealt with separately for convenience but the two are interdependent due to the interaction between the propulsion device and the flow around the hull. It is important that the ship has enough, but not too much, power for the speed required. A ship moves through water and air experiencing both water and air forces opposing that movement. The faster it moves the greater the resistance it experiences. The water and air masses may themselves be moving; the water due to currents and the air as a result of winds. These will, in general, be of different magnitudes and directions. The resistance is studied initially in still water with no wind. Separate allowances are made for wind and the resulting distance travelled corrected for water movements. Unless the winds are strong the water resistance will be the dominant factor in determining speed.

FLUID FLOW

Classical hydrodynamics leads to a flow pattern past a body of the type shown in Figure 7.1.

As the fluid moves past the body the spacing of the streamlines, and the velocity of flow, change because the mass flow within streamlines is constant. There are corresponding changes in pressure and Bernoulli's theorem applies. For a given streamline, if p, ρ, v and h are the pressure, density, velocity and height above a selected datum level, then:

$$\frac{p}{\rho} + \frac{v^2}{2} + gh = \text{constant}$$

Simple hydrodynamic theory deals with fluids without viscosity. In a non-viscous fluid a deeply submerged body experiences no resistance. Although the fluid is disturbed by the passage of the body, it returns to its original state of rest once the body has passed. There are local forces acting on the body, due to the pressure changes occasioned by the changing velocities in the flow, but these cancel each other out when integrated over the whole body surface.

Introduction to Naval Architecture. DOI: http://dx.doi.org/10.1016/B978-0-08-098237-3.00007-2

FIGURE 7.1 Streamlines round elliptic body.

In studying fluid dynamics it is useful to develop a number of non-dimensional parameters to characterise the flow and the forces, based on the fluid properties. The physical properties of interest in the present context are the density, ρ, viscosity, μ and the static pressure in the fluid, p. Taking R as the resistance, V as velocity and L as a typical length, dimensional analysis leads to an expression for resistance as a function of the variables:

$$R = f(L, V, \rho, \mu, g, p)$$

The quantities involved in this expression can all be expressed in terms of the fundamental dimensions of time, T, mass, M, and length, L. For instance resistance is a force and therefore has dimensions ML/T^2, ρ has dimensions M/L^3, μ has dimensions M/LT and g has dimensions L/T^2. Substituting these fundamental dimensions in the relationship above it will be seen that the expression for resistance can be written as:

$$R = \rho V^2 L^2 \left[f_1\left(\frac{\mu}{\rho VL}\right), f_2\left(\frac{gL}{V^2}\right), f_3\left(\frac{p}{\rho V^2}\right) \right]$$

Thus the analysis indicates that the following non-dimensional combinations are likely to be significant:

$$\frac{R}{\rho V^2 L^2}, \quad VL\frac{\rho}{\mu}, \quad \frac{V}{(gL)^{0.5}}, \quad \frac{p}{\rho V^2}$$

The first three ratios are termed, respectively, the *resistance coefficient, Reynolds' number*, and *Froude number*. The ratio μ/ρ is called the *kinematic viscosity* and is denoted by ν. The values of density and kinematic viscosity, for fresh and salt water, as recommended by the International Towing Tank Conference (ITTC), are given in Chapter 9. The fourth ratio is related to cavitation and is discussed later. In a wider analysis the speed of sound in water, α, and the surface tension, σ, can be introduced. These lead to non-dimensional quantities V/α, and $\sigma/g\ \rho L^2$ which are termed the *Mach number* and *Weber number*. These two numbers are not important in the context of this present book and are not considered further. At this stage it is assumed that these non-dimensional quantities are independent of each other. Ignoring cavitation the expression for the resistance can then be written as:

$$R = \rho V^2 L^2 \left[f_1 \left(\frac{v}{VL} \right) + f_2 \left(\frac{gL}{V^2} \right) \right]$$

Consider first f_2 which, as we shall see, is concerned with wave-making resistance. Take two geometrically similar ships or a ship and a geometrically similar model, denoted by subscripts 1 and 2.

$$R_{w1} = \rho_1 V_1^2 L_1^2 f_2 \left(\frac{gL_1}{V_1^2} \right) \quad \text{and} \quad R_{w2} = \rho_2 V_2^2 L_2^2 f_2 \left(\frac{gL_2}{V_2^2} \right)$$

Hence:

$$\frac{R_{w2}}{R_{w1}} = \frac{\rho_2}{\rho_1} \times \frac{V_2^2}{V_1^2} \times \frac{L_2^2}{L_1^2} \times \frac{f_2(gL_2/V_2^2)}{f_2(gL_1/V_1^2)}$$

The form of f_2 is unknown, but, whatever its form, provided $gL_1/V_1^2 = gL_2/V_2^2$ the values of f_2 will be the same. It follows that:

$$\frac{R_{w2}}{R_{w1}} = \frac{\rho_2 \, V_2^2 \, L_2^2}{\rho_1 \, V_1^2 \, L_1^2}$$

Since $L_1/V_1^2 = L_2/V_2^2$, this leads to:

$$\frac{R_{w2}}{R_{w1}} = \frac{\rho_2 L_2^3}{\rho_1 L_1^3} \quad \text{or} \quad \frac{R_{w2}}{R_{w1}} = \frac{\Delta_2}{\Delta_1}$$

For this relationship to hold $V_1/(gL_1)^{0.5} = V_2/(gL_2)^{0.5}$ assuming ρ is constant.

Putting this into words, the wave-making resistances of geometrically similar forms will be in the ratio of their displacements when their speeds are in the ratio of the square roots of their lengths. This is known as *Froude's law of comparison* and the quantity $V/(gL)^{0.5}$ is called the *Froude number*. In this form it is non-dimensional. If g is omitted from the Froude number, as it is in the presentation of some data, then it is dimensional and care must be taken with the units in which it is expressed. When two geometrically similar forms are run at the same Froude number they are said to be run at *corresponding speeds*. At corresponding speeds the wave patterns will look the same but be to a different scale.

The other function, f_1, in the total resistance equation determines the frictional resistance. Following an analysis similar to that for the wave-making resistance, it can be shown that the frictional resistance of geometrically similar forms will be the same if:

$$\frac{\nu_1}{V_1 L_1} = \frac{\nu_2}{V_2 L_2}$$

This is commonly known as *Rayleigh's law* and the quantity VL/ν is called the *Reynolds' number*. As the frictional resistance is proportional to the square of the length, it suggests that it will be proportional to the wetted surface of the hull.

For two geometrically similar forms, complete dynamic similarity can only be achieved if the Froude number and Reynolds' number are equal for the two bodies. This would require $V/(gL)^{0.5}$ and VL/ν to be the same for both bodies. This cannot be achieved for two bodies of different size running in the same fluid.

NUMERICAL METHODS AND COMPUTATIONAL FLUID DYNAMICS

For many years the only way in which a ship's resistance could be established with any accuracy was by measuring the resistance of a model and scaling that to the full size. It is these methods which are concentrated on in this chapter. With the increasing computer power becoming available the 1970s saw a big increase in theoretical work but it was in the 1980s that computational fluid dynamics (CFD), as applied to hydrodynamics, really took off with applications to fluid flow around the hull and resistance. Later it was applied to propellers, the interactions between propeller and hull, motions and manoeuvring. These numerical methods are now seen as essential tools for the designer but they require considerable experience for their correct application. Often consultants will be used to carry out such investigations.

These advanced methods cannot be covered in this book, but briefly they are based on an assumption of incompressible flow and the equations governing the motion are specified applying the concepts of continuity of mass and momentum. That is, the mass and momentum of the fluid entering each element of volume are equal to those of the fluid exiting the volume. The boundary conditions must be accurately defined. The equations cannot be solved directly because of turbulence effects and this is dealt with by averaging the flow over a period of time. This time period is short compared with overall ship motions but long compared to the turbulence fluctuations. The resulting equations are known as the Reynolds-averaged Navier–Stokes equations (RANSE). The numerical solution of the partial differential equations of flow requires that the continuous nature of the flow be represented in discrete form. The domain is divided into cells or elements to form a volume grid. The size and shape of the cells need careful selection to suit the problem, a finer grid being required in zones of rapid change. The equations are then expressed in discrete form at each point in the grid by using finite difference, finite volume or finite element methods.

The first approaches – the *boundary element methods (BEMs)* – used a boundary element (panel) method in which the integrals over the whole fluid domain are transformed to integrals at the boundaries of that domain, those

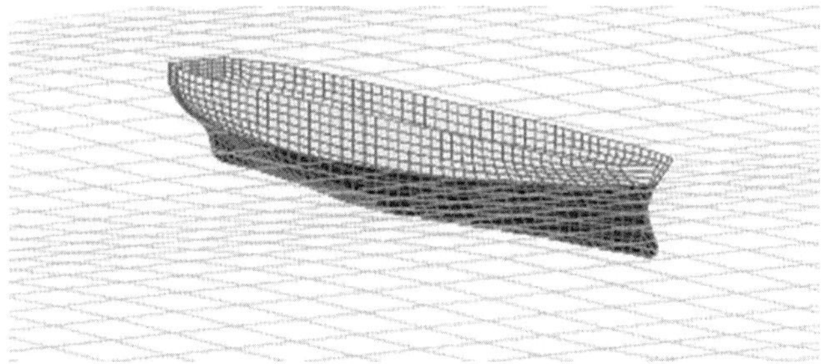

FIGURE 7.2 A grid for a ship in waves. *Courtesy of RINA.*

boundaries being divided into discrete elements (the panels). The equations are made to match the boundary conditions on the ship's hull, and part of the adjacent water surface. Figure 7.2 shows a grid constructed for a ship in waves. BEMs are widely used for flow and resistance studies. Later 'field' methods discretise the whole fluid domain, and include *finite difference methods (FDMs)* where the field equations are approximated by finite differences. This can lead to violation of the continuity conditions which can be overcome in *finite volume methods (FVMs)* which form the base of most present day commercial RANSE solvers. Grid generation is a more difficult task for the FDMs and FVMs but they can compute the pressure and velocity components and turbulence at the nodes or centroids of each grid cell. Grid generation accounts for the larger part of the time needed for a CFD investigation and the simpler representations will be used where they provide answers of sufficient accuracy.

In the marine field, CFD has been applied mainly to problems in hydrodynamics such as visualising the flow around a hull, flow into a propeller, propeller performance, resistance and motions, but it is also used for evaluating such features as air wake, ventilation flows and flow-induced noise. In ship motions, as an example, CFD can provide a significant improvement over the old strip theories. Application of CFD leads to the forces on a hull at any instance, which will determine the next movements of the ship. The same forces can be regarded as subjecting the ship to stresses deforming it. Thus, the whole dynamic system of a ship moving and flexing in the sea can be treated as one. As computations become cheaper one can expect the methods to be more widely applied. As with many other analysis methods there is a range of programs available differing in the accuracy of prediction. The simpler programs have more simplifying assumptions built in (e.g. ship treated as rigid rather than flexible) but they are quicker to run and require less computing capacity. Unfortunately, it is not always easy to determine the simplifying assumptions that have been made in specific applications. They must be used with care (ITTC 7.5-03-01-03).

Although numerical methods are used extensively in naval architecture today, they cannot, as yet, enable the designer to do away with older analysis methods and model testing. Nor do they provide such a clear understanding of the basic nature of a problem as some of the older methods of analysis. It is hoped that describing these older methods, many still widely used, will help the student grasp the essentials of the subject.

Bertram (2012) gives a good account of numerical methods as applied to resistance, propulsion, seakeeping, vibrations and manoeuvring. Molland et al. (2011) deals in more depth with resistance and propulsion. Both books set CFD in the context of the other assessment methods available.

TYPES OF RESISTANCE

When a moving body is near, or on, the free surface of the fluid, the pressure variations around it are manifested as waves on the surface. The energy needed to maintain these waves leads to a resistance. Also all practical fluids are viscous and movement through them causes tangential forces opposing the motion. The two resistances are known as the *wave-making resistance* and the *viscous* or *frictional resistance*. The viscosity modifies the flow around the hull, inhibiting the build-up of pressure around the after end which is predicted for a perfect fluid. This leads to what is sometimes termed *viscous pressure resistance*. The flow around the hull varies in velocity causing local variations in frictional resistance. Where the hull has sudden changes of section the flow may not be able to follow the lines exactly and 'breaks away'. This occurs at a transom stern. In breaking away, eddies are formed which absorb energy and thus cause a resistance. Because the flow variations and eddies are created by the particular ship form, this resistance is sometimes linked to the *form resistance*. Finally the ship has a number of appendages. Each has its own characteristic length and it is best to treat their resistances (they can generate each type of resistance associated with the hull) separately from that of the main hull. Collectively they form the *appendage resistance*.

Wave-making resistance arises from waves which are controlled by gravity, so it is to be expected that it will depend upon the Froude number. Frictional resistance being due to fluid viscosity will be related to Reynolds' number. As it is not possible to satisfy both the Froude number and the Reynolds' number in the model and the ship, the total resistance of the model cannot be scaled directly to the full scale. Indeed because of the different scaling it is not even possible to say that, if a model has less total resistance than another, a ship based on the first will have less total resistance than one based on the second. William Froude realised this and proposed that the model should be run at the corresponding Froude number to measure the total resistance, and that the frictional resistance of the model be calculated and subtracted from the total. The remainder, or *residuary resistance*, he scaled to full scale in proportion to the displacement of the ship to model. To the result he added an assessment of the skin friction resistance of the ship. The frictional resistance in each case was based on that of the equivalent flat plate. Although not theoretically correct this approach yields results which are sufficiently accurate for most purposes and it has provided the basis of ship-model correlations ever since.

Although the different resistance components are assumed independent of each other in the above non-dimensional analysis, in practice each type of resistance will interact with the others. Thus the waves created will change the wetted surface of the hull and the drag it experiences from frictional resistance.

Wave-Making Resistance

A body moving on an otherwise undisturbed water surface creates a varying pressure field which manifests itself as waves because the pressure at the surface must be equal to the local atmospheric pressure. When the body moves at a steady speed, the wave pattern appears to remain the same and move with the body. The energy for creating and maintaining this wave system must be provided by the ship's propulsive system; the waves cause a drag force on the ship which must be opposed by the propulsor. This drag force is the *wave-making resistance.*

A submerged body near the surface will also cause waves. A whale or a submarine may betray its presence in this way. The waves, and the associated resistance, decrease in magnitude quite quickly with increasing depth of the body until they become negligible at depths a little over half the body length.

The Wave Pattern

The nature of the wave system created by a ship is similar to that which Kelvin demonstrated for a moving pressure point. Kelvin showed that the wave pattern had two main features: diverging waves on each side of the pressure point with their crests inclined at an angle to the direction of motion and transverse waves with curved crests intersecting the ship's centreline at right angles. The angle the divergent waves, from a pressure point, make to the centreline is $\sin^{-1} 1/3$, that is just under 20°, Figure 7.3.

A similar pattern is clear if one looks down on a ship travelling in a calm sea. The diverging waves are readily apparent to anybody on board. The waves move with the ship so the length of the transverse waves must correspond to this speed, that is their length is $2\pi V^2/g$. Chapter 9 discusses wave characteristics.

The pressure field around the ship can be approximated by a moving pressure field close to the bow and a moving suction field near the stern. Each creates its own wave system (Figure 7.4). The after, suction, field

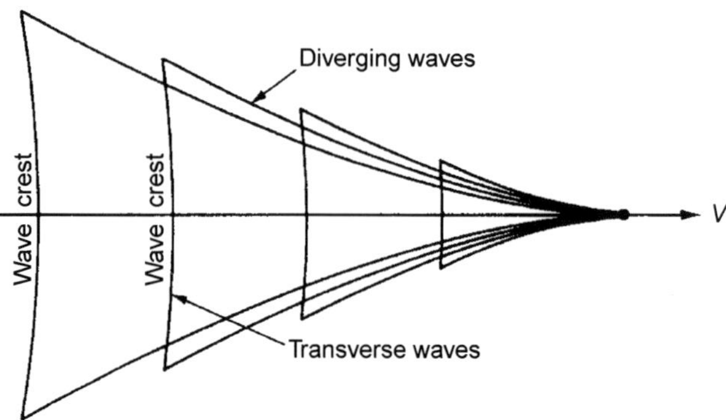

FIGURE 7.3 Pressure point wave pattern.

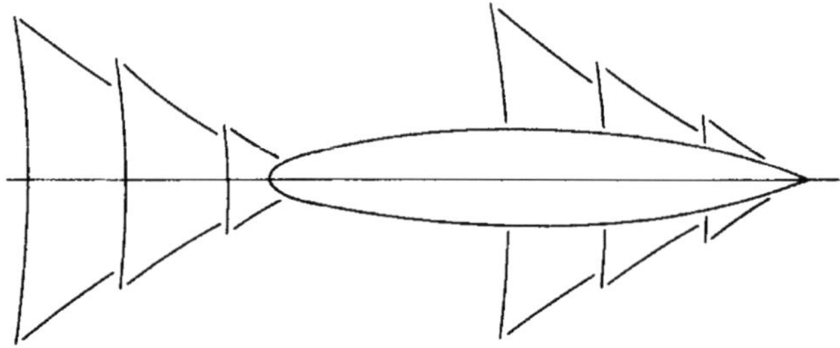

FIGURE 7.4 Bow and stern wave systems.

creates a trough near the stern instead of a crest as is created at the bow. The angle of the divergent waves to the centreline will not be exactly that of the Kelvin wave field. The maximum crest heights of the divergent waves do lie on a line at an angle to the centreline and the local crests at their maxima are at about twice this angle. The stern-generated waves are less clear, partly because they are weaker, but mainly because of the interference they suffer from the bow system.

Interference Effects

In addition to the waves created by the bow and stern others may be created by local discontinuities along the ship's length. However, the qualitative nature of the interference effects in wave-making resistance is illustrated by considering just the bow and stern systems. The transverse waves from the bow travel aft relative to the ship, reducing in height. When they reach the stern-generated waves they interact with them. If crests of the two systems coincide the resulting wave is of greater magnitude than either because their energies combine. If the crest of one coincides with a trough in the other

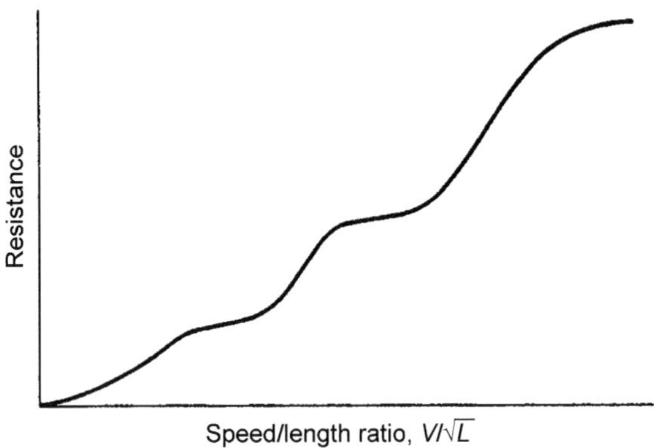

FIGURE 7.5 Humps and hollows in resistance curve.

the resultant energy will be less. Whilst it is convenient to picture two wave systems interacting, in fact the bow wave system modifies the pressure field around the stern so that the waves it generates are altered. Both wave systems are moving with the ship and will have the same lengths. As ship speed increases the wavelengths increase so there will be times when crests combine and others when crest and trough become coincident. The ship will suffer more or less resistance depending upon whether the two waves augment each other or partially cancel each other out. This leads to a series of 'humps and hollows' in the resistance curve as speed increases, relative to a smoothly increasing curve. Figure 7.5.

This effect was shown experimentally by Froude (1877) by testing models with varying lengths of parallel middle body but the same forward and after ends. Figure 7.6 illustrates some of these early results.

The distance between the two pressure systems is approximately 0.9 L. The condition therefore that a crest or trough from the bow system should coincide with the first stern trough is:

$$V^2/0.9\,L = g/N\pi$$

The troughs will coincide when N is an odd integer and for even values of N crests from the bow coincide with the stern trough. The most pronounced hump occurs when $N = 1$ and this hump is termed the *main hump*. The hump at $N = 3$ is often called the *prismatic hump* as it is greatly affected by the ship's prismatic coefficient.

Scaling Wave-Making Resistance

It has been shown that for geometrically similar bodies moving at corresponding speeds, the wave pattern generated is similar and the wave-making resistance can be taken as proportional to the displacements of the bodies

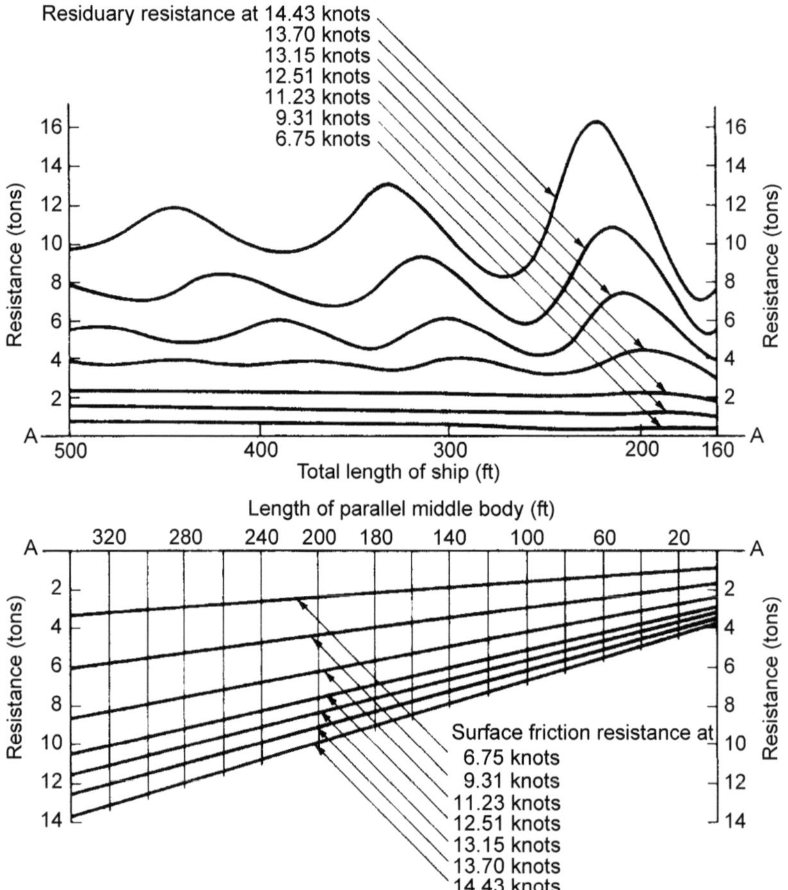

FIGURE 7.6 Resistance curves.

concerned. This assumes that wave making is unaffected by the viscosity, which is the usual assumption made in such studies. In fact there will be some viscosity effect but this will be confined largely to the boundary layer. To a first order then, the effect of viscosity on wave-making resistance can be regarded as that of modifying the hull shape in conformity with the boundary layer addition. These effects are relatively more pronounced at model scale than the full scale which means there is some scale effect on wave-making resistance. This is ignored in this book.

Frictional Resistance

Water is viscous and the conditions for dynamic similarity are geometric similarity and constancy of Reynolds' number. Due to the viscosity the particles immediately adjacent to the hull adhere to it and move at the speed of the ship. At a distance from the hull the water is at rest. There is a velocity gradient which is greatest close to the hull. The volume of water which moves with the body is known as the *boundary layer*. Its thickness is usually defined as the distance from the hull at which the water velocity is 1% of the ship speed.

Frictional resistance is associated with Reynolds because of the study he made of flow through pipes. He showed that there are two types of flow. In the first, *laminar flow*, each fluid particle follows its own streamlined path with no mass transfer between adjacent layers. It occurs only at relatively low Reynolds' numbers. At higher numbers the steady flow pattern breaks down and becomes confused and is called *turbulent flow*.

Reynolds showed that different laws of resistance applied to the two flow types. Further, if care was taken to ensure that the fluid entered the mouth of the pipe smoothly the flow began as laminar but at some distance along the tube changed to turbulent. This occurred at a critical velocity dependent upon the pipe diameter and the fluid viscosity. For different pipe diameters, d, the critical velocity, V_c, was such that $V_c d/v$ was constant. Below the critical velocity, resistance to flow was proportional to the velocity of flow. As velocity increased above the critical value there was an unstable region where the resistance appeared to obey no simple law. At higher velocity again the flow was fully turbulent and the resistance became proportional to V raised to the power 1.723.

Reynolds' work related to pipes but qualitatively the conclusions apply to ships. There are two flow regimes – laminar and turbulent. The change from one to the other depends on a *critical Reynolds' number* and different resistance laws apply.

Calculations have been made for laminar flow past a flat surface, length L and wetted surface area S, and these lead to a formula developed by Blassius as:

$$\textit{Specific resistance coefficient} = \frac{R_F}{\frac{1}{2}\rho S V^2} = 1.327\left(\frac{VL}{\nu}\right)^{-0.5}$$

Plotting the values of the specific resistance coefficient, C_F, against Reynolds' number together with results for turbulent flow past flat surfaces gives Figure 7.7.

In line with Reynolds' conclusions the resistance at higher numbers is turbulent and resistance is higher. The critical Reynolds' number at which breakdown of laminar flow occurs depends upon the smoothness of the surface and the initial turbulence present in the fluid. For a smooth flat plate it occurs at a Reynolds' number between 3×10^5 and 10^6. In turbulent flow the boundary layer still exists but in this case, besides the molecular friction force there is an interaction due to momentum transfer of fluid masses

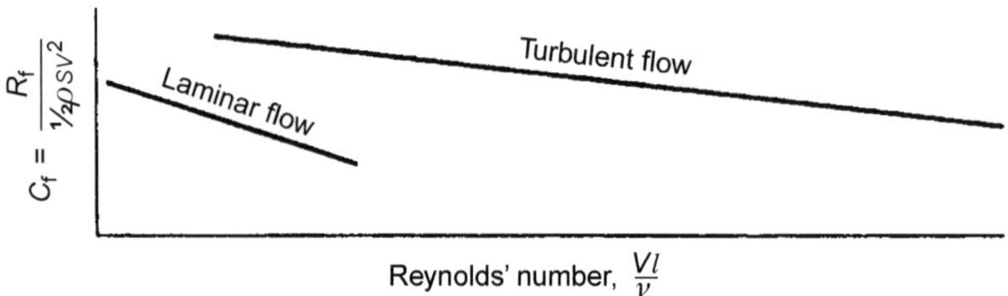

FIGURE 7.7 Laminar and turbulent flow.

between adjacent layers. The *transition* from one type of flow to the other is a matter of stability of flow. At low Reynolds' numbers, disturbances die out and the flow is stable. At the critical value the laminar flow becomes unstable and the slightest disturbance will create turbulence. The critical Reynolds' number for a flat plate is a function of the distance, l, from the leading edge and is given by:

$$\text{Critical Reynolds}' \text{ number} = Vl/\nu$$

Ahead of the point defined by l the flow is laminar. At l transition begins and after a *transition region* turbulence is fully established. For a flat plate the critical Reynolds' number is about 10^6. A curved surface is subject to a pressure gradient and this has a marked effect on transition; decreasing pressure delays transition. The thickness of the turbulent boundary layer, δx, is given by:

$$\frac{\delta x}{L} = 0.37\,(R_{\mathrm{L}})^{-0.2}$$

where L is the distance from the leading edge and R_{L} is the corresponding Reynolds' number.

Even in turbulent flow the fluid particles in contact with the surface are at rest relative to the surface. There exists a very thin *laminar sublayer*. Although thin, it is important as a body appears smooth if the surface roughness does not project through this sublayer. Such a body is said to be *hydraulically smooth*.

The existence of two flow regimes is important for model tests conducted to determine a ship's resistance. If the model is too small it may be running in the region of mixed flow. The ship will have turbulent flow over the whole hull. If the model flow was completely laminar this could be allowed for by calculation. However, this is unlikely and the small model would more probably have laminar flow forward turning to turbulent flow at some point along its length. To remove this possibility models are fitted with some form of *turbulence stimulation* at the bow. This may be a trip wire, a strip of sandpaper or a line of studs.

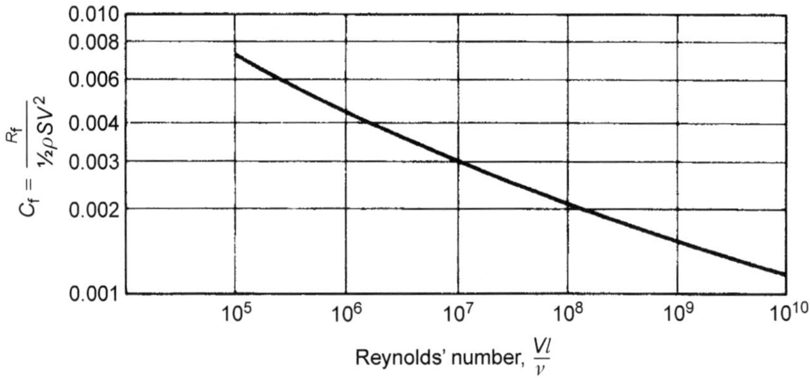

FIGURE 7.8 Schoenherr line.

Formulations of Frictional Resistance

Dimensional analysis shows that the resistance can be expressed as:

$$C_F = \frac{R_F}{0.5\rho S V^2} = F\left(\frac{\nu L}{V}\right) = \text{Function of Reynolds}' \text{ number}$$

The form the function of Reynolds' number takes has to be determined by experiment. Schoenherr (1932) developed a formula, based on all the available experimental data, in the form:

$$\frac{0.242}{(C_F)^{0.5}} = \log_{10}(R_n C_F)$$

from which Figure 7.8 is plotted.

In 1957, the ITTC (Hadler, 1958) adopted a *model-ship correlation line*, based on:

$$C_F = \frac{R_F}{0.5\rho S V^2} = \frac{0.075}{(\log_{10} R_n - 2)^2}$$

The term correlation line was used deliberately in recognition of the fact that the extrapolation from model to full scale is not governed solely by the variation in skin friction. C_F values from Schoenherr and the ITTC line are compared in Figure 7.9.

Eddy-Making Resistance

Where there are rapid changes of section the flow breaks away from the hull and eddies are created. The effects can be minimised by *streamlining* the body shape so that changes of section are more gradual. However, a typical

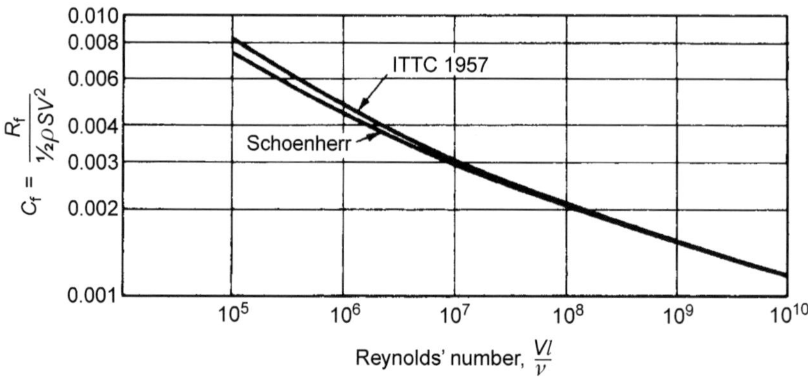

FIGURE 7.9 Comparison of Schoenherr and ITTC 1957 line.

ship has many features which are likely to generate eddies. Transom sterns, bilge keels and rudders are examples. Bilge keels are aligned with the smooth water flow lines, as determined in a circulating water channel, but at other loadings or when the ship is in waves, the bilge keels are likely to create eddies. In multi-shaft ships the shaft bracket arms are produced with streamlined sections and are aligned with the local flow. This is important for resistance and to improve the flow into the propellers.

Flow break away can occur on an apparently well-rounded form. This is due to the velocity and pressure distribution in the boundary layer. The velocity increases where the pressure decreases. Bearing in mind that the water is already moving slowly close to the hull, the pressure increase towards the stern can bring the water to a standstill or even cause a reverse flow to occur. That is the water begins to move ahead relative to the ship. Under these conditions separation occurs. The effect is more pronounced with steep pressure gradients associated with full forms.

Appendage Resistance

Appendages include rudders, bilge keels, shaft brackets and bossings, and stabilisers. Each appendage has its own characteristic length and therefore, if attached to the model, would be running at an effective Reynolds' number different from that of the main model. Although obeying the same scaling laws, the resistance of each would scale differently to the full scale so resistance models are run naked and allowance made for the resistance of appendages to give the total ship resistance. The allowances can be assessed or obtained by testing appendages separately and scaling to the ship. Fortunately the overall additions are generally relatively small, say 10–15% of the hull resistance, so errors in their assessment are not likely to be critical.

Wind Resistance

In conditions of no natural wind the air resistance is likely to be small in relation to the water resistance because water is so much denser than air. When a wind is blowing the fore and aft resistance force will depend upon its direction and speed and the shape of the ship exposed to the wind. If coming from directly ahead the relative velocity will be the sum of wind and ship speed. The resistance force will be proportional to the square of this relative velocity. Work at the National Physical Laboratory (Shearer and Lynn, 1959–1960) introduced the concept of an *ahead resistance coefficient* (ARC) defined by:

$$\text{ARC} = \frac{\text{fore and aft component of wind resistance}}{\frac{1}{2}\rho V_R^2 A_T}$$

where V_R is the relative velocity and A_T is the transverse cross-sectional area.

For a tanker, the ARC values ranged from 0.7 in the light condition to 0.85 in the loaded condition and were sensibly steady for winds from ahead and up to 50° off the bow. For winds astern and up to 40° off the stern the values were −0.6 to −0.7. Between 50° off the bow and 40° off the stern the ARC values varied approximately linearly. Two cargo ships showed similar trends but the ARC values were about 0.1 less. The figures allowed for the wind's velocity gradient with height. Because of this ARC values for small ships would be relatively greater and if the velocity was only due to ship speed they would also be greater.

CALCULATION OF RESISTANCE

The total resistance is assumed to be composed of two elements, each scaling differently. They are the residuary resistance and frictional resistance.

Residuary Resistance

This is the part of the resistance due to the shape and size of hull and comprises the:

- wave-making resistance;
- viscous pressure resistance, including:
  - *induced drag* due to the creation of vortices;
  - *boundary layer displacement drag* due to the thickening of the boundary layer towards the stern effectively modifying the hull shape; and
  - *separation of flow drag* due to the flow breaking away (separating) from the hull towards the stern. This and the boundary layer displacement drag are sometimes known collectively as the form drag which can be significant for full forms.
- *Frictional form resistance* which is the additional frictional resistance due to the fact that the hull is a three-dimensional shape rather than a flat plate.

The terminology used here is that commonly adopted by the UK Ministry of Defence.

The Calculation

Although CFD methods of calculation are used, the physical model, or data obtained from methodical model experiments, is still the principal method used for a final check. As proposed by Froude the ship resistance is obtained from that of the model by:

- Measuring the total model resistance by running it at the corresponding Froude number.
- Calculating the frictional resistance of the model and subtracting this from the total leaving the residuary resistance.
- Scaling the model residuary resistance to the full scale by multiplying by the ratio of the ship to model displacements.
- Adding a frictional resistance for the ship calculated on the basis of the resistance of a flat plate of equivalent surface area and roughness.
- Calculating, or measuring separately, the resistance of appendages.
- Making an allowance, if necessary, for air resistance.

It will be appreciated that residuary resistance is an artificial concept including a number of different physical phenomena. Also the flat plate assessment is an approximation. That is why a number of modifying factors are used by different authorities based on historical data from actual ship trials. Such factors can vary because they allow for differences in roughness of the ship and model. New hull coatings can reduce ship roughness appreciably.

ITTC Method (ITTC 7.5-02-03-01.4)

The resistance coefficient is taken as $C = (\text{Resistance})/0.5\ \rho SV^2$. Subscripts T, V, R and F denote the total, viscous, residual and frictional resistance components. Using subscripts M and S for the model and ship:

$$C_{\text{TM}} = R_{\text{TM}}/0.5\rho SV^2$$

The residual resistance is then

$$C_{\text{RM}} = C_{\text{TM}} - C_{\text{FM}}(1 + k)$$

where k is a form factor

Noting that $C_{\text{RM}} = C_{\text{RS}} = C_{\text{R}}$

The total resistance of the ship without bilge keels is:

$$C_{\text{TS}} = (1 + k)C_{\text{FS}} + C_{\text{R}} + \Delta C_{\text{F}} + C_{\text{A}} + C_{\text{AA}}$$

where ΔC_{F} is a roughness allowance, C_{A} is a correlation allowance and C_{AA} is the air resistance.

TABLE 7.1 Coefficients for the ITTC 1957 Model-Ship Correlation Line

Reynolds' Number	C_F	Reynolds' Number	C_F
10^5	0.008 333	10^8	0.002 083
5×10^5	0.005 482	5×10^8	0.001 671
10^6	0.004 688	10^9	0.001 531
5×10^6	0.003 397	5×10^9	0.001 265
10^7	0.003 000	10^{10}	0.001 172
5×10^7	0.002 309	5×10^{10}	0.000 991

The values of C_F are obtained from the ITTC model-ship correlation line for the appropriate Reynolds' number. See Table 7.1:

$$C_F = \frac{0.075}{(\log_{10} R_n - 2)^2}$$

k is determined from model tests at low speed and assumed to be independent of speed and scale. Molland et al. (2011) give a number of approximations to k one of which, due to Conn and Ferguson, is:

$$k = 18.7[C_B B/L]^2$$

The roughness allowance is calculated from:

$$\Delta C_F = 0.044[(k_s/L_{WL})^{1/3} - 10\,Re^{-1/3}] + 0.000125$$

where k_s is the roughness of hull, taken as 150×10^{-6} m, if no measured data is available.

$C_A = (5.68 - 0.6 \log Re) \times 10^{-3}$ and the air resistance is calculated from: $C_{AA} = C_{DA}(\rho_A A_{VS})/\rho_S S_S$ where A_{VS} is the projected area of the ship to the transverse plane, S_S is the ship wetted area, ρ_A is the air density. C_{DA} is the air drag coefficient of the ship above the waterline. It ranges from 0.5–1.0 but a default value of 0.8 can be taken.

If the ship is fitted with bilge keels, wetted area S_{BK} the total resistance is increased by the ratio $(S + S_{BK})/S$.

The method of extrapolating to the ship from the model is illustrated diagrammatically in Figure 7.10. It will be noted that if the friction lines used are displaced vertically but remain parallel, there will be no difference in the value of total resistance calculated for the ship. That is the actual

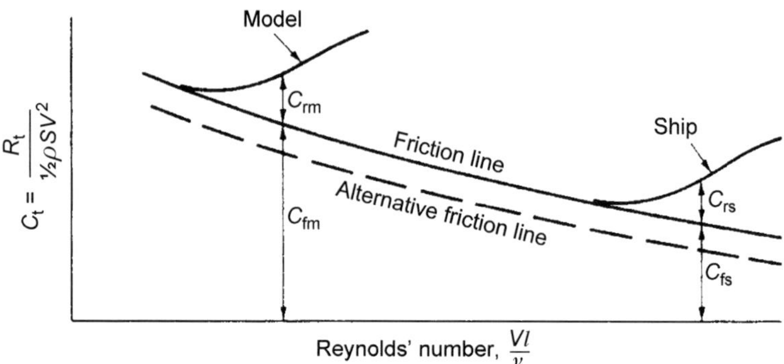

FIGURE 7.10 Extrapolation to ship.

frictional resistance taken is not critical as long as the error is the same for model and ship and all the elements making up the residuary resistance obey the Froude law of comparison. It is the slope of the skin friction line that is most important.

Notwithstanding this, the skin friction resistance should be calculated as accurately as possible so that an accurate wave-making resistance is obtained for comparing results between different forms and for comparing experimental results with theoretical calculations.

Wetted Surface Area

Frictional resistance depends on the wetted surface area of the hull. This can be found approximately from a plot the girths of the ship at various points along its length the area under the curve leading to the wetted surface area. Froude derived his circular S values by this method. A more accurate value is obtained by allowing for the inclination of the hull surface to the centreline plane, by assessing a mean hull surface length in each section as a correction to the girth readings. Alternatively an overall mean surface length can be found by averaging the distances round the waterline boundaries for a range of draughts. The wetted surface of the ship at rest is usually used.

Various approximate formulae are available for estimating wetted surface area from the principal hull parameters. With the usual notation and taking T as the draught, and Δ as the volume of displacement those proposed by various people have been

$$\text{Denny,} \quad S = L(C_B B + 1.7T)$$
$$\text{Taylor,} \quad S = C(\Delta L)^{0.5}$$

where C is a constant depending upon the breadth/draught ratio and the midship section coefficient.

Worked Example 7.1

To illustrate the use of a model in calculating ship resistance a worked example is given here. The ship is 140 m long, 19 m beam, 8.5 m draught and has a speed of 15 knots. Other details are:

Block coefficient = 0.65
Midship area coefficient = 0.98
Wetted surface area = 3300 m^2
Density of sea water = 1025 kg/m^3

Tests on a geometrically similar model 4.9 m long, run at corresponding speed, gave a total resistance of 19 N in fresh water whose density was 1000 kg/m^3.

Solution

$$\text{Speed of model} = 15\left(\frac{4.9}{140}\right)^{0.5} = 2.81 \text{ knots} = 1.44 \text{ m/s}$$

$$\text{Wetted surface of model} = 3300\left(\frac{4.9}{140}\right)^2 = 4.04 \text{ m}^2$$

$$\text{Speed of ship} = V_s = \frac{15 \times 1852}{3600} = 7.717 \text{ m/s}$$

If the kinematic viscosity for fresh water is 1.139×10^{-6} m^2/s and that for sea water is 1.188×10^{-6} m^2/s, the Reynolds' numbers can be calculated for model and ship.

$$\text{For model } R_n = \frac{4.9 \times 1.44}{1.139 \times 10^{-6}} = 6.195 \times 10^6$$

$$\text{For ship } R_n = \frac{140 \times 15 \times 1852}{3600 \times 1.188 \times 10^{-6}} = 9.094 \times 10^8$$

Schoenherr

The values of C_f for model and ship are 3.172×10^{-3} and 1.553×10^{-3}, respectively. Now:

$$C_{tm} = \frac{R_{tm}}{\frac{1}{2}\rho SV^2} = \frac{19}{\frac{1}{2} \times 1000 \times 4.04 \times 1.44^2} = 0.004536$$

$$C_{fm} = 0.003172$$
$$C_{wm} = C_{ws} = 0.001364$$
$$C_{fs} = 0.001553$$
$$C_{ts} = 0.002917$$

$$R_{ts} = \tfrac{1}{2}\rho SV^2 \times C_{ts} = \tfrac{1}{2} \times 1025 \times 3300 \times 7.717^2 \times 0.002917$$
$$= 293,800 \text{ N}$$

This makes no allowance for roughness. The usual addition for this to C_f is 0.0004. This would give a C_{ts} of 0.003317 and the resistance would be 334,100 N.

ITTC Correlation Line
This gives:

$$C_f = \frac{0.075}{(\log_{10} R_n - 2)^2}$$

which yields:

$$\text{For the model } C_{fm} = 0.003266$$
$$\text{For the ship } C_{fs} = 0.001549$$

Hence:

$$C_{wm} = C_{ws} = 0.004536 - 0.003266 = 0.001270$$
$$C_{ts} = 0.001270 + 0.001549 = 0.002819$$
$$R_{ts} = \frac{1}{2} \times 1025 \times 3300 \times 7.717^2 \times 0.002819 = 283,900 \text{ N}$$

Making the same allowance of 0.0004 for roughness, yielding $R_{ts} = 324,200$ N.

METHODICAL SERIES

Apart from tests of individual models a great deal of work has gone into ascertaining the influence of hull form changes on resistance. Tests start with a *parent form* and then a number of form parameters, which are likely to be significant, are varied systematically. Such a series of tests is called a *methodical* or *standard series*. The results show how resistance varies with the form parameters used and are useful in estimating power for new designs at an early stage. To cover n values of m variables would require m^n tests so the amount of work and time involved can be enormous. In planning a methodical series great care is needed in deciding the parameters and range of variables.

One methodical series is that carried out by Taylor. He took as variables the prismatic coefficient, displacement to length ratio and beam to draught ratio. With eight, five and two values of the variables, respectively, he tested 80 models. Taylor standardised his results on a ship length of 500 ft (152 m) and a wetted surface coefficient of 15.4. He plotted contours of R_F/Δ with $V/L^{0.5}$ and $\Delta/(L/100)^3$ as in Figure 7.11. R_F/Δ was in pounds per ton displacement. Taylor also presented correction factors for length and contours for wetted surface area correction. The residuary resistance, R_r, was plotted in a similar way but with prismatic coefficient in place of $V/L^{0.5}$ as abscissa, see Figure 7.12.

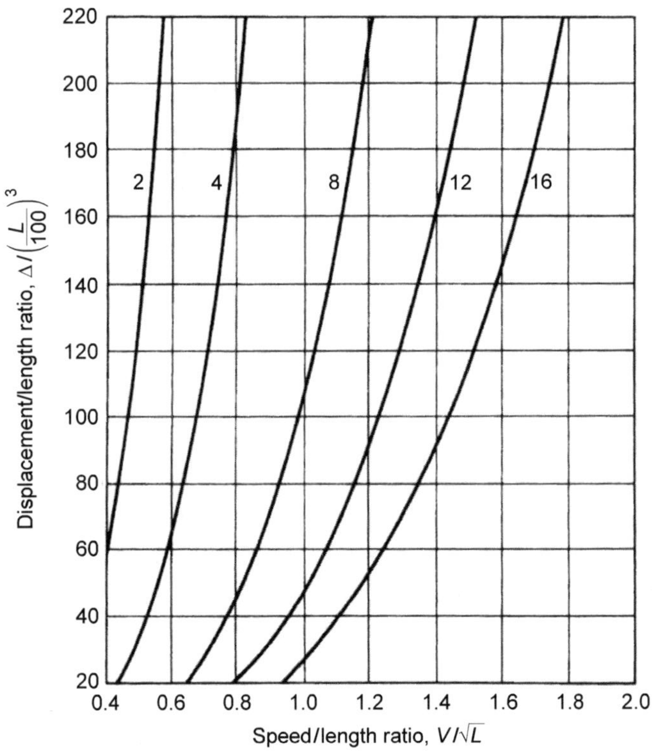

FIGURE 7.11 Contours of frictional resistance in pounds per ton displacement for 500-ft ship.

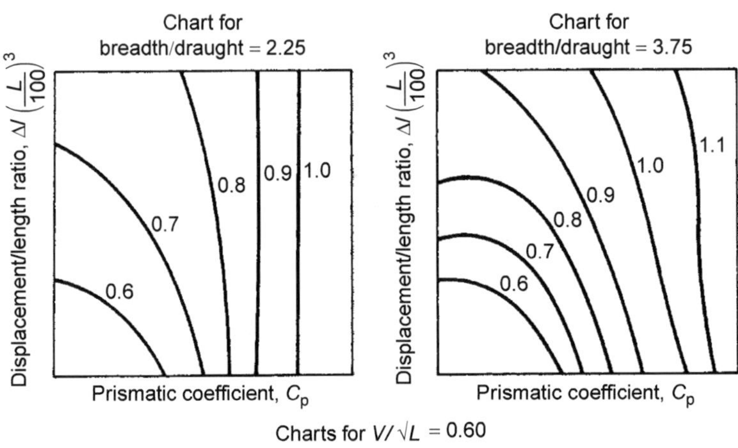

FIGURE 7.12 Taylor's contours of residuary resistance in pounds per ton displacement.

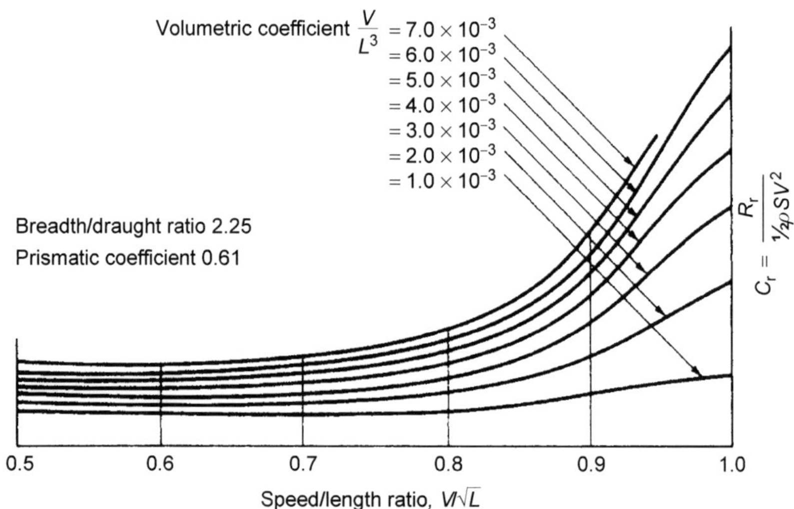

FIGURE 7.13 Typical chart from re-analysis of Taylor's data.

Taylor's data was re-analysed (Gertler, 1954) using C_F and C_r instead of resistance in pounds per ton of displacement. Frictional resistance was calculated from the Schoenherr formula rather than being based on the Froude data used by Taylor. A typical chart from the re-analysed data is given in Figure 7.13.

More recent methodical series for merchant ships have been by the then British Shipbuilding Research Association (BSRA), the David Taylor Model Basin (DTMB) and Wageningen (now MARIN). The BSRA tests varied block coefficient, length to displacement ratio, breadth to draught ratio and longitudinal position of the centre of buoyancy (LCB). Data was presented in circular *C* form to a base of block coefficient for various speeds. Correction factors are presented for the variation in the other parameters. The forms represent single screw ships with cruiser sterns. The DTMB data covers the same variables as the BSRA tests. Data is presented in circular *C* form and uses both the Froude skin friction correction and the ITTC 1957 ship-model correlation line.

A designer must consult the methodical series data directly in order to use it to estimate the resistance of a new design. Unless the new design is of the type and within the general range of the variables covered by the methodical series errors are likely. In this case other data may be available from which to deduce correction factors. The forms used in the methodical series are now somewhat dated and must be used with care. Use can also be made of methodical series data built up by numerical calculations (the mathematical ship tank) or direct estimates can be made. The final form needs to be checked by model tests.

ROUGHNESS

Apart from the wetted surface area and speed the major factor in determining the frictional resistance is the roughness of the hull. For slow ships the frictional resistance is the major part of the total and it is important to keep the hull as smooth as possible.

Owing to the increase in boundary layer thickness, the ratio of a given roughness amplitude to boundary layer thickness decreases along the length of the hull. Protrusions have less effect at the after end than forward. In the towing trials of HMS *Penelope*, the hull roughness, measured by a wall roughness gauge, was found to be 0.3 mm mean apparent amplitude per 50 mm. This mean apparent amplitude per 50 mm gauge length is the standard parameter used in the United Kingdom to represent hull roughness. Roughness can be considered under three headings:

- *Structural roughness*. This depends upon the design and method of construction. A riveted ship was rougher than a welded ship, but in welded hulls the plating exhibits a waviness between frames, particularly in thin plating. This is also a form of roughness.
- *Corrosion*. Steel corrodes in sea water creating a roughened surface. Modern painting systems are reasonably effective in reducing corrosion all the while the coating remains intact. If it is abraded in one area then corrosion is concentrated at that spot and pitting can be severe. This is bad from the structural point of view as well as for frictional resistance. To reduce corrosion, many ships are fitted with cathodic protection systems.
- *Fouling*. Marine organisms such as weed and barnacles can attach themselves to the hull. This would represent a very severe roughening if steps were not taken to prevent it. Traditionally the underwater hull has been coated with anti-fouling compositions. Early treatments contained toxic materials such as compounds of mercury or copper which leached out into the water and prevented the marine growth taking a hold on the hull. Unfortunately these compounds pollute the general ocean and other treatments are now used. Fouling is very dependent upon the time a ship spends in port relative to its time at sea, and the ocean areas in which it operates. Fouling increases more rapidly in port and in warmer waters. In the *Lucy Ashton* towing trials it was found that the frictional resistance increased by about 5% over 40 days, that is by about 1/8 of 1% per day. This was a common allowance made for time out of dock but with modern coatings a lower allowance is appropriate.

For an operator the deterioration of the hull surface with time results in a slower speed for a given power or more power being needed for a given speed. This increases running costs which must be set against the costs of docking, cleaning the underwater hull and applying new coatings.

The Schoenherr and ITTC resistance formulations were intended to apply to a perfectly smooth surface. This will not be true even for a newly completed ship. The usual allowance for roughness was to increase the frictional coefficient by 0.0004 for a new ship but with modern coatings a figure of 0.00025 would be more realistic. The ITTC-recommended allowance was given above. In addition an allowance must be made for the increase in frictional resistance with time out of dock. This has been estimated to be 10% and 40% of the frictional resistance per year, depending upon the service details.

RESISTANCE IN SHALLOW WATER

In shallow water the flow around the hull is modified and the pressures acting on the hull change. Also in water of depth h waves are propagated at a velocity, C, given by:

$$C^2/gh = (\lambda/2\pi h)[\tanh(2\pi\lambda)]$$

where tan$h$ is a trigonometrical function.

In deep water tan$h(2\pi\lambda)$ tends to unity and, when h/λ tends to zero, tan$h(2\pi\lambda)$ tends to $2\pi\lambda$ and $C^2 = gh$, no longer depending upon wavelength and waves of different length propagate at the same speed. The quantity $(gh)^{0.5}$ is known as the *critical speed.*

In shallow water, below the critical speed resistance is greater than in deep water and it increases rapidly as that speed is approached. If the ship has enough power to go faster than the critical speed the resistance begins to fall and may become a little less than that in deep water for the same speed.

FORM PARAMETERS AND RESISTANCE

There can be no absolutes in terms of optimum form and a designer makes many compromises. Even in terms of resistance one form may be better than another at one speed but inferior at another speed. It is important, therefore, to decide whether emphasis is on maximum or cruising speed, particularly in warships where the difference between the two can be marked. Another complication is the interdependence of many form factors, including those chosen for discussion below, and only generalised comments are possible.

Frictional resistance is directly related to the wetted surface area and any reduction in this will reduce skin friction resistance. This is not, however, a parameter that can be changed in isolation from others. Other form changes may have most effect on wave-making resistance but also affect frictional resistance because of consequential changes in surface area and flow around the hull.

Length

This is the most influential parameter. An increase increases frictional resistance but usually reduces wave-making resistance although this is complicated by the interaction of the bow and stern wave systems. Thus while fast ships (Froude number more than about 0.35) will benefit overall from being longer than slow ships (Froude number less than about 0.25), there will be bands of length in which the benefits will be greater or less.

Length to Beam Ratio

A higher value is beneficial for faster ships.

Prismatic Coefficient

The main effect, particularly close to the prismatic hump, is on wave-making resistance and choice of prismatic coefficient is not so important for slow ships where it is likely to be chosen to give better cargo-carrying capacity.

Fullness of Form

Fullness may be represented by the block or prismatic coefficient. For most ships resistance will increase as either coefficient increases as a full ship creates a greater disturbance as it moves through the water. There is evidence of optimum values of the coefficients on either side of which the resistance might be expected to rise. This optimum might be in the working range of high-speed ships but is usually well below practical values for slow ships. Generally the block coefficient should be reduced as the desired ship speed increases.

In moderate-speed ships, power can always be reduced by reducing block coefficient so that machinery and fuel weights can be reduced. However, for given overall dimensions, a lower block coefficient means less payload. A balance must be struck between payload and resistance based on a study of the economics of running the ship.

Slimness can be defined by the volumetric coefficient which is the volume of displacement divided by the cube of the length. For a given length, greater volume of displacement requires steeper angles of entrance and run for the waterplane endings. Increase in volumetric coefficient can be expected to lead to increased resistance. Generally in high-speed forms with low block coefficient, the displacement length ratio must be kept low to avoid excessive wave resistance. For slow ships this is not so important. Fast ships require larger length to beam ratios than slow ships.

Breadth to Draught Ratio

This ratio is generally dictated by stability requirements having a secondary effect on resistance. Generally resistance increases with increase in breadth to draught ratio within the normal working range of this variable. This can again be explained by the angles at the ends of the waterlines increasing and causing more disturbance. With very high values of beam to draught ratio the flow around the hull would tend to be in the vertical plane rather than the horizontal. This could lead to a reduction in resistance.

Longitudinal Distribution of Displacement

Even when the main hull parameters have been fixed it is possible to vary the distribution of displacement along the ship length. This distribution can be characterised by the LCB. For a given block coefficient the LCB position governs the fullness of the ends of the ship. As the LCB moves towards one end that end will become fuller and the other finer. There will be a position where the overall resistance will be minimised. This generally varies from just forward of amidships for slow ships to about 5–10% of the length aft of amidships for fast ships. The distribution of displacement along the length the curve of areas should be smooth. Sudden changes of curvature could cause waves or eddies.

Length of Parallel Middle Body

In high-speed ships with low block coefficient there is usually no parallel middle body. To get maximum capacity at minimum cost, high block coefficients are used with parallel middle body to avoid the ends becoming too full. For a given block coefficient, as the length of parallel middle body increases the ends become finer. There will be an optimum value of parallel middle body for a given block coefficient.

Section Shape

It is not possible to generalise on the shape of section to adopt but slow- to moderate-speed ships tend to have U-shaped sections in the fore body and V-shaped sections aft. It can be argued that the U-sections forward keep more of the ship's volume away from the waterline and so reduce wave making. The angle of entrance of the waterplane should be lower at higher speeds, the half angle reducing, say, from 20° at low speed to 10° at high speed.

Bulbous Bow

Originally the bulbous bow was sized, shaped and positioned so as to create a bow wave system partially cancelling out the ship's own bow wave system and reducing wave-making resistance. This applied only over a limited speed range and at the expense of resistance at other speeds. Many merchant ships operate at a steady speed for much of their lives so the bulb was designed for that speed. It was initially applied to moderate- to high-speed ships but has also been found to be beneficial in relatively slow ships such as bulk carriers, which now often have bulbous bows. The effectiveness of the bulb in the slower ships, where wave-making resistance is small, suggests the bulb reduces frictional resistance as well. This is due to the change in flow velocities over the hull and into the propeller, and the benefit is mainly in the ballast condition. Sometimes the bulb is sited well forward and it can extend beyond the fore perpendicular.

Transom Area Ratio

A transom stern can cause a high-resistance penalty at speeds below that at which flow breaks away. Transoms show reduced resistance at Froude numbers greater than about 0.38. Fast forms tend to trim by the stern at speed and the basic design should aim for a low immersed transom area. Transom flaps, to control stern trim, have been found beneficial in some applications.

Triplets

The designer cannot be sure of the change in resistance of a form, as a result of small changes, unless data is available for a similar form. However, changes are often necessary in the early design stages and it is desirable that their consequences are known. One approach is to run a set of three models. One is the base model and the other two have one parameter varied by a small amount. Typically the parameters changed would be beam and length and the variation would be a simple linear expansion of about 10% in the chosen direction. Only one parameter is varied at a time and the models are not geometrically similar. The variation in resistance, or its effective power, of the form can be expressed as:

$$\frac{\mathrm{d}R}{R} = \frac{a_1\,\mathrm{d}L}{L} + \frac{a_2\,\mathrm{d}B}{B} + \frac{a_3\,\mathrm{d}T}{T}$$

The values of a_1, a_2, etc. can be deduced from the results of the three experiments.

MODEL EXPERIMENTS

Full-scale resistance trials are difficult and expensive. Most data on ship resistance have been gained from model experiments. W. Froude was the pioneer of model testing and the towing tank he opened in Torquay in 1872 was the first of its kind. The tank was a channel about 85 m long, 11 m wide and 3 m deep. Over this channel ran a carriage, towed at uniform speed by an endless rope carrying a dynamometer. Models were attached to the carriage through the dynamometer and their resistances were measured by the extension of a spring. Models were made of paraffin wax which is easily shaped and altered. Since Froude's time great advances have been made in the design of tanks, their carriages and the recording equipment. However, the basic principles remain the same. Every maritime nation now has towing tanks and the ITTC provides a list of facilities on their website. An average good form can be improved by 3–5% by model tests, hence fuel savings pay for all the testing.

Early work on ship models was carried out in smooth water. Most resistance testing is still in this condition but now tanks are fitted with wavemakers so that the added resistance in waves can be studied. Wavemakers are fitted to one end of the tank and can generate regular or long crested irregular waves. For these experiments the model must be free to heave and pitch and these motions are recorded as well as the resistance. In towing tanks, testing is limited to head and following seas. Large seakeeping basins are discussed in the chapter on seakeeping and these can be used to determine model performance when manoeuvring in waves.

FULL-SCALE TRIALS

The final test of the accuracy of any prediction method based on extrapolation from models must be the resistance of the ship itself. This cannot be found from normal speed trials although the overall accuracy of power estimation can be checked by them. In special trials to measure a ship's resistance it is vital to ensure that the ship under test is running in open, smooth water and that the method of towing or propelling does not interfere with the flow of water around the test vessel. Towing has been the usual method adopted.

The earliest tests were conducted by Froude on HMS *Greyhound* in 1874. *Greyhound* was a screw sloop and was towed by HMS *Active*, a vessel of about 30.9 MN displacement, using a 58 m towrope attached to the end of a 13.7 m outrigger in *Active*. Tests were carried out with *Greyhound* at three displacements ranging from 11.57 MN to 9.35 MN, and over a speed range of 3–12.5 knots. The pull in the towrope was measured by dynamometer and speed by a log. Results were compared with those derived from a model of *Greyhound* and showed that the curve of resistance against speed was of the same character as that from the model but somewhat higher. This was attributed to the greater roughness of the ship surface than that assumed in the calculations. The experiment effectively verified Froude's law of comparison.

In the late 1940s, the British Ship Research Association carried out full-scale tests on the *Lucy Ashton*. The problems of towing were overcome by fitting the ship with four jet engines mounted high upon the ship and outboard of the hull to avoid the jet efflux impinging on the ship or its wake. Most of the tests were at a displacement of 3.9 MN with speeds ranging from 5 to 15 knots. Results were compared with tests on six geometrically similar models of lengths ranging from 2.7 to 9.1 m. Estimates of the ship resistance were made from each model using various skin friction formulae, including those of Froude and Schoenherr, and the results compared to the ship measurements. Generally the Schoenherr formulae gave better results. The trials showed that the full-scale resistance is sensitive to small roughness and the better outer bottom coatings gave about 5% less skin friction resistance. Fairing the seams gave a reduction of about 3% in total resistance. Forty days fouling on the bituminous aluminium hull increased skin frictional resistance by about 5%. The results indicated that the interference between skin friction and wave-making resistance was not significant over the range of the tests.

Later trials were conducted by the Admiralty Experiment Works, Haslar on the frigate HMS *Penelope* which was towed using a mile-long nylon rope. The main purpose of the trial was to measure radiated noise and vibration for a dead ship. Both propellers were removed and the wake pattern measured by a pitot fitted to one shaft. Propulsion data for *Penelope* were obtained from separate measured mile trials with three sets of propellers. Correlation of ship and model data showed the ship resistance to be some 14% higher than predicted over the speed range 12–13 knots. There appeared to be no significant wake scale effects. Propulsion data showed higher thrust, torque and efficiency than predicted.

EFFECTIVE POWER

The *effective power* at any speed is defined as the power needed to overcome the resistance of the naked hull at that speed. It is the power that would be expended if the ship were to be towed through the water without the flow around it being affected by the means of towing. Another, higher, effective power would apply if the ship were towed with its appendages fitted. The ratio of this power to that needed for the naked ship is known as the *appendage coefficient*. That is:

$$\text{the appendage coefficient} = \frac{\text{Effective power with appendages}}{\text{Effective power naked}}$$

Froude, because he dealt with imperial units, used the term *effective horsepower* or *ehp*. Even in mathematical equations the abbreviation ehp was used. The abbreviation now used is P_E.

For a given speed the effective power is the product of the total resistance and the speed. Thus returning to the earlier worked example, the effective powers for the two cases considered, would be:

- Using Schoenherr, total resistance = 334,100 N, allowing for roughness and effective power = 2578 kW
- Using the ITTC line, total resistance = 324,200 N and effective power = 2502 kW

This concept is developed in the next chapter to find the power required of the main machinery in driving the ship at the given speed.

SUMMARY

Resistance and propulsion are interdependent and the separation of the two is artificial although convenient. There is resistance to the passage of a ship through the water. The resistance of the naked hull measured in model tests can be considered as comprising two components, the frictional and the residuary resistance. These components scale differently in moving from the model to full scale. The residuary resistance, for geometrically similar hulls at corresponding speeds, scales as the ratio of the displacements. The frictional resistance component is estimated from experimental data and scaled in relation to Reynolds' number. The naked hull resistance must take account of surface roughness and be increased to allow for appendages. Where necessary an allowance can be made for the resistance of the above water form due to its passage through the air although in the absence of a natural wind this is likely to be small.

Model tests may be of an individual ship or a methodical series. A few full-scale towing tests have been carried out to validate model predictions. The effective power of a hull is the power needed to move the naked hull at a given speed.

Chapter 8

Propulsion

INTRODUCTION

The concept of effective power is the starting point for discussing the propulsion of the ship. This chapter discusses the means of producing the driving force together with the interaction between the propulsor and the flow around the hull. It is convenient to study the propulsor performance in open still water and then the change in that performance when placed close behind a ship. The situation is more complex with the ship in waves. Concerns related to the pollution of the atmosphere by funnel gases are leading to more attention being paid to reducing installed power and to the use of cleaner fuels.

GENERAL PRINCIPLES

When a propulsor is introduced it modifies the flow at the stern causing an augmentation of the resistance experienced by the hull and affecting the average velocity of water through the propulsor. This will not be the same as the ship speed through the water. These two effects are taken together as a measure of hull efficiency. The other effect of the combined hull and propulsor is that the flow through the propulsor is not uniform and generally not along the propulsor axis. The ratio of the propulsor efficiency in open water to that behind the ship is termed the relative rotative efficiency. There will also be losses in the transmission of power between the main machinery and the propulsor.

Extension of the Effective Power Concept

If the installed power is the *shaft power* (P_S) then the *overall propulsive efficiency* is determined by the *propulsive coefficient*, where

$$\text{Propulsive coefficient (PC)} = \frac{P_E}{P_S}$$

Introduction to Naval Architecture. DOI: http://dx.doi.org/10.1016/B978-0-08-098237-3.00008-4

The intermediate stages in moving from the effective to the shaft power are usually taken as:

Effective power for a hull with appendages $= P'_E$
Thrust power developed by propulsors $= P_T$
Power delivered by propulsors when propelling ship $= P_D$
Power delivered by propulsors when in open water $= P'_D$.

With this notation the overall propulsive efficiency can be written:

$$\text{PC} = \frac{P_E}{P_S} = \frac{P_E}{P'_E} \times \frac{P'_E}{P_T} \times \frac{P_T}{P'_D} \times \frac{P'_D}{P_D} \times \frac{P_D}{P_S}$$

The term P_E/P'_E is the inverse of the *appendage coefficient*. The other terms in the expression are a series of efficiencies which are termed, and defined, as follows:

$P'_E/P_T =$ hull efficiency $= \eta_H$
$P_T/P'_D =$ propulsor efficiency in open water $= \eta_O$
$P'_D/P_D =$ relative rotative efficiency $= \eta_R$
$P_D/P_S =$ shaft transmission efficiency.

This can be written:

$$\text{PC} = \left(\frac{\eta_H \times \eta_O \times \eta_R}{\text{appendage coefficient}}\right) \times \text{transmission efficiency}$$

The expression in brackets is termed the *quasi-propulsive coefficient* (QPC) and is denoted by η_D. The QPC is obtained from model experiments and to allow for errors in applying this to the full scale an additional factor is needed. Some authorities use a *QPC factor* (QPCF) which is the ratio of the propulsive coefficient determined from a ship trial to the QPC obtained from the corresponding model. Others use a *load factor*, where

$$\text{Load factor} = (1 + x) = \frac{\text{transmission efficiency}}{\text{QPC factor} \times \text{appendage coefficient}}$$

In this expression the *overload fraction*, x, is meant to allow for hull roughness, fouling and weather conditions on trial.

Some authorities use P_{EA} for the effective power of the hull with appendages. More importantly some use the term propulsive coefficient as the ratio P'_E/P_S. It is important in using data from any source to check the definitions used.

It remains to establish how the hull, propulsor and relative rotative efficiencies can be determined. This is dealt with later.

PROPULSORS

Propulsion devices take many forms, all relying upon imparting momentum to a mass of fluid which causes a force to act on the ship. In the case of air cushion vehicles the fluid is air but usually it is water. The most common device is the propeller. This may take various forms, but attention in this chapter is focused on the fixed pitch propeller. (Pitch is constant but may change with radius.) Before defining such a propeller it is instructive to consider the general case of a simple actuator disc imparting momentum to water.

Momentum Theory

The propeller is replaced by an actuator disc, area A, which is assumed to be working in an ideal fluid. The disc imparts an axial acceleration to the water which, in accordance with Bernoulli's principle, requires a change in pressure at the disc (Figure 8.1).

It is assumed that the water is initially, and finally, at pressure p_O. At the actuator disc it receives an incremental pressure increase d_p. The water, initially at rest, achieves a velocity aV_a at the disc, goes on accelerating and finally has a velocity bV_a at infinity behind the disc. The disc is moving at a velocity V_a relative to the still water. Assuming the velocity

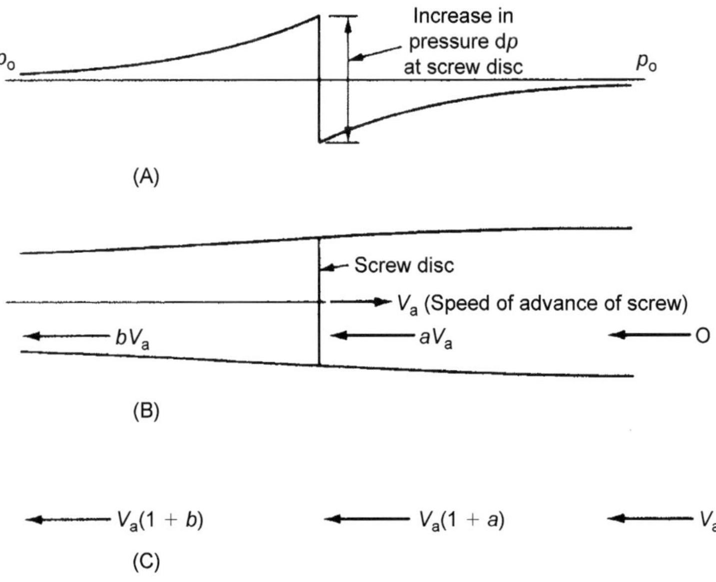

FIGURE 8.1 (A) Pressure; (B) absolute velocity; and (C) velocity of water relative to screw.

increment is uniform across the disc and only the column of water passing through the disc is affected:

$$\text{Velocity of water relative to the disc} = V_a(1 + a)$$

where a is termed the *axial inflow factor*, and:

$$\text{Mass of water acted on in unit time (passing through the disc)} = \rho A V_a(1 + a)$$

Since this mass finally achieves a velocity bV_a, the change of momentum in unit time is

$$\rho A V_a(1 + a)bV_a$$

Equating this to the thrust generated by the disc:

$$T = \rho A V_a^2(1 + a)b$$

The work done by the thrust on the water is

$$TaV_a = \rho A V_a^3(1 + a)ab$$

This is equal to the kinetic energy in the water column:

$$\frac{\rho A V_a(1 + a)(bV_a)^2}{2}$$

Equating this to the work done by the thrust:

$$\rho A V_a^3(1 + a)ab = \frac{\rho A V_a^3(1 + a)b^2}{2} \quad \text{and} \quad a = \frac{b}{2}$$

That is, half the velocity ultimately reached is acquired by the time the water reaches the disc. Thus the effect of a propulsor on the flow around the hull, and therefore the hull's resistance, extends both ahead and astern of the propulsor.

The useful work done by the propeller is equal to the thrust multiplied by its forward velocity. The total work done is this plus the work done in accelerating the water so:

$$\text{Total work} = \rho A V_a^3(1 + a)ab + \rho A V_a^3(1 + a)b$$

The efficiency of the disc as a propulsor is the ratio of the useful work to the total work. That is:

$$\text{Efficiency} = \frac{\rho A V_a^3(1 + a)b}{\rho A V_a^3[(1 + a)ab + (1 + a)b]} = \frac{1}{1 + a}$$

This is termed the *ideal efficiency*. For good efficiency a must be small and for a given speed and thrust the propulsor disc must be large. The larger the disc area the less the velocity that has to be imparted to the water for a given thrust. A lower race velocity means less energy in the race and more energy usefully employed in driving the ship.

So far it has been assumed that only an axial velocity is imparted to the water. In a real propeller, because of the rotation of the blades, the water will also have rotational motion imparted to it. Allowing for this it can be shown (Carlton, 2007) that the overall efficiency becomes:

$$\eta = \frac{1 - a'}{1 + a}$$

where a' is the *rotational inflow factor*. Thus the effect of imparting rotational velocity to the water is to reduce efficiency further.

THE SCREW PROPELLER

A screw propeller may be regarded as part of a helicoidal surface which 'screws' its way through the water.

A Helicoidal Surface

Consider a line AB, perpendicular to line AA', rotating at uniform angular velocity about AA' and moving along AA' at uniform velocity (Figure 8.2). AB sweeps out a helicoidal surface. The *pitch* of the surface is the distance travelled along AA' in making one complete revolution. A propeller with a flat face and constant pitch could be regarded as having its face trace out the helicoidal surface. If AB rotates at N revolutions per unit time, the circumferential velocity of a point, distant r from AA', is $2\pi Nr$ and the axial velocity is NP. The point travels in a direction inclined at θ to AA' such that:

$$\tan\theta = \frac{2\pi Nr}{NP} = \frac{2\pi r}{P}$$

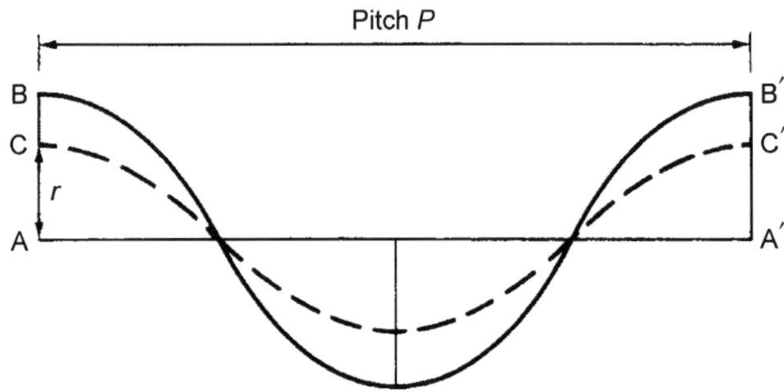

FIGURE 8.2 Creating helicoidal surface.

If the path is unwrapped and laid out flat the point will move along a straight line as shown in Figure 8.3.

Propellers can have any number of blades but three, four and five are most common in marine propellers, although special reduced noise designs often have more blades. Each blade can be regarded as part of a different helicoidal surface. In modern propellers the pitch of the blade varies with radius so that sections at different radii are not on the same helicoidal surface.

Propeller Features

The *diameter* of a propeller (Figure 8.4) is that of a circle passing tangentially through the tips of the blades. At their inner ends the blades are attached to a boss, the diameter of which is kept as small as possible consistent with strength. Blades and boss are often one casting for fixed

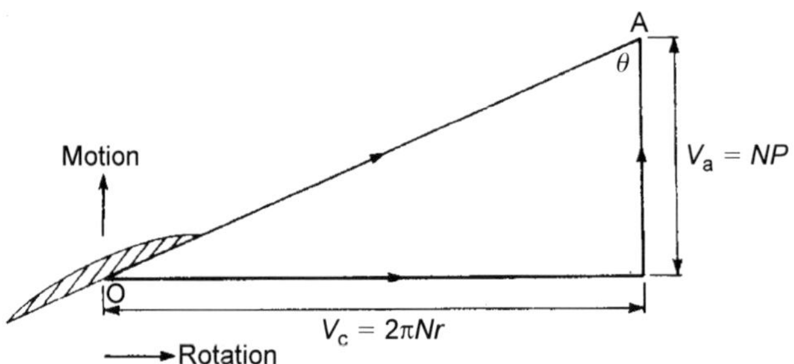

FIGURE 8.3 Point movement.

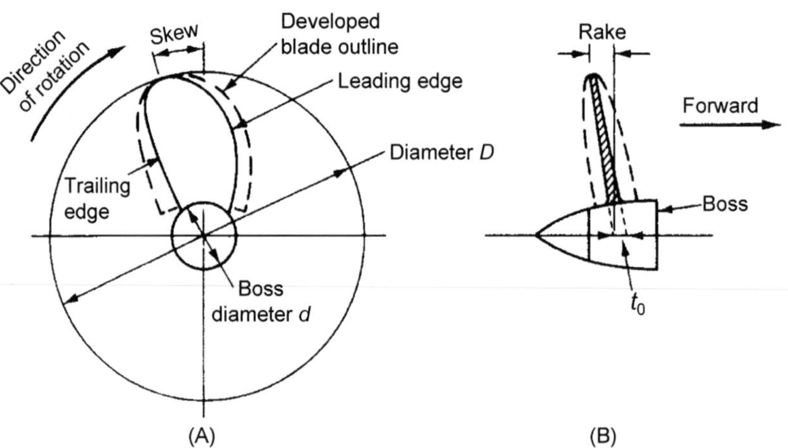

FIGURE 8.4 (A) View along shaft axis and (B) Side elevation.

pitch propellers. The boss diameter is usually expressed as a fraction of the propeller diameter.

The blade outline can be defined by its projection on to a plane normal to the shaft. This is the *projected outline*. The *developed outline* is the outline obtained if the circumferential chord of the blade, i.e. the circumferential distance across the blade at a given radius, is set out against radius. The shape is often symmetrical about a radial line called the *median*. In some propellers the median is curved back relative to the rotation of the blade. Such a propeller is said to have *skew back*. Skew is expressed in terms of the circumferential displacement of the blade tip. Skew back can be advantageous where the propeller is operating in a flow with marked circumferential variation. In some propellers the face in profile is not normal to the axis and the propeller is said to be *raked*. It may be raked forward or backward, but generally the latter to improve the clearance between the blade tip and the hull. Rake is usually expressed as a percentage of the propeller diameter.

Blade Sections

A section (Figure 8.5) is a cut through the blade at a given radius, i.e. it is the intersection between the blade and a circular cylinder. The section can be laid out flat. Early propellers had a flat face and a back in the form of a circular arc. Such a section was completely defined by the blade width and maximum thickness.

Modern propellers use aerofoil sections. The *median or camber line* is the line through the mid-thickness of the blade. The *camber* is the maximum distance between the camber line and the *chord* which is the line joining the forward and trailing edges. The camber and the maximum thickness are usually expressed as percentages of the chord length. The maximum

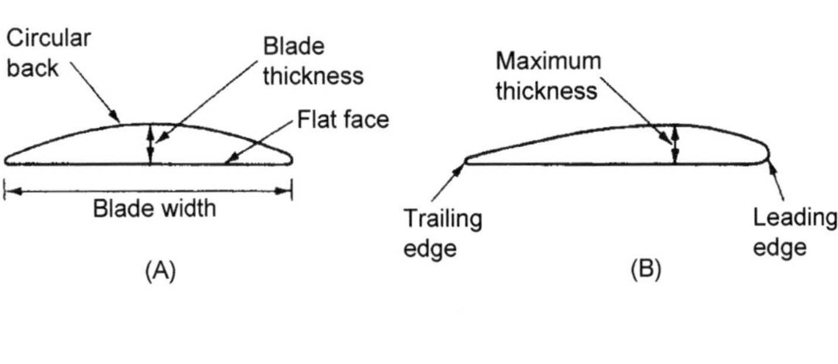

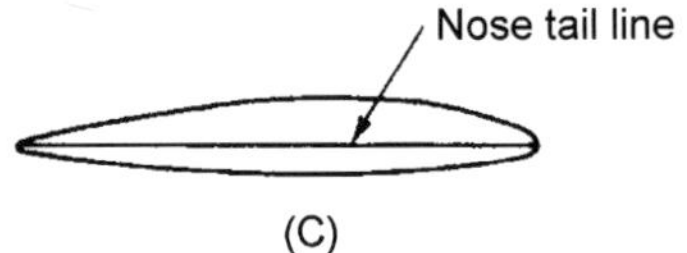

FIGURE 8.5 (A) Flat face, circular back; (B) aerofoil; and (C) cambered face.

thickness is usually forward of the mid-chord point. In a flat face circular back section the camber ratio is half the thickness ratio. For a symmetrical section the camber line ratio would be zero. For an aerofoil section the section must be defined by the ordinates of the face and back as measured from the chord line.

The maximum thickness of blade sections decreases towards the tips of the blade. The thickness is dictated by strength requirements and does not necessarily vary in a simple way with radius. In simple, small, propellers thickness may reduce linearly with radius. This distribution gives a value of thickness that would apply at the propeller axis were it not for the boss. The ratio of this thickness, t_o, to the propeller diameter is termed the *blade thickness fraction*.

Pitch Ratio

The ratio of the pitch to diameter is called the *pitch ratio*. When pitch varies with radius, that variation must be defined. For simplicity a nominal pitch is quoted being that at a certain radius. A radius of 70% of the maximum is often used for this purpose.

Blade Area

Blade area is defined as a ratio of the total area of the propeller disc.

The usual form is

$$\text{Developed blade area ratio} = \frac{\text{developed blade area}}{\text{disc area}}$$

In some earlier work, the developed blade area was increased to allow for a nominal area within the boss. The allowance varied with different authorities and care is necessary in using such data. Sometimes the projected blade area is used, leading to a *projected blade area ratio*.

Handing of Propellers

If, when viewed from aft, a propeller turns clockwise to produce ahead thrust it is said to be right handed. If it turns anti-clockwise for ahead thrust it is said to be left handed. In twin screw ships the starboard propeller is usually right handed and the port propeller left handed. In that case the propellers are said to be outward turning. Should the reverse apply they are said to be inward turning. With normal ship forms inward turning propellers sometimes introduce manoeuvring problems which can be solved by fitting outward turning screws. Tunnel stern designs can benefit from inward turning screws.

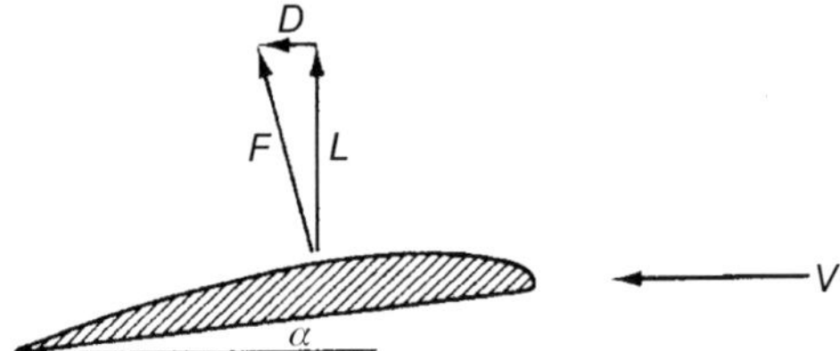

FIGURE 8.6 Forces on blade section.

Forces on a Blade Section

From dimensional analysis it can be shown that the force experienced by an aerofoil can be expressed in terms of its area, A, chord, c, and its velocity, V, as

$$\frac{F}{\rho A V^2} = f\left(\frac{v}{Vc}\right) = f(R_n)$$

Another factor affecting the force is the attitude of the aerofoil to the velocity of flow past it. This is the *angle of incidence* or *angle of attack*. Denoting this angle by α, the expression for the force becomes

$$\frac{F}{\rho A V^2} = f(R_n, \alpha)$$

This resultant force F (Figure 8.6) can be resolved into two components. That normal to the direction of flow is termed the *lift*, L, and the other in the direction of the flow is termed the *drag*, D. These two forces are expressed non-dimensionally as

$$C_L = \frac{L}{\frac{1}{2}\rho A V^2} \quad \text{and} \quad C_D = \frac{D}{\frac{1}{2}\rho A V^2}$$

Each of these coefficients will be a function of the angle of incidence and Reynolds number. For a given Reynolds number they depend on the angle of incidence only and a typical plot of lift and drag coefficients against angle of incidence is presented in Figure 8.7.

Initially the curve for the lift coefficient is practically a straight line starting from a small negative angle of incidence called the *no lift angle*. As the angle of incidence increases further the curve reduces in slope and then the coefficient begins to decrease. A steep drop occurs when the angle of incidence reaches the *stall angle* and the flow around the aerofoil breaks down. The drag coefficient has a minimum value near the zero angle of incidence, rises slowly at first and then more steeply as the angle of incidence increases.

Lift Generation

Hydrodynamic theory shows the flow round an infinitely long circular cylinder in a non-viscous fluid as shown in Figure 8.8.

At points A and B the velocity is zero and these are called *stagnation points*. The resultant force on the cylinder is zero. This flow can be transformed into the flow around an aerofoil as shown in Figure 8.9, when the

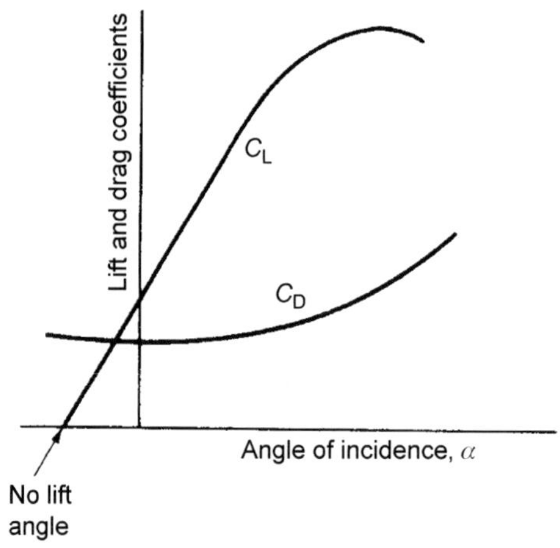

FIGURE 8.7 Lift and drag curves.

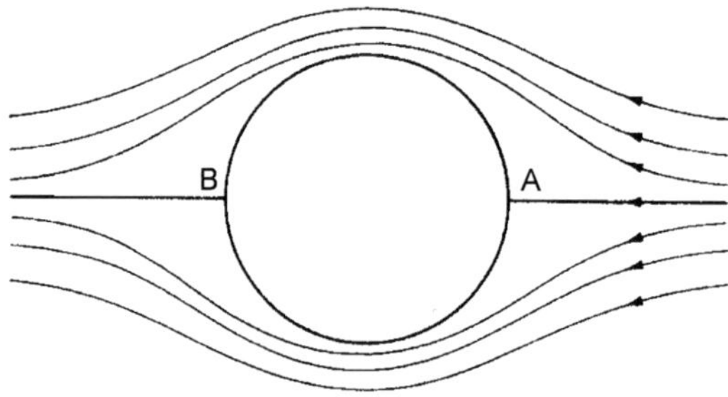

FIGURE 8.8 Flow round circular cylinder.

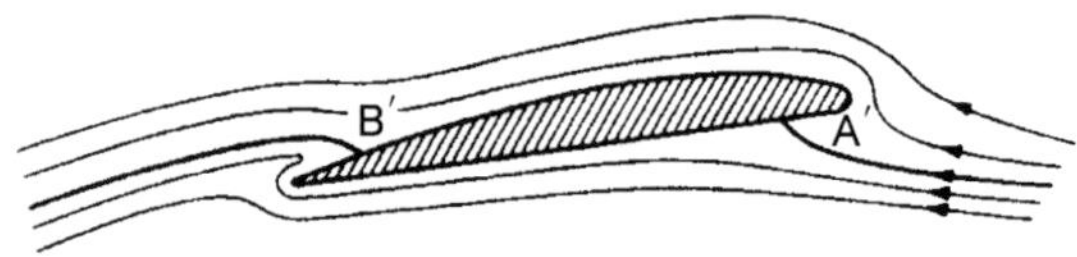

FIGURE 8.9 Flow round aerofoil without circulation.

stagnation points moving to A′ and B′. The force on the aerofoil in these conditions is also zero.

In a viscous fluid the very high velocities at the trailing edge produce an unstable situation due to shear stresses. The potential flow pattern breaks down and a stable pattern develops with one of the stagnation points at the trailing edge (Figure 8.10).

The new pattern is the original pattern with a *vortex* superimposed upon it. The vortex is centred on the aerofoil and the strength of its circulation depends upon the shape of the section and its angle of incidence. Its strength is such as to move B′ to the trailing edge. It can be shown that the lift on the aerofoil, for a given strength of circulation, τ, is

$$\text{Lift} = L = \rho V \tau$$

The fluid viscosity introduces a small drag force but has little influence on the lift generated.

The simple approach assumes an aerofoil of infinite span over which the flow is two-dimensional. The lift force is generated by the difference in pressures on the face and back of the foil. In practice an aerofoil will be finite in span and there will be a tendency for the pressures on the face and back to try to equalise at the tips by a flow around the ends of the span reducing the lift in these areas. Some lifting surfaces have plates fitted at the ends to prevent this. The effect is relatively greater the less the span in relation to the chord. This ratio of span to chord is termed the *aspect ratio*. As aspect ratio increases the lift characteristics approach more closely those of two-dimensional flow.

Pressure Distribution Around an Aerofoil

The effect of the flow past, and circulation round, the aerofoil is to increase the velocity over the back and reduce it over the face. By Bernoulli's principle there will be corresponding decreases in pressure over the back and increases over the face. Both pressure distributions contribute to the total lift, the reduced pressure over the back making the greater contribution as shown in Figure 8.11.

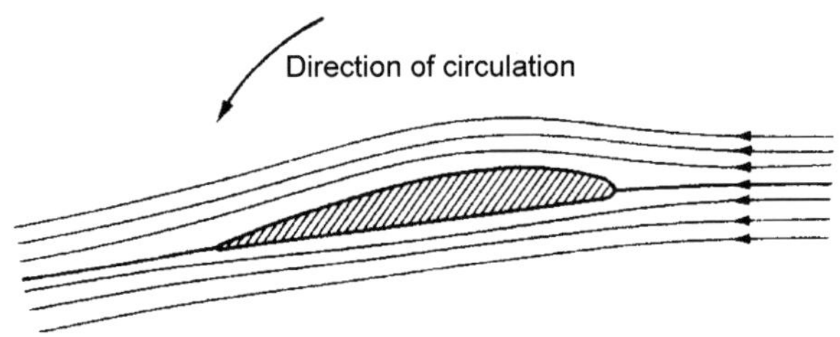

FIGURE 8.10 Flow round aerofoil with circulation.

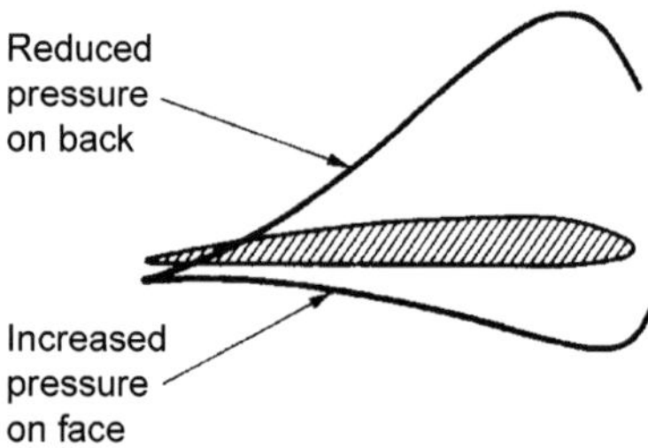

FIGURE 8.11 Pressure distribution on aerofoil.

The maximum reduction in pressure occurs at a point between the mid-chord and the leading edge. If the reduction is too great in relation to the ambient pressure in a fluid like water, bubbles form filled with air and water vapour. The bubbles are swept towards the trailing edge and they collapse as they enter an area of higher pressure. This is known as *cavitation* and is bad for noise and efficiency. The forces generated when the bubbles collapse can cause physical damage to the propeller.

PROPELLER THRUST AND TORQUE

To deal with thrust and torque the actuator disc used in the momentum theory is replaced by a screw with a large number of blades.

Blade Element Theory

Consider the forces on a radial section of a propeller blade taking account of the axial and rotational velocities at the blade as deduced from the momentum theory. The flow conditions can be represented diagrammatically as shown in Figure 8.12.

At a radial section, r, from the axis, with revolutions of N per unit time the rotational velocity is $2\pi Nr$. If the blade was a screw rotating in a solid it would advance axially at a speed NP, where P is the pitch of the blade. As water is not solid the screw actually advances at a lesser speed, V_a. The ratio V_a/ND is termed the *advance coefficient* and is denoted by J. Alternatively the propeller can be considered as having 'slipped' by an amount $NP - V_a$. The *slip* or *slip ratio* is

$$\text{Slip} = (NP - V_a)/NP = 1 - J/p$$

where p is the *pitch ratio* $= P/D$.

In Figure 8.12 the line OB represents the direction of motion of the blade relative to still water. Allowing for the axial and rotational inflow velocities,

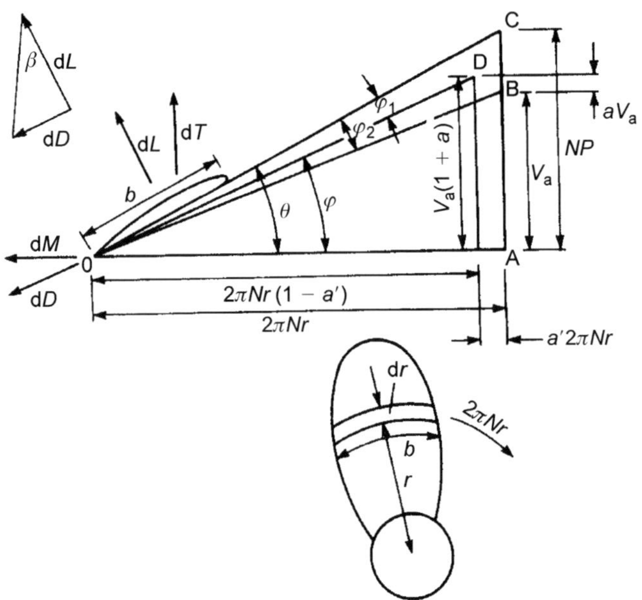

FIGURE 8.12 Forces on blade element.

the flow is along OD. The lift and drag forces on the blade element, area dA, shown are

$$\mathrm{d}L = \frac{1}{2}\rho V_1^2 C_\mathrm{L} \quad \mathrm{d}A = \frac{1}{2}\rho C_\mathrm{L}\left[V_\mathrm{a}^2(1+a)^2 + 4\pi^2 r^2(1-a')^2\right] b\,\mathrm{d}r$$

where

$$V_1^2 = V_\mathrm{a}^2(1+a)^2 + 4\pi^2 r^2(1-a')^2$$

$$\mathrm{d}D = \frac{1}{2}\rho V_1^2 C_\mathrm{D} \quad \mathrm{d}A = \frac{1}{2}\rho C_\mathrm{D}\left[V_\mathrm{a}^2(1+a)^2 + 4\pi^2 r^2(1-a')^2\right] b\,\mathrm{d}r$$

The contributions of these elemental forces to the thrust, T, on the blade follow as

$$\mathrm{d}T = \mathrm{d}L\cos\varphi - \mathrm{d}D\sin\varphi = \mathrm{d}L\left(\cos\varphi - \frac{\mathrm{d}D}{\mathrm{d}L}\sin\varphi\right)$$

$$= \frac{1}{2}\rho V_1^2 C_\mathrm{L}(\cos\varphi - \tan\beta\sin\varphi)b\,\mathrm{d}r$$

where

$$\tan\beta = \mathrm{d}D/dL = C_\mathrm{D}/C_\mathrm{L}$$

$$= \frac{1}{2}\rho V_1^2 C_L \frac{\cos(\varphi+\beta)}{\cos\beta} b\,\mathrm{d}r$$

Since

$$V_1 = V_\mathrm{a}(1+a)/\sin\varphi$$

$$\mathrm{d}T = \frac{1}{2}\rho C_\mathrm{L}\frac{V_\mathrm{a}^2(1+a)^2\cos(\varphi+\beta)}{\sin^2\varphi\cos\beta} b\,\mathrm{d}r$$

The total thrust acting is obtained by integrating this expression from the hub to the tip of the blade. Similarly, the transverse force acting on the blade element is given by

$$\mathrm{d}M = \mathrm{d}L\sin\varphi + \mathrm{d}D\cos\varphi = \mathrm{d}L\left(\sin\varphi + \frac{\mathrm{d}D}{\mathrm{d}L}\cos\varphi\right)$$

$$= \frac{1}{2}\rho V_1^2 C_\mathrm{L}\frac{\sin(\varphi+\beta)}{\cos\varphi} b\,\mathrm{d}r$$

Continuing as before, substituting for V_1 and multiplying by r to give torque:

$$\mathrm{d}Q = r\,\mathrm{d}M = \frac{1}{2}\rho C_\mathrm{L}\frac{V_\mathrm{a}^2(1+a)^2\sin(\varphi+\beta)}{\sin^2\varphi\cos\beta} br\,\mathrm{d}r$$

The total torque is obtained by integration from the hub to the tip of the blade.

The thrust power of the propeller will be proportional to TV_a and the shaft power to $2\pi NQ$. So the propeller efficiency will be $TV_\mathrm{a}/2\pi NQ$. Correspondingly there is an efficiency associated with the blade element in the ratio of the thrust to torque on the element. This is

$$\text{Blade element efficiency} = \frac{V_\mathrm{a}}{2\pi Nr}\times\frac{1}{\tan(\varphi+\beta)}$$

But from Figure 8.12,

$$\frac{V_\mathrm{a}}{2\pi Nr} = \frac{V_\mathrm{a}(1+a)}{2\pi Nr(1+a')}\times\frac{1-a'}{1+a} = \frac{1-a'}{1+a}\tan\varphi$$

This gives a blade element efficiency:

$$\frac{1-a'}{1+a}\times\frac{\tan\varphi}{\tan(\varphi+\beta)}$$

So the efficiency of the blade element is governed by the 'momentum factor' and the blade section characteristics in the form of the angles φ and β, the latter representing the ratio of the drag to lift coefficients. If β were zero the blade efficiency reduces to the ideal efficiency deduced from the momentum theory. Thus the drag on the blade leads to an additional loss of efficiency.

The simple analysis ignores many factors which have to be taken into account in more comprehensive theories. These include:

- The finite number of blades and the variation in the axial and rotational inflow factors.
- Interference effects between blades.
- The flow around the tip from face to back of the blade which produces a tip vortex modifying the lift and drag for that region of the blade.

It is not possible to cover adequately the more advanced propeller theories here and the reader should refer to a more specialist treatise (Bertram, 2012; Carlton, 2007). Theory has developed greatly in recent years due mainly to the increasing power of modern computers. As research has continued the computations have been applied to more complex propeller geometries. So that the reader is familiar with the terminology mention can be made of:

- *Lifting line models*. In these the aerofoil blade element is replaced with a single bound vortex at the radius concerned. The strength of the vortices varies with radius and the line in the radial direction about which they act is called the *lifting line*.
- *Lifting surface models*. In these the aerofoil is represented by an infinitely thin bound vortex sheet. The vortices in the sheet are adjusted to give the lifting characteristics of the blade. That is they are such as to generate the required circulation at each radial section. In some models the thickness of the sections is represented by source-sink distributions to provide the pressure distribution across the section. Pressures are needed for studying cavitation.
- *Surface vorticity models*. In this case rather than being arranged on a sheet the vortices are arranged around the section. Thus they can represent the section thickness as well as the lift characteristics.
- *Vortex lattice models*. In such models the surface of the blade and its properties are represented by a system of vortex panels.

PRESENTATION OF PROPELLER DATA

Dimensional analysis can be applied to propulsion data.

Thrust and Torque

The thrust, T, and the torque, Q, developed by a propeller depend upon:

- size as represented by its diameter, D;
- rate of revolutions, N;

- speed of advance, V_a;
- the viscosity and density of the fluid it is operating in;
- gravity.

Performance generally also depends upon the static pressure in the fluid, but this affects cavitation and is discussed later. The thrust and torque can be expressed in terms of the above variables and the fundamental dimensions of time, length and mass substituted in each. Equating the indices of the fundamental dimensions leads to a relationship:

$$T = \rho V^2 D^2 \left[f_1 \left(\frac{ND}{V_a} \right), f_2 \left(\frac{\nu}{V_a D} \right), f_3 \left(\frac{gD}{V_a^2} \right) \right]$$

This gives thrust in the units of force and the various expressions in brackets are non-dimensional. f_1 is a function of advance coefficient and is likely to be important. f_2 is a function of Reynolds number. The drag on the blades due to viscosity is likely to be small in comparison with other dynamic forces acting. Reynolds number is therefore neglected at this stage. f_3 is a function of Froude number and is concerned with gravity effects. Unless the propeller is acting close to a free surface where waves may be created, or is being tested behind a hull, it too can be ignored.

Hence for deeply immersed propellers in the non-cavitating condition, the expression for thrust reduces to

$$T = \rho V_a^2 D^2 \times f_T \left(\frac{ND}{V_a} \right)$$

For two geometrically similar propellers, operating at the same advance coefficient the expression in the brackets will be the same for both. Hence using subscripts 1 and 2 to denote the two propellers:

$$\frac{T_1}{T_2} = \frac{\rho_1}{\rho_2} \times \frac{V_{a1}^2}{V_{a2}^2} \times \frac{D_1^2}{D_2^2}$$

If it is necessary to take Froude number into account:

$$\frac{gD_1}{V_{a1}^2} = \frac{gD_2}{V_{a2}^2}$$

To satisfy both Froude number and advance coefficient:

$$\frac{T_1}{T_2} = \frac{\rho_1}{\rho_2} \times \frac{D_1^3}{D_2^3} = \frac{\rho_1}{\rho_2} \lambda^3$$

where λ is the ratio of the linear dimensions. Since ND/V_a is constant:

$$\frac{N_1}{N_2} = \frac{V_{a1}}{V_{a2}} \times \frac{D_2}{D_1} = \frac{1}{\lambda^{0.5}}$$

Thus for dynamic similarity the model propeller must rotate faster than the corresponding ship propeller in the inverse ratio of the square root of the linear dimensions.

The thrust power is the product of thrust and velocity and for the same Froude number:

$$\frac{P_{T1}}{P_{T2}} = \frac{\rho_1}{\rho_2}\lambda^{3.5}$$

Correspondingly for torque it can be shown that:

$$Q = \rho V_a^2 D^3 \times f_Q\left(\frac{ND}{V_a}\right)$$

The ratio of torques for geometrically similar propellers at the same advance coefficient and Froude number will be as the fourth power of the linear dimensions. That is

$$\frac{Q_1}{Q_2} = \frac{\rho_1}{\rho_2}\lambda^4$$

Coefficients for Presenting Data

It has been shown that

$$T = \rho V_a^2 D^2[f_T(J)] \quad \text{and} \quad Q = \rho V_a^2 D^3[f_Q(J)]$$

Substituting $V_a = NDJ$ in these expressions:

$$T = \rho N^2 D^4 J^2[f_T(J)] \quad \text{and} \quad Q = \rho N^2 D^5 J^2[f_Q(J)]$$

$J^2[f(J)]$ is a new function of J, say $F(J)$, and thus

$$T = \rho N^2 D^4 F_T(J) \quad \text{and} \quad Q = \rho N^2 D^5 F_Q(J)$$

Non-dimensional coefficients for thrust and torque are

$$K_T = T/\rho N^2 D^4 = F_T(J) \quad \text{and} \quad K_Q = Q/\rho N^2 D^5 = F_Q(J)$$

The other parameter of concern is the *propeller efficiency* which can be defined as the ratio of output to the input power. Thus:

$$\eta_o = \frac{TV_a}{2\pi QN} = \frac{K_T}{K_Q} \times \frac{J}{2\pi}$$

Thrust and torque coefficients and efficiency when plotted against advance coefficient produce plots as shown in Figure 8.13. Both thrust and

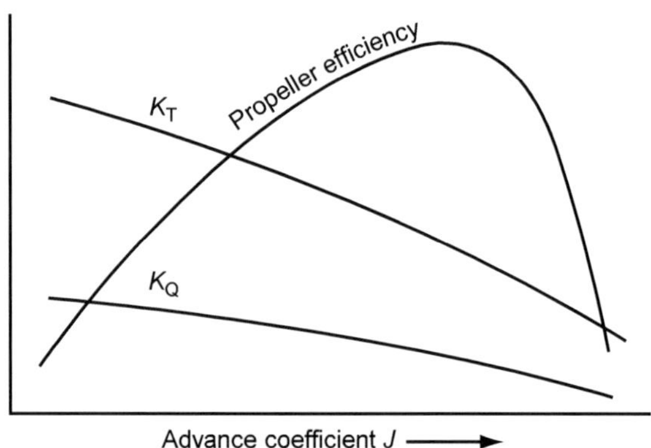

FIGURE 8.13 Thrust, torque and efficiency curves.

torque coefficients decrease with increasing advance coefficient, whereas efficiency rises to a maximum and then falls off steeply.

This format is good for presenting the data for a given propeller but not very useful for design purposes. In design the problem is usually to find the diameter and pitch of a propeller to provide the desired power at set revolutions and speed. The thrust power, P_T, is the product of thrust and speed.

$$\text{Thrust power} = TV_a = \rho V_a^3 D^2 f_T(J) = \frac{\rho V_a^5}{N^2 J^2} f_T(J)$$

That is, $P_T(N^2/\rho V_a^5) = G(J)$, where G is a new function of J.

Taylor used U to denote thrust power and using seawater as the fluid, dropped ρ and took the square root of the left-hand side of the above equation to give a coefficient B_U. He used a corresponding coefficient, B_P, for shaft power which he designated P. That is

$$B_U = \frac{NU^{0.5}}{V_a^{2.5}} \quad \text{and} \quad B_P = \frac{NP^{0.5}}{V_a^{2.5}}$$

For a series of propellers in which the only parameter varied was pitch ratio, Taylor plotted B_U or B_P against pitch ratio in the form of contours for constant δ values, δ being the reciprocal of the advance coefficient. A typical plot is shown in Figure 8.14.

To use the plot the designer decides upon a value of revolutions for a given power and advance coefficient. This gives B_U or B_P. Erecting an ordinate at this value gives a choice of values of δ from which the diameter is obtained. Associated with each diameter is a value of pitch ratio. For a given B_P the maximum efficiency that can be obtained is that defined by the efficiency contour which is tangential to the ordinate at that B_P. In other words a line of maximum efficiency can be drawn through the points where

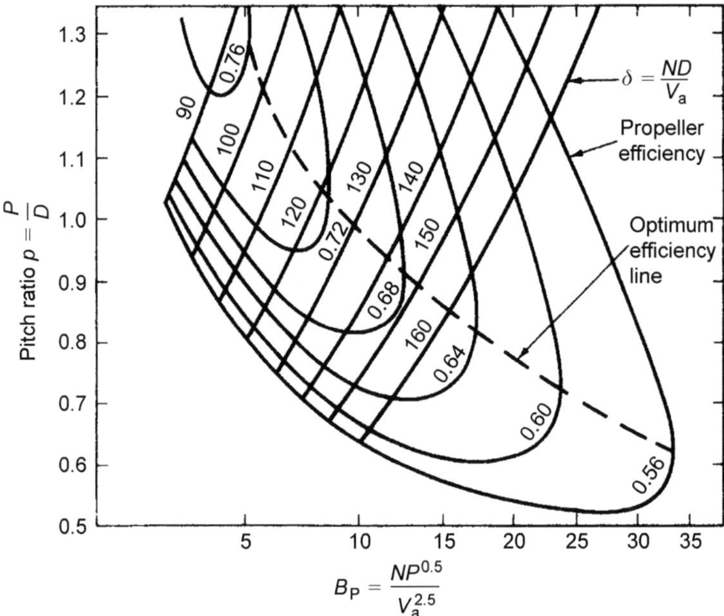

FIGURE 8.14 Typical Taylor plot.

the efficiency contours are vertical. Such a line is shown in Figure 8.14. The intersection of this line with the designer's B_P value establishes the pitch and diameter of the most efficient propeller.

Taylor used as units the horsepower, speed in knots, N in revolutions per minute and diameter in feet. With these units:

$$B_P = 33.08\left(\frac{K_Q}{J^5}\right)^{0.5} \quad \text{and} \quad \delta = \frac{ND}{V_a} = \frac{101.27}{J}$$

Keeping speed in knots and N in revolutions per minute, but putting diameter in metres and power in kilowatts:

$$B_P = 1.158\frac{NP^{0.5}}{V_a^{2.5}} \quad \text{and} \quad \delta = 3.2808\frac{ND}{V_a}$$

The Taylor method of presentation is widely used for plotting model propeller data for design purposes.

Open Water Tests

Open water tests of propellers are used in conjunction with tests behind models to determine the wake and relative rotative efficiency. Also methodical propeller testing is carried out in a towing tank. The propeller

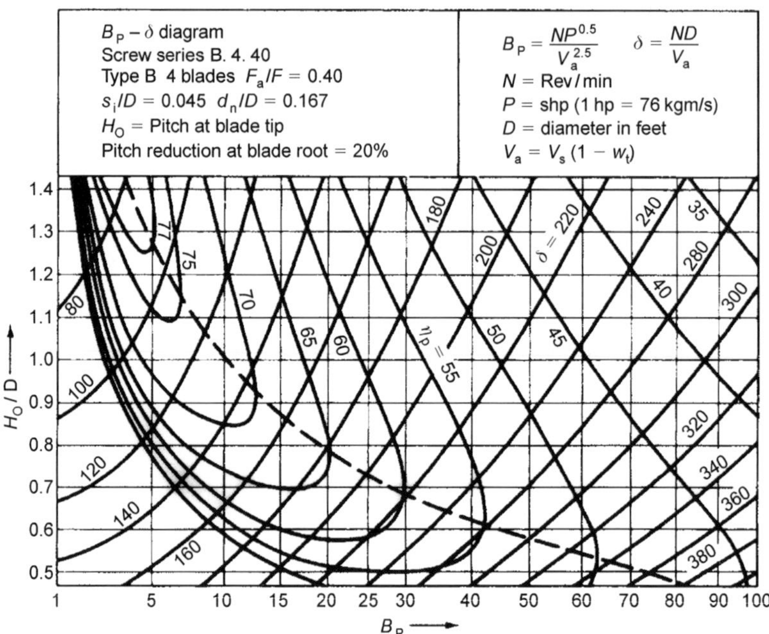

FIGURE 8.15 Propeller diagram.

is powered from the carriage through a streamlined housing. It is pushed along the tank with the propeller ahead of the housing so that the propeller is effectively in undisturbed water. Records of thrust and torque are taken for a range of carriage speeds and propeller revolutions (advance coefficient). Such tests eliminate cavitation and provide data on propeller in uniform flow. This methodical series data can be used by the designer, making allowance for the actual flow conditions a specific design is likely to experience behind the hull it is to drive.

There have been many methodical series. Those by Froude, Taylor, Gawn, Troost and van Lammeren are worthy of mention. The reader should refer to published data if it is wished to make use of these series. See Molland et al. (2011). A typical plot for a four bladed propeller from Troost's series is presented in Figure 8.15.

HULL EFFICIENCY ELEMENTS

The Propeller Behind the Ship

So far the resistance of the ship and the propeller performance have been treated in isolation. When the two are brought together there will be interaction effects.

Wake

The presence of the ship modifies the flow conditions into the propeller. The water locally will have a velocity relative to the ship and due to this *wake* the average speed of advance of the propeller through the local water will differ from the ship speed. The wake comprises three main elements:

- The velocity of the water as it passes round the hull varies, being less than average at the ends.
- Due to viscous effects the hull drags a volume of water along with it creating a boundary layer.
- The water particles in the waves created by the passage of the ship move in circular orbits.

The first two reduce the velocity of flow into the propeller. The last reduces or increases the velocity depending upon whether there is a crest or trough at the propeller position. If the net result is that the water is moving in the same direction as the ship the wake is said to be positive. This is the case for most ships but for high-speed ships, with a large wave-making component in the wake, it can become negative. The wake will vary across the propeller disc area, being higher close to the hull or behind a structural element such as a shaft bracket arm. Thus the blades operate in a changing velocity field as the propeller rotates leading to a variable angle of incidence. The pitch cannot be constantly varied to optimise the angle and an average value has to be chosen. The design of each blade section is based on the mean wake at any radius.

Model tests in a towing tank can be used to study the wake, but it must be remembered that the boundary layer thickness will be less relatively in the ship. Model data has to be modified to take account of full-scale measurements as discussed later.

In preliminary propeller design, before the detailed wake pattern is known, an average speed of flow over the whole disc is taken. This is usually expressed as a fraction of the speed of advance of the propeller or the ship speed. It is termed the *wake fraction* or the *wake factor*. Froude used the speed of advance and Taylor the ship speed in deriving the wake fraction, so that if the difference in ship and local water speed is V_{w}:

$$\text{Froude wake fraction, } \omega_{\mathrm{F}} = \frac{V_{\mathrm{w}}}{V_{\mathrm{a}}}$$

$$\text{Taylor wake fraction, } \omega_{\mathrm{T}} = \frac{V_{\mathrm{w}}}{V_{\mathrm{s}}} \quad \text{where } V_{\mathrm{w}} = V_{\mathrm{s}} - V_{\mathrm{a}}$$

Generally the wake fraction has been found to be little affected by ship speed although for ships where the wave-making component of the wake is large there will be some speed effect due to the changing wave pattern. The full-scale towing trials of HMS *Penelope* indicated no significant scale effect on the wake. (Canham, 1974)

The wake varies with the after end shape and the relative propeller position. The wake fraction can be expected to be higher for a single screw ship than for twin screws. In the former the Taylor wake fraction may be as high as 0.25–0.30.

Relative Rotative Efficiency (RRE)

The wake fraction was based on the average wake velocity across the propeller disc. As the flow varies over the disc and in general will be at an angle to the shaft line, the propeller operating in these flow conditions will have a different efficiency to that it would have in uniform flow. The ratio of the two efficiencies is called the *relative rotative efficiency*. This ratio is usually close to unity and is often taken as such in design calculations.

Augment of Resistance, Thrust Deduction

In the simple momentum theory of propeller action it was seen that the water velocity begins to increase ahead of the propeller disc. This causes a change in velocity of flow past the hull. The action of the propeller also modifies the pressure field at the stern. If a model is towed in a tank and a propeller is run behind it in the correct relative position, but run independently of the model, the resistance of the model is greater than that measured without the propeller. The propeller causes an augment in the resistance. The thrust, T, required from a propeller will be greater than the towrope resistance, R. The propeller–hull interaction effect can be regarded as an augment of resistance or a reduction in thrust. This leads to two expressions for the same phenomenon.

$$\text{Augment of resistance, } a = \frac{T - R}{R}$$

and

$$\text{Thrust deduction factor, } t = \frac{T - R}{T}$$

Hull Efficiency

Using the thrust deduction factor and Froude's notation:

$$T(1 - t) = R \quad \text{and} \quad TV_s(1 - t) = RV_s = TV_a(1 + \omega_F)(1 - t)$$

Now TV_a is the thrust power of the propeller and RV_s is the effective power for driving the ship, with appendages, at V_s. Thus:

$$P'_E = (P_T)(1 + \omega_F)(1 - t)$$

Using Taylor's notation, $P'_E = (P_T)(1 - t)/(1 - \omega_T)$.

In terms of augment of resistance $(1-t)$ can be replaced by $1/(1 + a)$.

The ratio of P'_E to P_T is called the *hull efficiency* and for most ships is a little greater than unity. This is because the propeller gains from the energy already imparted to the water by the hull. Augment and wake are functions of Reynolds number as they arise from viscous effects. The variation between model and ship is usually ignored and the error this introduces is corrected by applying a factor obtained from ship trials.

The factors augment, wake and relative rotative efficiency are collectively known as the *hull efficiency elements*.

Quasi-Propulsive Coefficient

This coefficient is obtained by dividing the product of the hull, propeller and relative rotative efficiencies by the appendage coefficient. If the overall *propulsive coefficient* is the ratio of the naked model effective power to the shaft power:

$$\text{The propulsive coefficient} = \text{QPC} \times \text{transmission efficiency}$$

The transmission efficiency can be taken as 0.98 for ships with machinery aft and 0.97 for ships with machinery amidships. The difference is due to the greater length of shafting in the latter.

Since the energy of rotation in the propeller wake represents a loss of efficiency various ways have been proposed for recovering this energy. The drive to reduce ship emissions and create a 'greener' ship makes it more important to improve propulsive efficiency. One way is to introduce a stator behind the propeller to straighten out the flow. This is done as the final stage of the integrated unit known as a pump jet. In other applications a stator is mounted separately on the forward side of the fixed rudder support; the hull is shaped to impart a pre-swirl to the water entering the propeller or the rudder itself is designed so that the wake is diverted in slightly different directions above and below the propeller axis. Modern computational fluid dynamics methods make it possible to design such features with some accuracy instead of relying on tunnel tests or trial and error.

Determining Hull Efficiency Elements

Having outlined the elements involved in propulsion it remains to quantify them. This can be done in a series of model tests. The model is fitted with propellers which are driven through a dynamometer which registers the shaft

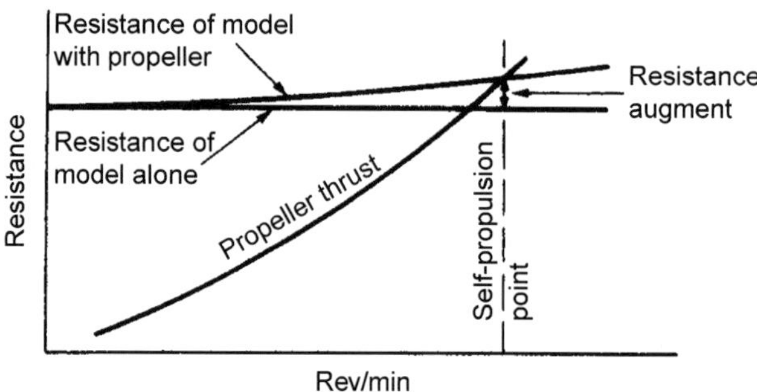

FIGURE 8.16 Wake and thrust deduction.

thrust, torque and revolutions. With the model being towed along the tank at its corresponding speed for the ship speed under study, the propellers are run at a range of revolutions straddling the self-propulsion point for the model. The model would already have been run without propellers to find its resistance. Data from the test can be plotted as shown in Figure 8.16.

The *self-propulsion point* for the model is the point at which the propeller thrust equals the model resistance with propellers fitted. The difference between this resistance, or thrust, and the resistance of the model alone, is the augment of resistance or thrust deduction.

The propeller is now run in open water and the value of advance coefficient corresponding to the thrust needed to drive the model is determined. This leads to the average flow velocity through the propeller which can be compared to the ship speed corresponding to the self-propulsion point. The difference between the two speeds is the wake assuming a uniform distribution across the propeller disc. The difference in performance due to the wake variation across the disc is given by relative rotative efficiency which is the ratio of the torques needed to drive the propeller in open water and behind the model at the revolutions for self-propulsion.

Although the propellers used in these experiments are made as representative as possible of the actual design, they are small. The thrust and torque obtained are not accurate enough to use directly. The hull efficiency elements obtained are used with methodical series data or specific cavitation tunnel tests to produce the propeller design.

THE ITTC PERFORMANCE PREDICTION METHOD

The International Towing Tank Conference (ITTC) introduced this method to help provide consistency between different research establishments. Apart from the results from model resistance tests, which were dealt with in the previous chapter, the method uses data from self-propulsion and propeller open water tests.

The thrust (T) and torque (Q) are represented by the non-dimensional parameters:

$$K_{TM} = \frac{T}{\rho D^4 n^2} \quad \text{and} \quad K_{QM} = \frac{Q}{\rho D^5 n^2}$$

From the plots of the model propeller characteristics, K_{TM} enables J_{TM} and K_{QTM} to be read off. Then:

the wake fraction, $w_{TM} = 1-(J_{TM}D_M)/V_M$, where V_M is the model speed;
the relative rotative efficiency $= \eta_R = K_{QTM}/K_{QM}$.

The thrust deduction factor is given by

$$t = \frac{(T + F_D - R_C)}{T}$$

where F_D is the towing force applied in the propulsion test

$$F_D = 0.5\rho_M S_M V_M^2[C_{FM} - (C_{FS} + \Delta C_F)]$$

and R_C is the resistance corrected for the temperature difference between the resistance and self-propulsion tests. That is:

$$R_C = R_{TM}[(1 + k)C_{FMC} + C_R]/[(1 + k)C_{FM} + C_R]$$

where
C_{FMC} is the corrected frictional resistance.

Full-Scale Prediction

We deduced in Chapter 7 that for the ship fitted for bilge keels:

$$[S + S_{BK}]/S[(1 + k)C_{FS} + \Delta C_F] + C_R + C_A + C_{AA}$$

The full-scale propeller characteristics are now calculated from

$$K_{TS} = K_{TM} - \Delta K_T \quad \text{and} \quad K_{QS} = K_{QM} - \Delta K_Q$$

where

$$\Delta K_T = -0.3\Delta C_D(P/D)(cZ/D)$$

and

$$\Delta K_Q = 0.25\Delta C_D(cZ/D)$$

The difference in drag coefficient, $\Delta C_D = C_{DM}-C_{DS}$.
where

$$C_{DM} = 2(1 + 2t/c)\{[0.044/(Re_\infty)^{1/6}] - [5/(Re_\infty)^{2/3}]\}$$
$$C_{DS} = 2(1 + 2t/c)\{1.89 + 1.62\log(c/k_p)\}^{-2.5}$$

In these formulae c is the chord length, t is the maximum thickness, Z is the number of blades, P/D is pitch ratio and Re_∞ is the local Reynolds number at three quarter radius. The blade roughness, k_p is taken as 30×10^{-6} m. Re_∞ must not be less than 2×10^5 in the open water test.

The full-scale wake is calculated as follows:

$$w_{TS} = (t + w_R) + (w_{TM} - t - w_R)\{[(1 + k)C_{FS} + \Delta C_F]/[(1 + k)C_{FM}]\}$$

where

w_{TM} = the model wake
t is the model thrust deduction
the factor w_R allows for rudder effects and is taken as 0.04 if no other estimate is available.

The load of the ship propeller is given by

$$K_T/J^2 = [S/2D^2][C_{TS}/(1 - t)(1 - w_{TS})^2]$$

Using this value of K_T/J^2 the ship advance coefficient and torque coefficient are read off from the full-scale propeller characteristics. The following can then be determined for each propeller:

- Revolutions: $n_S = (1 - w_{TS})V_S/J_{TS}D$ r/s
- Delivered power: $P_{DS} = 2\pi\rho D^5 n_S^3[K_{QTS}/\eta_R] \cdot 10^{-3}$ kW
- Propeller thrust: $T_S = (K_T/J^2)J_{TS}^2\rho D^4 n_S^2$ N
- Propeller torque: $Q_S = (K_{QTS})/\eta_R)\rho D^5 n_S^2$ Nm
- Effective power: $P_E = C_{TS}[0.5]\rho V_S^3 S \cdot 10^3$ kW
- Total efficiency: $\eta_D = N_p P_{DS}/P_E$, where N_p is the number of propellers
- Hull efficiency: $\eta_H = (1 - t)/(1 - w_{TS})$.

CAVITATION

The lift force on a propeller blade is generated by increased pressure on the face and reduced pressure on the back, the latter making the greater contribution (Figure 8.11). If the reduction in pressure on the back is great enough cavities form and fill up with air coming out of solution and by water vapour. Thus local pressures in the water, and their variation as the propeller rotates, are important to the study of propellers. In deriving non-dimensional parameters that might be used to characterise fluid flow, it can be shown that the parameter associated with the pressure, p, in the fluid is $p/\rho V^2$. There is always an 'ambient' pressure in water at rest due to atmospheric pressure acting on the surface plus a pressure due to the water column above the point considered. If the water is moving with a velocity V then the pressure reduces to, say, p_V, from this ambient value, p_O, according to Bernoulli's principle.

Inflow velocities (magnitude and direction) vary across the propeller disc due to the ship's wake so the angle of attack of individual blades is constantly varying and cannot be optimum. Also the pressure varies with depth below the water surface causing variations in cavitation.

Comparing Ship and Model Under Cavitating Conditions

For dynamic similarity of ship and model conditions the non-dimensional quantity must be the same for both. That is, using subscripts m and s for model and ship:

$$\frac{p_{\mathrm{m}}}{\rho_{\mathrm{m}} V_{\mathrm{m}}^2} \text{ must equal } \frac{p_{\mathrm{s}}}{\rho_{\mathrm{s}} V_{\mathrm{s}}^2}$$

If the propellers are to operate at the same Froude number, as they would need to if the propeller–hull combination is to be used for propulsion tests:

$$V_{\mathrm{m}} = \frac{V_{\mathrm{s}}}{(\lambda)^{0.5}}$$

where λ is the ratio of the linear dimensions. That is

$$p_{\mathrm{m}} = \frac{\rho_{\mathrm{m}}}{\rho_{\mathrm{s}}} \times \frac{p_{\mathrm{s}}}{\lambda}$$

Assuming water is the medium in which both model and ship are run, the difference in density values will be small. For dynamic similarity the pressure must be scaled down in the ratio of the linear dimensions. This can be arranged for the water pressure head, but the atmospheric pressure requires special action. The only way in which this can be scaled is to run the model in an enclosed space in which the pressure can be reduced. This can be done by reducing the air pressure over a ship tank and running a model with propellers fitted at the correctly scaled pressure as is done in a special *depressurised towing tank* facility at MARIN in the Netherlands. The tank is 240 m long, 18 m wide with a water depth of 8 m. The pressure in the air above the water can be reduced to 0.03 bar. The more usual approach is to use a *cavitation tunnel*.

Cavitation Number

The value $(p_{\mathrm{O}} - p_{\mathrm{V}})/\rho V^2$ or $(p_{\mathrm{O}} - p_{\mathrm{V}})/0.5\rho V^2$ is called the *cavitation number*. Water contains dissolved air and at low pressures this air will come out of solution and below a certain pressure, the *vapour pressure of water*, water vapour forms. Hence, as the pressure on the propeller blade drops, bubbles form. This *cavitation* occurs at a cavitation number given by:

$$\text{Cavitation number, } \sigma = (p_O - e)/\tfrac{1}{2}\rho V^2$$

where e is water vapour pressure.

The actual velocity experienced and the value of p_O vary with position on the blade. For a standard, a representative velocity is taken as speed of advance of the propeller through the water and p_O is taken at the centre of the propeller hub. For a local cavitation number the actual velocity at the point concerned, including rotational velocity and any wake effects, and the corresponding p_O for the depth of the point at the time must be taken. Blade elements experience different cavitation numbers as the propeller rotates and cavitation can come and go.

Occurrence and Effects of Cavitation

Since cavitation number reduces with increasing velocity, cavitation is most likely to occur towards the blade tips where the rotational component of velocity is highest. It can also occur near the roots, where the blade joins the hub, as the angle of incidence can be high there. The greatest pressure reduction on the back of the blade occurs between the mid-chord and the leading edge so bubbles are likely to form there first. They will then be swept towards the trailing edge and as they enter a region of higher pressure they will collapse. The collapse of the bubbles generates very high local forces and these can damage the blade material causing it to be 'eaten away'. This phenomenon is called *erosion*.

Water temperature, dissolved air or other gases, and the presence of nuclei to provide an initiation point for bubbles, all affect the pressure at which cavitation first occurs. Face cavitation usually appears first near the leading edge of the section. It results from an effective negative angle of incidence where the wake velocity is low. This face cavitation disappears as the propeller revolutions and slip increase. Tip vortex cavitation is next to appear, resulting from the low pressure within the tip vortex. As the pressure on the back of the blade falls further the cavitation extends from the leading edge across the back until there is a sheet of cavitation. When the sheet covers the whole of the back of the blade the propeller is said to be fully cavitating or *super-cavitating*. Propellers working in this range do not experience erosion on the back and the drag due to the frictional resistance to flow over the back disappears. Thus when fairly severe cavitation is likely to occur anyway there is some point in going to the super-cavitation condition as the design aim. *Super-cavitating propellers* are sometimes used for fast motor boats.

Flat faced, circular back sections tend to have a less peaky pressure distribution than aerofoil sections. For this reason they have often been used for heavily loaded propellers. However, aerofoil sections can be designed to

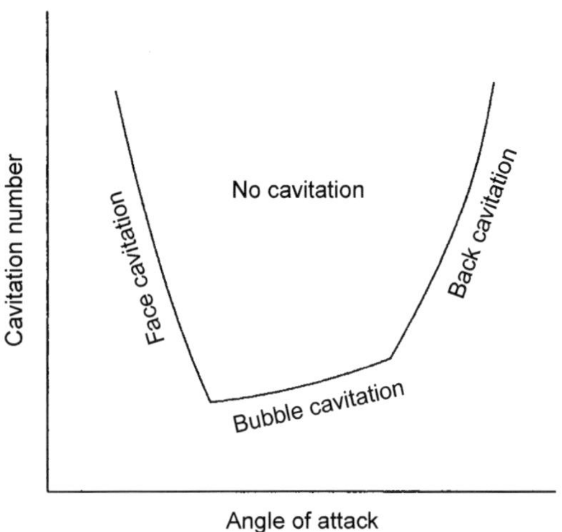

FIGURE 8.17 Cavitation bucket.

have a more uniform pressure distribution and this approach is to be preferred. For a given thrust, more blades and greater blade area will reduce the average pressures and therefore the peaks. It will be found that heavily loaded propellers have much broader blades than lightly loaded ones. However, increased blade area leads to greater torque and merchant ships usually go to smaller areas to improve propeller efficiency. In warships, where the noise from cavitation can betray a ship's presence, a lower efficiency is accepted. Thus the speeds at which cavitation begins – the *cavitation inception speed* – can approach 15 knots in a frigate. For merchant ships it may be only about a third of this.

A useful presentation for a designer is the *bucket diagram*. This shows (Figure 8.17) for the propeller, the combinations of cavitation number and angle of attack or advance coefficient for which cavitation can be expected. There will be no cavitation as long as the design operates within the bucket. The wider the bucket, the greater the range of angle of attack or advance coefficient for cavitation free operation at a given cavitation number.

The Cavitation Tunnel

A cavitation tunnel is a closed channel in the vertical plane as shown in Figure 8.18. Water is circulated by means of an impeller in the lower horizontal limb. The extra pressure here removes the risk of the impeller cavitating. The model propeller under test is placed in a working section in the upper horizontal limb which is provided with glass viewing ports and is designed to give uniform flow across the test section. The water circulates in such a way that it meets the model propeller before passing over its drive

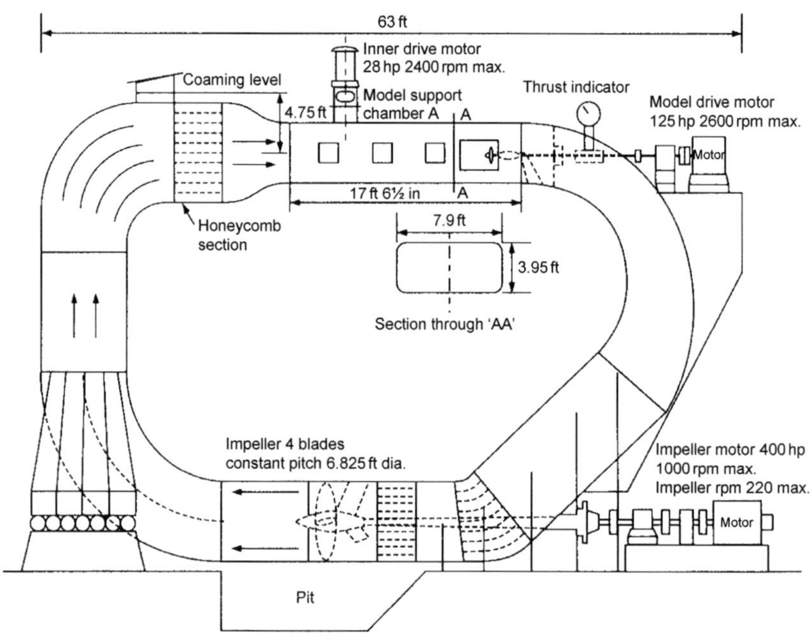

FIGURE 8.18 Large cavitation tunnel. *Courtesy of RINA.*

shaft. That is the propeller is effectively tested in open water. A vacuum pump reduces the pressure in the tunnel and usually some form of de-aerator is fitted to reduce the amount of dissolved air and gas in the tunnel water. Usually the model is tested with the water flow along its axis, but there is often provision for angling the drive shaft to take measurements in an inclined flow. This is the more typical ship condition.

A limitation of straight tunnel tests is that the ship wake variations are not reproduced in the model test. If the tunnel section is large enough this is overcome by fitting a model hull in the tunnel modified to reproduce the correctly scaled boundary layer at the test position. In these cases the flow to the propeller must be past the hull. An alternative is to create an artificial wake by fixing a grid ahead of the model propeller. The grid would be designed so that it reduced the water velocities differentially to produce the correctly scaled wake pattern for the hull to which the propeller is to be fitted.

Cavitation Tunnel Tests

Experiments are usually conducted as follows:

- The water speed is made as high as possible to keep Reynolds number high and reduce scaling effects due to friction on the blades. Since wave

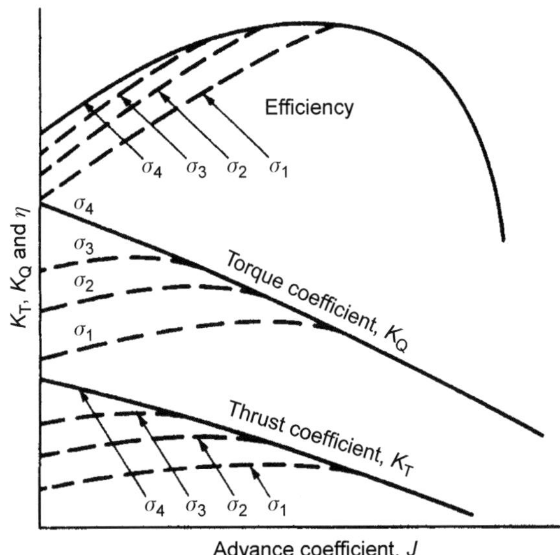

FIGURE 8.19 Propeller curves with cavitation.

effects are not present and the hull itself is not under test the Froude number can be varied.

- The model is made to the largest possible scale consistent with avoiding tunnel wall interference effects.
- The shaft revolutions are adjusted to give the correct advance coefficient.
- The tunnel pressure is adjusted to give the desired cavitation number at the propeller axis.
- A series of runs are made over a range of shaft revolutions, that being a variable which is easy to change. This gives a range of advance coefficients. Tests can then be repeated for other cavitation numbers.

Figure 8.19 shows typical curves of thrust and torque coefficient and efficiency to a base of advance coefficient for a range of cavitation number. Compared with non-cavitating conditions, the values of all three parameters fall off at low advance coefficient, the loss being greater the greater the cavitation number.

When cavitation is present the propeller can be viewed using a stroboscopic light set at a frequency which makes the propeller seem stationary to the eye and photographs can be taken. A similar technique is used in propeller viewing trials at sea when the operation of the propeller is observed through special glass viewing ports fitted in the shell plating, or a type of periscope.

The propeller, particularly when cavitating, is a significant source of noise. It would be useful to be able to take noise measurements in a cavitation tunnel. This is not possible in most tunnels because of the background noise levels, but a few tunnels have been built for acoustical measurements.

Propeller Singing

A new propeller may generate a high pitched note and this is known as *singing*. This is caused by the frequency of the trailing edge vortices exciting the structural frequency of vibration of that area of the propeller blade. It is audible in the range 10–1200 Hz. The cure usually adopted is to cut back the trailing edge slightly leaving a flat trailing edge with sharp corners. Sometimes blade damping is modified by special treatments.

OTHER PROPULSOR TYPES

Besides the fixed pitch screw propeller other propulsors are outlined below.

Controllable Pitch Propeller

The machinery must develop enough torque to turn the propeller at the revolutions appropriate to the power being developed or the machinery will *lock up*. This matching is not always possible with fixed blades and some ships are fitted with propellers in which the blades can be rotated about axes normal to the drive shaft. These are termed *controllable pitch propellers* (CPPs). The pitch can be altered to satisfy a range of operating conditions which is useful in tugs and trawlers where there is a great difference in propeller loading when towing or trawling and when running free. The machinery can be run at constant speed so that full power can be developed over the range of operating conditions.

The pitch of the blades is changed by gear fitted in the hub and controlled by linkages passing down the shaft. The larger boss limits the blade area ratio to about 0.8 which affects cavitation performance. A CPP is mechanically complex which limits the total power transmitted. By reversing the pitch an astern thrust can be produced thus eliminating the need for a reversing gear box. Variation in thrust for manoeuvring can be more rapid as it only involves changing blade angle rather than shaft revolutions, but for maximum acceleration or deceleration there will be an optimum rate of change of blade angle.

A CPP should not be confused with a *variable pitch propeller*. In the latter pitch varies with radius, the blades themselves being fixed.

Self-Pitching Propellers

A propeller which has found favour for auxiliary yachts and motorsailers is the self-pitching propeller (Miles et al., 1993). The blades are free to rotate through 360° about an axis approximately at right angles to the drive shaft. The angle the blades take up, and therefore their pitch, is dictated solely by the hydrodynamic and centrifugal forces acting.

Surface Piercing Propellers

Those ships which have a large draught difference between the loaded and light condition may run, in the latter, with the propeller only partially immersed. However, the true surface piercing propeller is one that is designed to operate in this condition in order to gain certain advantages, usually in high-speed craft. At the design condition the waterline passes through the hub. Advantages claimed are ability to operate in shallow water, improved efficiency, reduced appendage drag and avoidance of cavitation effects through ventilation.

Shrouded or Ducted Propellers

The propeller is surrounded by a shroud or duct as depicted in Figure 8.20 to improve efficiency, avoid erosion of banks in confined waterways and shield noise generated on the blades.

The duct can be designed to contribute to ahead thrust offsetting the drag of the shroud and its supports. Early applications were to ships with heavily loaded propellers like tugs. Its use is now being extended and it is considered suitable for large tankers.

Pump Jets

This is an advanced variant of the ducted propeller for use in warships, particularly submarines, where noise reduction is important. A rotor with a large number of blades operates between sets of stator blades, the whole being surrounded by a specially shaped duct. The rotational losses in the wake

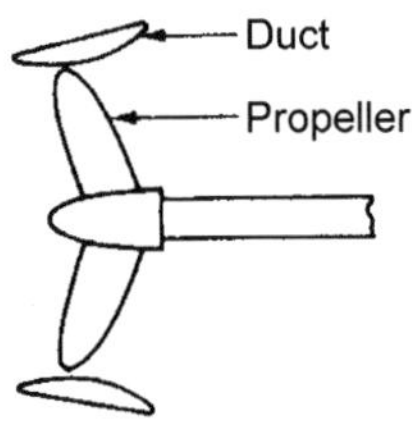

FIGURE 8.20 Shrouded propeller.

are eliminated, cavitation is avoided and there is no resultant heeling torque acting on the vessel (significant for single screw submarines).

Contra-Rotating Propellers

Another way of eliminating the net heeling torque, and reduce rotational losses in the race, is to use two propellers on the one shaft line rotating in opposite directions. It has been concluded (Glover, 1966–1967) that they can be useful in large tankers where by using slow running contra-rotating propellers (CRPs) the QPC can be increased by up to 20%. In high-speed dry cargo ships, where propeller diameter may be restricted by draught, propeller efficiency may be increased by 12%. Like CPPs, CRPs introduce mechanical complications.

One CRP system of about 60 MW, intended for a large 25 knot container ship, claimed:

- fuel savings of 12% compared with a conventional propeller;
- a diameter reduction of 10% because the thrust is shared between two propellers;
- reductions in cavitation and hull vibrations.

Azimuthing Propellers

These are propellers mounted in a housing, or pod, which can rotate through a full circle to give thrust in any direction. Early applications, typically to tugs which require good manoeuvrability, relied on mechanical transmission of power to the pod which limited the power of the installation. Nowadays a number of large ships, including cruise ships and Ro–Ro ferries, are fitted with diesel electric propulsion for a range of reasons. This gives them the opportunity to use large pods which are essentially motors driving a propeller. Units transmitting 5–8 MW of power are typical but units of about 15 MW have been fitted in large cruise ships. Fast, large container ships are also candidates for high-power azimuthing pods. The propeller is mounted on the motor's rotor and in some units two propellers are mounted on the rotor, one forward and one aft of the pod. This permits a greater blade area with an associated increase in efficiency. In yet others the two propellers are mounted independently, using two motors in the pod. It is claimed that such ships are cheaper to run and can be smaller to carry a given payload, besides having the advantages of good manoeuvrability. Separate stern thrusters are not needed in such ships.

Some advantages of pod propulsion are greatly improved manoeuvrability, less space required within the main hull and greater freedom in siting the propeller and in developing the hull form for greater efficiency.

Vertical Axis Propeller

This is essentially a horizontal disc, rotating about a vertical axis, carrying a series of vertical blades which can rotate about their own vertical axes. The individual vertical blades have aerofoil sections and generate lift forces. By controlling the angle of the blades as the horizontal disc turns, a thrust can be produced in any desired direction. Vertical axis propellers are fitted in tugs and ferries for good manoeuvrability. Drive is usually through bevel gears which limits the power.

Water Jets

This type of propulsion has become more common in recent years particularly for high-speed craft although they can be beneficial in medium-speed operations. They are used for some wind farm support catamarans. Water is drawn into the ship and then pushed out at the stern to develop thrust. The ejecting unit can be steerable to give a varying thrust direction. It is attractive for craft where it is desired to have no moving parts outside the hull. For this reason many applications are for craft operating in shallow water. The water jet can be discharged either above or below water. Some hydrofoil craft use the system, discharging above water.

Paddle Wheels

A paddle wheel is a ring of paddles rotating about a horizontal transverse axis. In very simple craft the paddles are fixed but in craft requiring greater efficiency their angle is changed as the wheel rotates. When fitted either side of a ship they can exert a large turning moment on the ship by being run one ahead and the other astern. Unfortunately this leads to a wide vessel. For use in narrow waterways the paddle wheel is mounted at the stern giving rise to the *stern wheeler* in the United States.

Wind

The wind was the only means, apart from oars, of propelling ships for many centuries. It has always been popular for pleasure craft. The rise in fuel costs and public concern with conserving energy sources has rekindled interest. Some ships have sails to use in place of their engines when wind conditions are suitable. Other applications have harnessed modern technology to use the old idea of rotating cylinders, the *Flettner rotor* concept, more effectively.

SHIP TRIALS

A complete range of trials is carried out on a ship when complete to confirm that the ship meets its specification. Amongst these is a speed trial which has the following uses:

- To demonstrate that the desired speed is attained. There are usually penalties imposed if a ship fails to meet the specified speed. It would be uneconomic to provide too much power and a designer must predict resistance and powering accurately in the design stages.
- To provide a feedback on the effectiveness of prediction methods and provide factors to be applied to compensate for simplifying assumptions in the methods.
- To provide data on the relationships between shaft revolutions, ship speed and power for the master.

To meet the last two aims it is desirable to gather data at a range of speeds. Therefore trials are run at progressively higher speeds up to the maximum and are often called *progressive speed trials*. The engine designer may wish to take readings of a wide range of variables concerned with the performance of the machinery itself. The naval architect, however, is concerned with the shaft revolutions, thrust, torque and speed achieved relative to the water. Thrust is not always measured. It can be measured by a special thrust meter but more commonly by a series of electrical resistance strain gauges fitted to the shaft. Torque is measured by the twist experienced over an accurately known length of shaft. It remains to determine the speed of the ship.

Speed Measurement

Ships are provided with a log for speed measurement. This is not accurate enough for speed trial purposes. Indeed the speed trial is often used to calibrate the log.

Traditionally a ship has been taken to a *measured mile* for speed trials although nowadays use can be made of accurate position fixing systems which are available in many areas. The measured mile (Figure 8.21) comprises a number of posts set up on land at known distances apart. These distances are not necessarily exactly one nautical mile but it simplifies analysis if they are. The posts are in parallel pairs clearly visible from the sea. There may be two pairs as in the figure, or three pairs to give a double reading on each run. By noting the time the ship takes to transit the distance between adjacent pairs of posts, the speed relative to land is obtained. For accuracy a number of precautions are needed:

- The ship must be travelling at right angles to the line of posts and the ship must have reached a steady speed for the power used by the time it passes the first pair of posts.

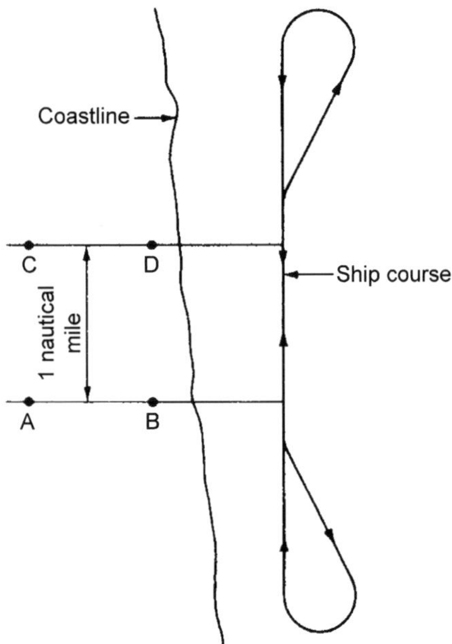

FIGURE 8.21 Measured mile.

- The depth of water must be adequate to avoid the speed being affected due to squat and trim. A minimum depth is defined by a *Froude water depth number*, F_H, where
  $F_H = 0.164V/(H)^{0.5}$, where V is in knots and water depth, H, in metres.
- Avoidance of wind and waves.
- The ship should be newly out of dock, with a clean bottom. If this condition is not met some allowance may be needed for the increased resistance due to time out of dock.
- After passing the last pair of posts the ship must continue on for some way and then turn for the return run, reaching a steady speed before passing the first set of posts. This may involve a run on of several miles and an easy turn to minimise the drop in speed associated with turning.
- The displacement must be accurately obtained by measuring the ship's draughts and the density of the water.

If there were no wind, current or tide, one run at each power setting would theoretically be enough and the speed through the water would be the same as that relative to land. In any practical situation a number of runs are needed in each direction so that the results can be analysed to remove current and tidal effects.

Determining Speed Through the Water

It is usually assumed that the current and tide effects will vary with time in accordance with an equation of the type

$$V_T = a_0 + a_1 t + a_2 t^2$$

where a_0, a_1 and a_2 are constants.

It is the component of tide along the ship's line of transit on the measure mile that is of concern and is used here. Suppose four runs are made, two in each direction. Two will be with the tide and two against. Using subscripts to denote the speeds recorded on the runs:

$$\begin{aligned} V_1 &= V + a_0 \\ V_2 &= V - a_0 - a_1 t_1 - a_2 t_1^2 \\ V_3 &= V + a_0 + a_1 t_2 + a_2 t_2^2 \\ V_4 &= V - a_0 - a_1 t_3 - a_2 t_3^2 \end{aligned}$$

where V is the speed through the water and the runs are at times zero and t_1, t_2 and t_3.

The equations can be solved and the speed relative to the water found.

If the tide varied linearly with time three runs would be enough. A higher order equation for tide can be used if more runs are made. Usually four runs are adequate.

Trial Condition

Ideally trials would be carried out for each of the likely operating conditions. This would be expensive and time consuming. The key condition is that for which the contract speed is defined, which is usually the deep load condition. If this level of loading cannot be achieved some lesser load is specified with a correspondingly higher speed to be obtained. In some ships the load condition can be achieved by water ballasting. The trial is carried out in calm conditions which are easy to define for contract purposes but are not representative of the average conditions a ship will meet in service. Increasingly it is realised that it is this speed that is of real interest and this has led to a lot of effort being devoted to obtaining and analysing voyage data.

Plotting Trials Data

The results from the ship trial can be plotted as shown in Figure 8.22. The revolutions will be found to plot as a virtually straight line against speed.

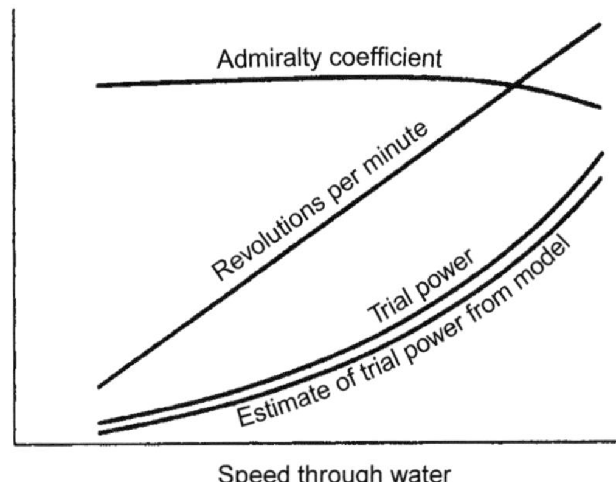

FIGURE 8.22 Trials data.

Power increases rapidly with speed. If enough readings are available the humps and hollows due to the interaction of bow and stern wave systems will be detectable. The figure shows a plot of *Admiralty coefficient*. This coefficient, or constant, is effectively the inverse of circular C and is given by

$$\frac{V^3 \Delta^{\frac{2}{3}}}{\text{Power}}$$

A comparison of the power measured on trial and that estimated from model tests gives a ship–model correlation factor to be used for future similar ships.

Wake Fraction from Ship Trials

If shaft torque is measured a torque coefficient can be calculated from the shaft revolutions and propeller diameter. The advance coefficient can be found from the ship speed and a plot made as shown in Figure 8.23.

From open water propeller tests the value of advance coefficient corresponding to any given torque coefficient can be found. This yields a value of V_a. The wake is the difference between the ship speed and V_a. This is the mean wake through the propeller disc. In the absence of open water model tests methodical series data can be used but with less accuracy.

Modern GPS systems allow the path of a ship, relative to the ground, to be established accurately. A speed trial could, therefore, in principle be carried out anywhere at sea avoiding the need for a measured mile and the possibility of shallow water effects. However, it leaves the problem of finding out the ship's speed relative to the water and runs in different

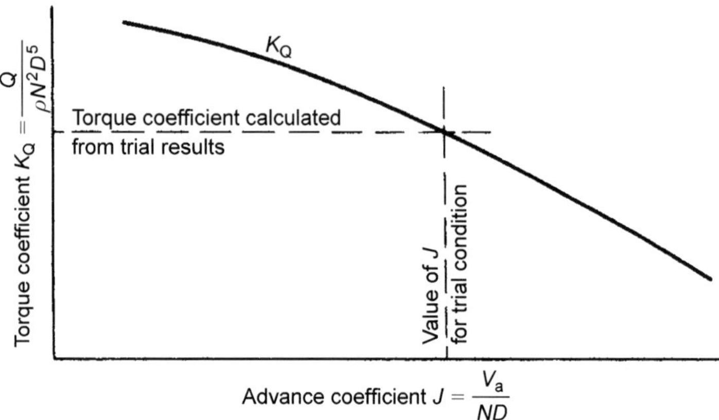

FIGURE 8.23 Wake fraction.

directions will be needed to enable tidal effects to be eliminated. What is very useful is the ability to establish average speeds in various sea conditions during typical voyages.

MAIN MACHINERY POWER

The objectives of the resistance and propulsion testing have been to develop an efficient hull form and propulsor design and to establish the main machinery power needed to drive the ship at the design speed. The point has been reached in the analysis where the last aim can be met.

The general principles involved were outlined at the beginning of this chapter. In the previous chapter an example was given illustrating the calculation of a hull's effective power. This same ship can be used to calculate the machinery power needed to propel it at the 15 knots for which the effective power was 2502 kW allowing for roughness.

Continuing:

P_E for rough hull = 2502 kW
Appendage allowance, say 5% = 125
P'_E in smooth water = 2627

If the hull efficiency elements and the QPC determined from experiment were Taylor wake fraction = 0.27, hull efficiency = 1.15, QPC = 0.75 and relative rotative efficiency = 1.00, then:

Required delivered power = 2627/0.75 = 3503 kW.
Transmission loss at say 2% = 70
Required installed power = 3573 kW

This is the power for calm conditions. If 20% is allowed for average service conditions the installed power to maintain 15 knots in these conditions is 4288 kW.

The actual power to be fitted will depend upon the powers of the machinery sets available. For the present example it is assumed that the closest power available is 4275 kW and that the slight difference is accepted by the designer. It follows that:

Power at propeller = 4190 kW
Speed of advance of propeller = 15(1−0.27) = 10.95 knots

The choice of propeller revolutions is generally a compromise between propeller performance and machinery characteristics. Propellers are more efficient at low revolutions and machinery is lighter, for a given power, at high revolutions. Reduction gear can be fitted to bridge the gap, but the cost and weight must be set against the advantages gained. It is assumed initially that propeller revolutions are to be 100.

$$B_{\mathrm{p}} = \frac{1.158NP^{0.5}}{V_{\mathrm{a}}^{2.5}}$$

$$= \frac{1.158 \times 100 \times (4190)^{0.5}}{(10.95)^{2.5}}$$

$$= 18.89$$

From the propeller curves presented in Figure 8.15, which are for a four bladed propeller of 0.4 blade area ratio:

$\delta = 3.2808ND/V_{\mathrm{a}} = 178$
Pitch ratio $= p = 0.8$
Efficiency $= \eta = 0.655$

These are the values for maximum efficiency. Since $D = (\delta \times V_{\mathrm{a}})/3.2808N = 5.94$ m.

$$Pitch = pD = 4.75\ m$$

$$QPC = (hull\ efficiency) \times (propeller\ open\ \eta) \times (RRE)$$

$$= 1.15 \times 0.655 \times 1.00 = 0.75$$

This QPC happens to be the same as that assumed in the calculation of power. Had it differed significantly then a repeat calculation would have been needed using the new value. The process can be repeated for other propeller revolutions to see how the propeller dimension and QPC would vary. For $N = 110$,

$$B_P = \frac{1.158 \times 110 \times (4190)^{0.5}}{(10.95)^{2.5}} = 20.78$$

From Figure 8.15 $\delta = 183$, $p = 0.78$, $\eta = 0.645$. Hence,

$$\text{Diameter} = \frac{183 \times 10.95}{110 \times 3.2808} = 5.55 \text{ m}$$

$$\text{Pitch} = 0.78 \times 5.55 = 4.33 \text{ m}$$

$$\text{QPC} = 1.15 \times 0.645 \times 1.00 = 0.74$$

For $N = 120$,

$$B_P = \frac{1.158 \times 120 \times (4190)^{0.5}}{(10.95)^{2.5}} = 22.67$$

and

Diameter = 5.29
Pitch = 3.97
QPC = 0.73

These results confirm that as expected a higher revving propeller is smaller in diameter and is less efficient.

Figure 8.15 did not allow for cavitation and should cavitation be a problem curves from cavitation tunnel tests should be used.

SUMMARY

Fitting a propulsor modifies the flow around the hull causing an augment in the resistance the hull experiences and modifying the wake in which the propulsor must generate its thrust. The flow through the propulsor is not uniform so the efficiency will vary from that found in open water tests. Taking all these factors into account the power to be delivered by the propulsor for a given ship speed can be calculated. The power required of the main propulsion machinery follows after making allowance for transmission losses.

This analysis process, including the assessment of resistance, is illustrated in Figure 8.24 and leads to the power needed in calm seas with no natural wind. This is usually the condition for which the required ship speed is set

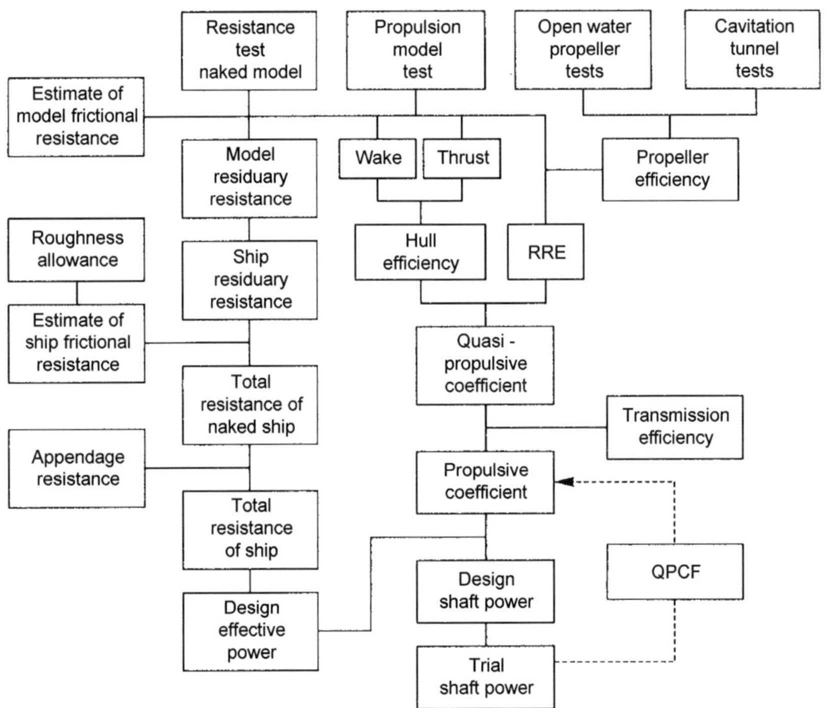

FIGURE 8.24 Power estimation.

out in the contract and which is aimed for in the speed trial conducted on completion. In service the ship will seldom be in these conditions. For more realistic powers and speeds allowance must be made for the wind resistance on the above water form and the effects of waves on the hull resistance and propulsor performance. This involves assessing the average conditions a ship is likely to meet or the range of conditions and their probability of occurrence.

This page intentionally left blank

Chapter 9

The Ship Environments

INTRODUCTION

There are two environments to consider, the external (ocean) and internal. The naval architect is concerned with the ocean environment because it presents significant risks to the safe operation of ships due to waves, wind and possibly ice. Also important is the range of temperatures experienced, of both air and water. The ship pollutes the environment in which it operates and the acceptable levels of pollution have been reduced significantly and this trend continues. The internal environment determines the well-being of crew and passengers and sometimes needs to be modified to suit the cargo carried. There is a two-way interaction between the two environments.

BS 6349-1: 2000 gives much useful data on the ocean environment.

THE OCEAN ENVIRONMENT

Water and Air – General Properties

Water is effectively incompressible so its density does not vary with depth although it does with temperature and salinity. The kinematic viscosity of water is affected by the same factors. The variations are given in Table 9.1, based on saltwater of standard salinity (3.5%). The figures are those recommended by the International Towing Tank Conference (ITTC). See ITTC 7.5-02-01-03.

Naval architects use standard figures in calculations, including a mass density of freshwater of 1.000 tonnes/m^3 and of seawater of 1.025 tonnes/m^3. For air at standard barometric pressure and temperature, with 70% humidity mass of 1.28 kg/m^3 is used.

The salinity and density of seawater vary locally over the ocean depending primarily on the temperature but are also affected by estuarial flows into the sea. The salinity of seawater together with oxygen in the atmosphere makes the sea a highly corrosive environment.

The ambient temperatures a ship is likely to meet at sea determine the amount of air-conditioning and insulation to be provided besides affecting the power produced by machinery. Extreme air temperatures of 52°C in the tropics

Introduction to Naval Architecture. DOI: http://dx.doi.org/10.1016/B978-0-08-098237-3.00009-6

TABLE 9.1 Water Properties

Temperature (°C)	Density (kgs^2/m^4)		Kinematic Viscosity ($m^2/s \times 10^6$)	
	Freshwater	Saltwater	Freshwater	Saltwater
0	101.95	104.83	1.787	1.828
10	101.93	104.71	1.306	1.354
20	101.78	104.49	1.004	1.054
30	101.52	104.18	0.801	0.849

TABLE 9.2 Design Temperatures

Area of World	Average Max. Summer Temperature (°C)			Average Min. Winter Temperature (°C)		
	Air		Sea	Air		Sea
	DB	WB		DB	WB	
Extreme tropic	34.5	30	33			
Tropics	31	27	30			
Temperate	30	24	29			
Temperate winter				−4	–	2
Sub-Arctic winter				−10	–	1
Arctic/Antarctic winter				−29	–	−2

Water temperatures are measured near the surface in deep water. Dry and wet bulb temperatures are quoted for the air.

in harbour and 38°C at sea have been recorded: also −40°C in the Arctic in harbour and −30°C at sea. Less extreme values are taken for design purposes and typical design figures for warships, in degrees Celsius, are as given in Table 9.2.

THE ATMOSPHERE

Winds

The air is seldom still and winds may be steady in strength or gusting. They may have a predominant direction or swirl around. Strong winds add

TABLE 9.3 The Beaufort Scale

Beaufort Wind Scale	Limits of Wind Speed (knots)	Wind Descriptive Terms	Probable Max. Wave Height (m)	Sea State	Sea State Description Terms	Significant Wave Height (m)
0	<1	Calm	–	0	Calm (glassy)	0.00
1	1–3	Light air	0.1	1	Calm (rippled)	0.00–0.10
2	4–6	Light breeze	0.3	2	Smooth wavelets	0.10–0.50
3	7–10	Gentle breeze	1.0	3	Slight	0.50–1.25
4	11–16	Moderate breeze	1.5	3–4	Slight to moderate	
5	17–21	Fresh breeze	2.5	4	Moderate	1.25–2.50
6	22–27	Strong breeze	4.0	5	Rough	2.50–4.00
7	28–33	Near gale	5.5	5–6	Rough to very rough	
				6	Very rough	4.00–6.00
8	34–40	Gale	7.5	6–7	Very rough to high	
9	41–47	Severe gale	10.0	7	High	6.00–9.00
10	48–55	Storm	12.5	8	Very high	9.00–14.00
11	56–63	Violent storm	16.0	8	Very high	
12	>64	Hurricane	–	9	Phenomenal	>14.00

to the resistance a ship experiences, make it heel and make manoeuvring difficult. Masters of sailing ships dreaded being caught close to a rocky lee shore. Winds create waves. Their strength is classified in broad terms by the *Beaufort Scale* (Table 9.3). The higher the wind speed the less likely it is to be exceeded. In the North Atlantic, for instance, a wind speed of 10 knots is likely to be exceeded for 60% of the time, 20 knots for 30% and 30 knots for only 10% of the time.

The term *significant wave height* denotes the average of the third highest waves present in a sea. This form of the sea state code is that most commonly met but other codes are used.

Wind speeds given by, and associated with, the Beaufort Wind Scale and sea states are usually for heights of 6 m above the sea surface. These are the values generally used for stability calculations. In some cases the variation of wind speed with height is allowed for. This variation (based on 100 knots at 6 m height) is illustrated in Figure 9.1.

The density of air increases with entrained water content due to rain. With 10%, 20% and 30% entrained water air densities are 2.25, 3.25 and 4.35 kg/m^3, respectively.

Winds are caused by the circulation of air initiated by the different air temperatures over the surface of the earth which in turn creates the prevailing global winds. Between the equator and 30° latitudes north and south there are the trade winds and between 30° north and 60° north and between 30° south and 60° south there are the prevailing westerlies. North of latitude 60° north and south of latitude 60° south are the prevailing easterlies.

Local winds occur over relatively small regions due to high and low pressure in the atmosphere. These create cyclones and anticyclones which result

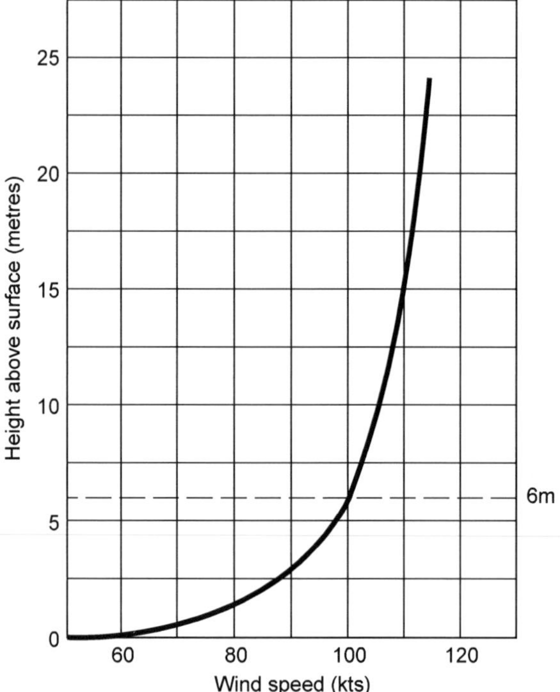

FIGURE 9.1 Variation of wind speed with height.

in severe storms and sea states. Hurricanes, or typhoons, are whirling winds of 75 miles per hour or more that form over the ocean in tropical regions then move into the more temperate regions, reaching wind speeds of 150 miles per hour.

THE OCEAN

Apart from the properties discussed above the ocean has tides, currents and waves.

Tides and Currents

Tides are created by the earth's rotation and the gravitational pull of the moon and sun on the body of water forming the oceans. Most areas experience two high and two low tides each day, but some only one. The height variations may determine whether a ship can enter a port at a given time.

Global currents in the sea move like large rivers caused by wind forces on the sea surface with their direction of flow dictated by the earth's rotation. Major currents in the Northern hemisphere are the Gulf Stream and the Californian Current. In the Southern Hemisphere are the Humbolt Current, the Brazil Current, the Benguela Current, the Agulhas Current and the West and East Australian Currents. Some of these currents can create extremely large and dangerous waves when in opposition to strong winds and when contacting with continental shelves.

Operating in Shallow Water

This book deals mainly with operations in deep water. Shallow water effects are discussed separately.

Waves

The characteristics of wind generated waves depend upon the wind's *strength*, the time for which it acts, its *duration*, and the distance over which it acts, its *fetch*. Although used in a general sense to denote the mass of water in which a ship moves the term *sea* is more strictly applied to waves generated locally by a wind. When waves have travelled out of the generation area they tend to be longer and more regular and are termed *swell*. The wave form depends upon depth of water, currents and local geographical features. Unless otherwise specified the waves referred to in this book are fully developed waves in deep water.

To an observer the sea surface can appear very confused. For many years it defied attempts at mathematical definition. The essential nature of this

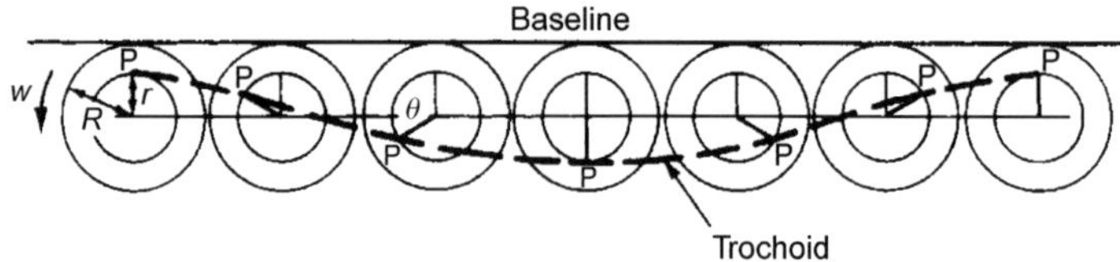

FIGURE 9.2 Trochoidal wave.

apparently random surface was understood by Froude (1905) who postulated that irregular wave systems are a compound of a large number of regular systems, individually of comparatively small amplitude, covering a range of periods and directions. Further he stated that the effect of such a compound wave system on a ship would be 'more or less the compound of the effects proper to the individual units composing it'. This is the basis for all modern studies of waves and ship motion. The mathematics to apply this theory had to wait until the 1950s. As individual wave components are assumed to be regular we first study the properties of regular waves and then combine them to create irregular seas.

Regular Waves

A uni-directional regular wave would appear constant in shape with time and resemble a sheet of corrugated iron of infinite width. As it passed a fixed point a height recorder would record a variation with time that would be constantly repeated. Two wave shapes are of particular significance to naval architects, the *trochoidal wave* and the *sinusoidal wave*.

The Trochoidal Wave

By observation the crests of ocean waves are sharper than the troughs. This is a characteristic of trochoidal waves and they were taken as an approximation to ocean waves in early calculations of longitudinal strength. The section of the wave is generated by a fixed point within a circle when that circle rolls along and under a straight line (Figure 9.2).

The crest of the wave occurs when the point is closest to the straight line. The wavelength, λ, is equal to the distance the centre of the circle moves in making one complete rotation, i.e. $\lambda = 2\pi R$. The wave height is $2r = h_W$. Consider the x-axis horizontal and passing through the centre of the circle, and the z-axis downwards with origin at the initial position of the centre of the circle. If the circle rolls through θ, its centre will move $R\theta$ and the wave generating point, P, has co-ordinates

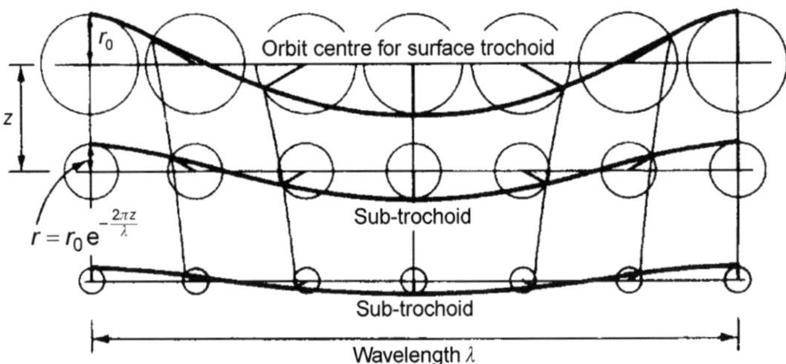

FIGURE 9.3 Sub-trochoids.

$$x = R\theta - r\sin\theta$$

$$z = r\cos\theta$$

From Figure 9.3 the following mathematical relationships can be derived:

- In deep water the wave velocity is C, where $C^2 = g\lambda/2\pi$.
- The still water surface will be at $z = (r_0^2/2R)$ reflecting the fact that the crests are sharper than the troughs.
- Particles in the wave move in circular orbits.
- Surfaces of equal pressure below the wave surface are trochoidal. These subsurface amplitudes reduce with depth so that, at z below the surface, the amplitude is

$$r = r_0 \exp\frac{-z}{R} = r_0 \exp\frac{-(2\pi z)}{\lambda}$$

- This exponential decay is very rapid and there is little movement at depths of more than about half the wavelength.

Wave Pressure Correction

The water pressure at the surface of the wave is zero and at a reasonable depth, planes of equal pressure will be horizontal. Hence the pressure variation with depth within the wave cannot be uniform along the length of the wave. The variation is due to the wave particles moving in circular orbits. It is a dynamic effect, not one due to density variations. It can be shown that the pressure at a point z below the wave surface (Figure 9.4) is the same as the hydrostatic pressure at a depth z', where z' is the distance between the mean, still water, axis of the surface trochoid and that for the subsurface trochoid through the point considered.

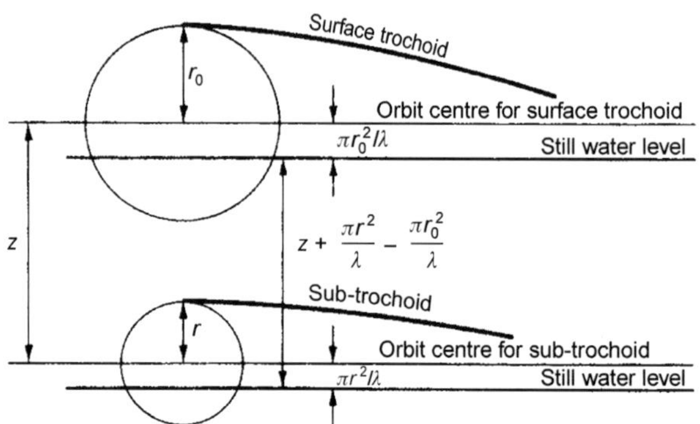

FIGURE 9.4 Pressure in wave.

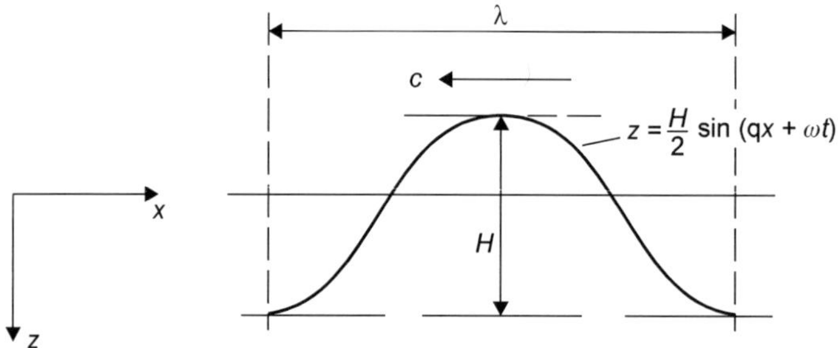

FIGURE 9.5 Profile of sinusoidal wave.

$$\text{Now} \quad z' = z - \frac{\pi}{\lambda}\left[r_0^2 - r^2\right]$$

$$= z - \frac{r_0^2}{2R}\left[1 - r^2/r_0^2\right] = z - \frac{r_0^2}{2R}\left[1 - \exp(-2z/R)\right]$$

To obtain the forces acting on the ship in the wave the usual hydrostatic pressure based on depth should strictly be corrected to allow for this. This correction is generally known as the *Smith effect*. Its effect is to increase pressure below the trough and reduce it below the crest for a given absolute depth. A correction used to be made for this effect when balancing a ship on a standard wave in longitudinal strength calculations. This is no longer done as such calculations are merely compared with results from similar ships.

The Sinusoidal Wave

Trochoidal waveforms are difficult to manipulate mathematically and irregular waves are analysed for their sinusoidal components. Taking the x-axis in the still water surface, the same as the mid-height of the wave, and z-axis vertically down, the wave surface height at x and time t can be written as (Figure 9.5):

$$z = \frac{H\sin(qx + \omega t)}{2}$$

In this equation $q = 2\pi/\lambda$ is termed the *wave number* and $\omega = 2\pi/T$ is known as the *wave frequency*. T is the *wave period*. The principal characteristics of the wave, including the *wave velocity*, C, are

$$C = \frac{\lambda}{T} = \frac{\omega}{q};\ T^2 = \frac{2\pi\lambda}{g};\ \omega^2 = \frac{2\pi g}{\lambda}\ \text{and}\ C^2 = \frac{g\lambda}{2\pi}$$

As with trochoidal waves water particles in the wave move in circular orbits, the radii of which decrease with depth in accordance with

$$r = \tfrac{1}{2}H\exp(-qz)$$

From this it is seen that for depth $\lambda/2$ the orbit radius is only $0.02H$ which can normally be ignored.

The average total energy per unit area of wave system is $\rho gH^2/8$, the potential and kinetic energies each being half of this figure. The energy of the wave system is transmitted at half the speed of advance of the waves. The front of the wave system moves at the speed of energy transmission so the component waves, travelling at twice this speed, will 'disappear' through the wave front.

For more information on sinusoidal waves, including proofs of the above relationships, the reader should refer to a standard text on hydrodynamics.

Irregular Waves

The irregular wave surface is taken as the compound of a large number of small waves, each component wave having its own length, height and direction. If they all travel in the same direction the irregular pattern is constant across the breadth of the wave, extending to infinity in each direction. Such a sea is said to be a *long crested irregular system* and is referred to as *one-dimensional*, the one dimension being frequency (length). In the more general case the component waves travel in a different direction and the sea surface resembles a series of humps and hollows with any apparent crests being short. Such a system is said to be a *short crested irregular wave system* or a *two-dimensional system*, the dimensions being frequency and direction. Only the simpler, long crested system is considered here and it will be called an irregular wave system. For a discussion of the effects of the spread of wave directions, see Lloyd (1998).

Evidence, based on both measured and visual data at a number of widely separated locations over the North Atlantic, indicates (Hogben, 1995) that mean wave heights increased over some 30 years or more at a rate of the order of about 1.5% per annum. It is not clear whether this is a general trend or part of a cycle. Extreme wave heights may also have increased slightly but the evidence for this is not conclusive. One possible cause for the increase in the mean height is increasing storm frequency giving waves less time to decay between storms. The fresh winds then act upon a surface with swell already present. This has important implications where a new design is based upon comparison with existing ships. The present data does not allow for this increase.

Describing an Irregular Wave System

A typical wave profile, as recorded at a fixed point, is shown in Figure 9.6. The wave heights could be taken as vertical distances between successive crests and troughs, and the wavelength measured between successive crests, as shown.

If λ_a and T_a are the average distance and time interval in seconds between crests, it has been found that, approximately:

$$\lambda_a = 2gT_a^2/6\pi = 1.04\,T_a^2 \text{ m};$$
$$T_a = 0.285V_w \text{ in seconds if } V_w \text{ is wind speed in knots.}$$

If the wave heights measured are arranged in order of reducing magnitude the mean height of the highest third of the waves is called the *significant wave height* and an observer tends to assess the height of a set of waves as being close to this figure. A general description of a sea state, related to significant wave height, is given by the *sea state code* (Table 9.4), which is quite widely accepted although an earlier code may be encountered.

The wave height data from Figure 9.6 can be plotted as a histogram showing the frequency of occurrence of wave heights within selected bands, as shown in Figure 9.7. A similar plot could be produced for wavelength. In such plots the number of records in each interval is usually expressed as a percentage of the total number in the record so that the total area under the curve is unity.

A distribution curve can be fitted to the histogram as shown (see also the note on statistics in Appendix A). For long duration records or for samples taken over a period of time a *normal* or *Gaussian distribution* is found to give a good approximation. The curve is expressed as

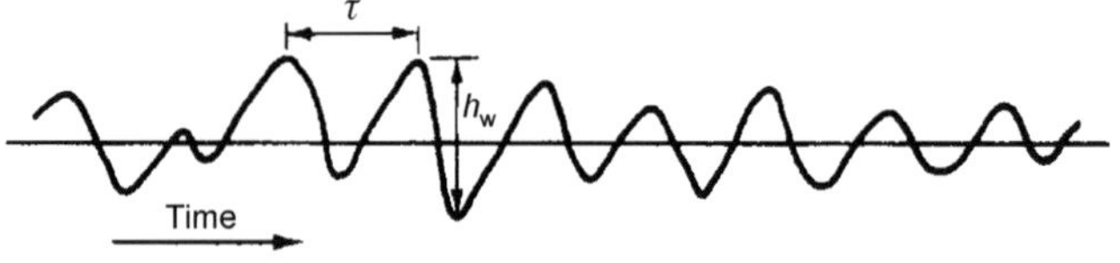

FIGURE 9.6 Wave record.

$$p(h) = \sigma^{-1}(2\pi)^{-0.5} \exp \frac{-(h-\overline{h})^2}{2\sigma^2}$$

where

$p(h)$ = the height of curve, the frequency of occurrence
h = wave height
$\overline{h}$ = mean wave height from record
σ = standard deviation.

TABLE 9.4 Sea State Code

Code	Description of Sea	Significant Wave Height (m)
0	Calm (glassy)	0
1	Calm (rippled)	0– 0.10
2	Smooth (wavelets)	0.10–0.50
3	Slight	0.50–1.25
4	Moderate	1.25–2.50
5	Rough	2.50–4.00
6	Very rough	4.00–6.00
7	High	6.00–9.00
8	Very high	9.00–14.00
9	Phenomenal	Over 14

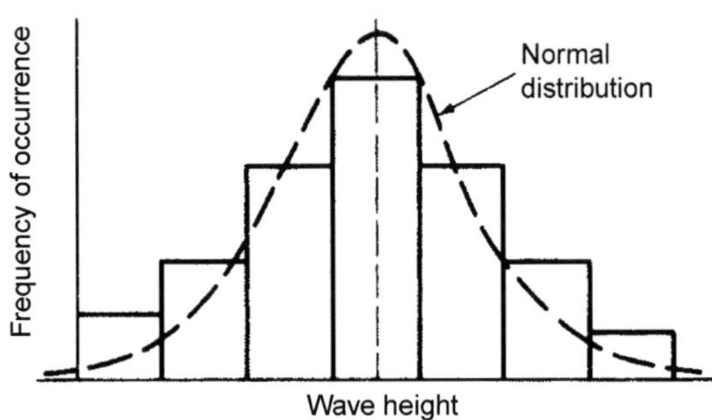

FIGURE 9.7 Histogram of wave height.

Where data are from a record of say 30 min duration, during which time conditions remain reasonably steady, a *Rayleigh distribution* is found to be a better fit. The equation for this type of distribution is

$$p(h) = \frac{2h}{E} \exp\frac{-h^2}{E}$$

where

$$E = \frac{1}{N}\sum h^2 = \text{mean value of } h^2,\ N \text{ being the total number of observations}$$

In these equations $p(h)$ is a probability density. The area under the curve is unity as it is certain that the variable will take some value of h. The area under the curve between two values of h represents the probability that the wave height will have a value within that range. Integrating the curve leads to a *cumulative probability distribution*. The ordinate at some value h on this curve represents the probability that the wave height will have a value less than or equal to h.

Energy Spectra

One of the most powerful means of representing an irregular sea (and a ship's responses) is the energy spectrum. The components of the sea can be found by Fourier analysis and the elevation of the sea surface at any point and time can be represented by

$$h = \Sigma h_n \cos(\omega_n + \varepsilon_n)$$

where h_n, ω_n and ε_n are the height, circular frequency and arbitrary phase angle of the nth wave component.

The energy per unit area of surface of a regular wave system is proportional to half the square of the wave height. The energy, therefore, of the nth component will be proportional to $h_n^2/2$, and the total energy of the composite system is given by

$$\text{Total energy} \ \alpha \frac{\Sigma h_0^2}{2}$$

Within a small interval, $\delta\omega$, the energy in the waves can be represented by half the square of the mean surface elevation in that interval. Plotting this against ω gives what is termed an *energy spectrum* (Figure 9.8). The ordinate of the spectrum is usually denoted by $S(\omega)$. Since the ordinate represents the energy in an interval whose units are 1/s its units will be (height)^2 (seconds). $S(\omega)$ is called the *spectral density*.

Some interesting general wave characteristics can be deduced from the area under the spectrum. If this is m_o, and the distribution of wave amplitude

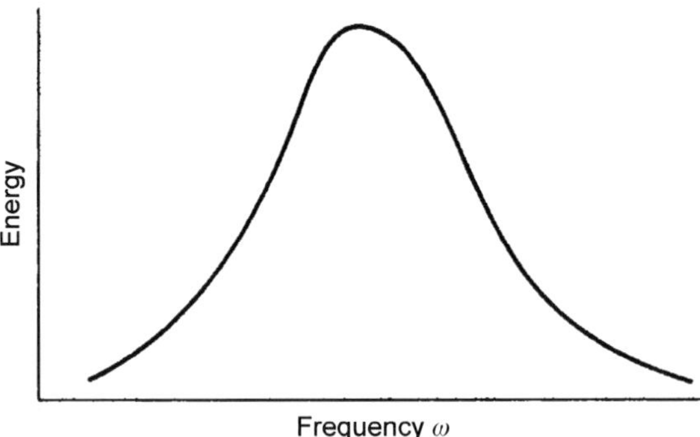

FIGURE 9.8 Energy spectrum.

is Gaussian, then the probability that the magnitude of the wave amplitude at a random instant will exceed some value ζ is

$$p(\zeta) = 1 - \operatorname{erf}\frac{\zeta}{(2m_o)^{0.5}}$$

In this expression erf is the *error function* which will be found in standard mathematical tables.

However, wave observations show that generally the Rayleigh distribution is better at representing the sea surface. In this case it can be shown that:

The most frequent wave amplitude $= 0.707(2m_o)^{0.5} = (m_o)^{0.5}$
Average wave amplitude $= 1.25(m_o)^{0.5}$
Average amplitude of 1/3 highest waves $= 2(m_o)^{0.5}$
Average amplitude of 1/10 highest waves $= 2.55(m_o)^{0.5}$.

Shapes of Wave Spectra

Even in deep water, a wave system will only become fully developed if the duration and fetch are long enough. The wave components produced first are those of shorter length, higher frequency. With time the longer length components appear so that the shape of the spectrum develops as shown in Figure 9.9. A similar progression would be found for increasing wind speed. As the wind abates and the waves die down, the spectrum reduces, the longer waves disappearing first because they travel faster, leaving the storm area.

So far we have discussed how a specific wave system can be represented once it has been measured. The naval architect needs to know the expected shape a fully developed spectrum can be expected to take for a given wind speed, which a ship is likely to meet in service. Early formulae attempted to define the spectrum purely in terms of wind speed and the Pierson and Moskowitz's formula is

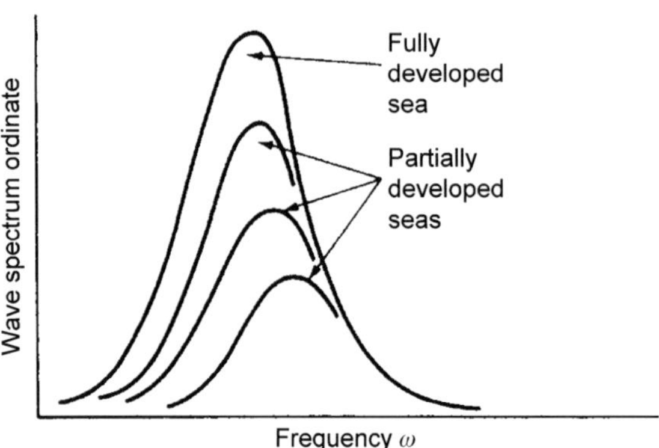

FIGURE 9.9 Developing spectra.

$$S(\omega) = \frac{8.1 \times 10^{-3} g^2}{\omega^5} \exp\left[-0.74\left(\frac{g}{V\omega}\right)^4\right] \text{ m}^2\text{s}$$

where g and V are in ms^{-2} and ms^{-1} units, respectively, V being the wind speed.

The spectrum recommended by the ITTC (2005) and widely adopted is based on the *Bretschneider spectrum* for open ocean conditions. It is known as the ITTC two-parameter spectrum and takes the form

$$S(\omega) = \frac{A}{\omega^5} \exp \frac{-B}{\omega^4} \text{ m}^2/(\text{rad/s})$$

where A and B are constants.

When both the significant wave height $\zeta_{1/3}$ and the characteristic period T_1 are known:

$$A = \frac{172.75\zeta_{\frac{1}{3}}^2}{T_1^4}, \text{ m}^2/\text{s}^4 \quad B = \frac{691}{T_1^4}, \text{ s}^{-4} \quad T_1 = \frac{2\pi m_o}{m_1}$$

where m_1 is the first moment of area of the energy spectrum about the axis $\omega = 0$. Although not part of the ITTC recommendation, when only the significant wave height is known, $S(\omega)$ can be represented approximately by

$$A = 8.10 \times 10^{-3}\ \text{g}^2 \quad \text{and} \quad B = 3.11/\zeta_{\frac{1}{3}}^2$$

It can be shown that:

The area under the spectrum $= m_o = A/4B = 0.0625\zeta_{\frac{1}{3}}^2$
The frequency of peak, $\omega_o = 4.849/T_1$ rad/s

The formula applies to the open ocean. In areas where there is limited fetch the ITTC recommend:

$S_J(\omega) = 155\{((\zeta_w)^2{}_{1/3})/(T_1{}^4\omega^5)\}\exp[(-944)/(T_1{}^4\omega^4)]3.3^\gamma \text{m}^2/(\text{rad/s})$, where

$\gamma = \exp\{-(0.191\omega T_1 - 1)^2/2\sigma^2\}$
$\sigma = 0.07$ if $\omega < 5.24/T_1$ or 0.09 if $\omega > 5.24/T_1$.

Wave Statistics

Having established a wave spectrum to use, the designer still needs to know the severity of waves a new design is likely to meet in service. For this, recourse is had to *ocean wave statistics*. See *IACS; REC 034 (2001) Standard wave data*. Over the years wave data have been obtained from observations and measurements. Although they must be somewhat subjective, visual observations are available for large ocean areas, particularly the main shipping routes, and they have been successfully integrated with measured data. Measurements can be at fixed points in the ocean using buoys, taken by shipboard recorders or by satellite. On board recorders need careful calibration to remove the influence of the ship on the wave system being recorded. Generally buoys deployed locally by the ship are preferred for trials. A suitably arranged group of buoys can give information on the dominant wave direction as well as on height and period.

The concept of using a satellite radar altimeter was established by *Skylab* in 1973. The higher the waves in the footprint of the satellite radar, the more spread out is the time of arrival of the return pulse. Adjusting the height of the return pulse to a constant value, the slope of the leading edge gives a measure of the significant wave height. Wind speed is indicated by the backscatter of the signal. Early radars did not permit the wave period to be measured but later synthetic aperture radars should fill this gap.

Statistical data on the probability of occurrence of various sea conditions at different times of the year with the predominant wave direction are available from a number of sources including the British Oceanographic Centre and an Atlas of Ocean Waves produced as a result of an EU initiative. Typically they show the number of observations within selected wave height and wave period bands. They show a spread of period for a given height and of height for a given period. This 'scatter' is not due to inaccuracies of observation but to the fact that the sea states observed are at various stages of development and includes swell as well as sea components.

The data can be combined in many ways. They can, for example, be averaged over the North Atlantic or world wide. Doing this confirms the popular impression that the Atlantic is one of the roughest areas: 21.4% of waves there can be expected to exceed 4 m whereas the corresponding percentage worldwide is 16.8.

Freak Waves

Between 1993 and 1997 more than 582 ships were lost, totalling some 4.5 million tonnes. Possibly a third of these were lost in bad weather. Each year some 1200 seafarers are lost. As Faulkner (2003) reported, for centuries mariners had spoken of 'walls of water', 'holes in the sea' or 'waves from nowhere'. Usually they were not believed, being thought guilty of gross exaggeration. Freak waves, according to the traditional view of oceanographers and meteorologists, could not exist; their mathematical models of the ocean did not permit them. The theory on which these models were based was a linear one and in real life things are often non-linear – markedly so as events become more extreme. The old theories failed to predict adequately extreme wave heights and their frequency of occurrence. Investigations following the loss of *MV Derbyshire* and subsequent research showed that freak waves do occur and that, although not frequent, they are not rare.

Faulkner discusses four types of abnormal waves apart from Tsunamis which do no harm to ships well out to sea. They are described below.

Extreme Waves in Normal Stationary Seas

These follow the usual laws and occur because waves of different frequencies are superimposed and at times their peaks will coincide. In a sample of 1500 waves a wave k times the significant height of the system will have decreasing probabilities of occurrence as k increases:

k	2.0	2.2	2.4	2.9
% probability	40	9.0	1.5	0.07

The longer a record the greater chance there is of finding an extreme wave.

Opposing Waves and Surface Currents

That these conditions lead to high waves has been known for a long time. The usually quoted example is that of the Agulhas Current off SE Africa. A 4 knot current opposing a 15 s wave increases its height by about 90%.

Standing Waves

These are transient waves which appear and disappear. They occur at the centres of tropical storms, in crossing seas and near steep coastlines.

Progressive Abnormal Waves

These arise from a number of causes including:

- accretion of waves within a group,
- coalescing independent wave groups,
- energy transfer to the faster bigger waves in a group until they become unstable,
- non-linear interactions between colliding waves.

Bateman (2010) discusses the development of wave theories to allow for non-linearity. He points out that to accurately predict waves and their statistical rates of occurrence, account must be taken of the non-linear, unsteady and directional nature of the ocean surface. Allowing for directionality and third order non-linearties it can be shown that significant internal energy transfers can occur. Larger waves are formed before breaking and 'walls of water' can form as reported by seafarers. The exact nature of energy movement is not known but more energy becomes aligned with the dominant wave direction. Compared with linear theories waves 20% higher and 200 m wider are predicted.

Damage Due to Abnormal Waves

Faulkner quotes a number of cases of damage showing that the problems are significant:

- Five container ships (4000–4500 twenty-foot equivalent units (TEU)) caught in a North Pacific storm which lost 700 containers overboard. Some reported rolling to 35–40°.
- A 29.5 m Agulhas wave that smashed a crows nest window 18 m above the waterline on a 256,000 dwt tanker.
- The liner *Michael Angelo* had its bridge, 0.26 l from forward, completely smashed, as were windows 24 m above the waterline.

Satellite images now give us better data on extreme waves, their heights, where they occur and their frequency. Fortunately many occur in the southern oceans, away from the most used shipping routes. The theories are now based on non-linear models and give a better feel for the physics involved.

ICE AND COLD WEATHER

General

Ice is a hazard for the mariner as illustrated by the loss of the *Titanic* in 1912. Cold conditions pose problems for ships in a variety of ways. For instance:

- ice may make ports inaccessible or only usable with the aid of an icebreaker;
- ice at sea may necessitate a reduction in speed;
- ice can form on rigging and superstructures, reducing stability;
- ships need heating to create an acceptable internal environment;
- exposed equipment can freeze up, requiring special heating arrangements;
- work in exposed areas is difficult and dangerous for the crew.

The exploitation of oil and gas fields in the Arctic has led to an increasing number of ice breakers, cargo vessels and service support vessels being designed for operation in Polar ice-infested waters.

Sea Ice

The mechanical properties and form of ice vary greatly and the requirements for the structure and power of vessels in ice are somewhat empirical relying on service experience for calibration of theoretical methods. Some steels are brittle at low temperatures, typically about $-10°C$ and less. Brittle fractures can propagate extremely fast leading to catastrophic failure.

Sea Ice Formation and Properties

Sea ice has a crystalline microstructure that grows vertically resulting in different mechanical properties in different directions. It is extremely brittle, weak in tension but strong in compression. The principal mode of failure is in tensile flexure. Ice breakers impose bending moments on consolidated ice to induce tensile stresses in its upper surface to break it. The mechanical properties increase with age, with reduction in temperature or salinity, and with increasing applied strain rate.

The following mechanical properties of sea ice are applicable to the short duration and rate of loading associated with ship/ice impacts.

- Tensile strength = 0.80–1.0 N/mm^2
- Flexural strength = 1.0–1.2 N/mm^2
- Shear strength = 0.80–1.0 N/mm^2
- Uni-axial compression = 5.0–10.0 N/mm^2
- E modulus = 2.1×10^3–5.1×10^3 N/mm^2.

Polar Ice

Level, multi-year, ice thickness in the Arctic typically varies from about 2 to 4 m with an average of about 2.90 m with a standard deviation of 1.80 m. Antarctic ice thicknesses are generally less and the area of multi-year ice is much less. Multi-year ice is typically up to 2.70 m thick extending from 15 to 20 nm from shore.

Non-Polar Ice

Ice in non-Polar waters is always first year ice. Typical thicknesses encountered range from 0.6 to 1.0 m.

Navigation in Ice

Ice nomenclature is laid down by the World Meteorological Organisation (WMO). Their thicknesses are used by IACS in defining Polar Ice Classes for ships intended for independent operation. These range from *Ice class PC 1* (year-round operation in all Polar waters, thickness > 3.0 m) to *Ice class PC 7* (summer/autumn operation in thin first year ice, thickness < 0.70 m). See IACS UR I (2011) Polar Class.

Operating in Extreme Cold

In addition to sea ice ships encounter low air and water temperatures, extreme and rapidly changing weather conditions with ice on deck and superstructures. There are reduced navigational aids, charts, meteorological and hydrographical information and limited search and rescue facilities in Polar waters. Antarctic waters are those south of latitude 60° south. Arctic waters vary from north of 56° north to 67° north depending on the proximity of the land mass. Due to the severity of the conditions, maximum allowable work times are laid down for able-bodied personnel together with the required protective garments. Below −30°C no outdoor work is permitted unless essential from a safety or operational perspective. All ships systems must be capable of functioning effectively and safely.

In ice, inadequate power can result in a ship being locked-in. The minimum speed for satisfactory progress in the most severe intended ice condition is 3 knots. Safe navigation requires information on ice conditions and data is gained from satellites, ground-based observations, Polar stations and ice breakers.

Winterisation of Ships

In addition to ice class notation some classification societies provide 'winterisation' notation.

Ships must be adapted for safe operation at a low outdoor air temperature of down to −30°C, including the hull structural material, deck equipment, lifting appliances, main engine cooling systems and propeller material. Propeller immersion must be enough to reduce interaction with ice. Ships must have sufficient intact and damage stability, taking account of icing up and adequate stability for ice breaking operations.

OTHER EXTREME ENVIRONMENTS

There are other extreme conditions the design of the ship and equipment may need to allow for. Most relate to equipment on exposed decks and requirements are usually laid down in general specifications such as British Standards or Mil Specs which define suitable tests and should be consulted by the designer. These other factors include the following (with typical standards shown):

- *Rain*. Equipment should operate with no degradation of performance in a rainfall intensity of 0.8 mm/min at 24°C with wind speeds up to 8 m/s for 10 min and 3 mm/min for up to 2 min. The test condition is 180 mm/h for 1 h. Driving sea spray can be more corrosive than rain.
- *Hail*. Equipment should be able to withstand hailstones of 6–25 mm striking at 14–25 m/s for a period of 7 min.

- *Green sea loading*. Exposed equipment and structure should be designed to withstand a pulse loading of 70 kPa acting for 350 ms with transients to 140 kPa for 15 ms.
- *Winds*. Equipment should be able to operate at full efficiency in a steady relative wind speed of 30 m/s with gusts of 1 min duration of 40 m/s. They should remain operational, with some lowering of performance, in wind speeds of up to 36 m/s with gusts of 1 min duration at 54 m/s.
- *Solar radiation*. This can be important in the tropics. Equipment should be able to cope with a heat flux of 1120 W/m^2 acting for 4 h causing a temperature rise of about 20°C on exposed surfaces.
- *Mould*. The fungi associated with mould feed by breaking down certain organic compounds. It occurs when the relative humidity is high – say 65% or higher – and in a temperature range of 0–50°C. The most rapid growth is in a stagnant atmosphere of 95% relative humidity and temperature range 20–35°C. Tests specify the ability to maintain operational capability for a period of days (typically 28 days) when exposed to a mould growth environment.

MARINE POLLUTION

General

A ship affects its environment and the pollution of the world's oceans and atmosphere has become a cause for increasing international concern. In 1972 the United Nations held a Conference on the Human Environment in Stockholm. This conference recommended that ocean dumping anywhere should be controlled and led to the Convention on the Prevention of Marine Pollution by Dumping of Wastes and Other Matter (the London Convention), covering dumping from any source. In 1996, the 'London Protocol' was adopted to modernise the London Convention and replacing it in 2006. Under the Protocol all dumping is prohibited, with limited exceptions.

Over many years IMO has, through the Marine Environment Protection Committee, proposed a wide range of measures to prevent and control pollution caused by ships and to mitigate the effects of any damage that may occur as a result of maritime operations and accidents. These have been shown to be successful in reducing vessel-sourced pollution.

MARPOL covers survey, port state control and certification relating to construction, equipment and operation of machinery spaces of all vessels, and the cargo areas of oil tankers. It covers noxious liquid substances, sewage systems (with discharge of sewage and reception facilities), disposal of garbage and the control of emissions from ships of ozone-depleting substances (ODS), nitrogen oxides (NO_x), sulphur oxides (SO_x) and volatile organic compounds (VOCs).

IMO recognises the importance of training and issues a number of guidance manuals.

The Ocean

Shipping is, statistically, the least environmentally damaging mode of transport. In comparison with land-based industry, shipping is a comparatively minor contributor to overall marine pollution from human activities. That is, however, no reason for complacency.

Wastes dumped from ships amount to about 10% of the pollutants that enter the sea annually. Land sources account for 44%, the atmosphere for 33% (most originating from the land), 12% come from maritime transportation and 1% from offshore production. Thus, the naval architect is concerned directly with about a quarter of the total problem. The Convention bans the dumping of some materials, limits others and controls the location and method of disposal.

Oil Pollution

In terms of quantity, oil is the most important pollutant and the harmful effects of large oil spillages receive wide publicity. Most incidents occur during the loading or discharge of oil at a terminal but far greater quantities of oil enter the sea as a result of normal tanker operations such as the cleaning of cargo residues.

In recent years the international maritime transport of oil has grown dramatically. Between 1983 and 2002 world seaborne trade rose from around 12 billion tonne miles to some 23 billion tonne miles. The carriage of oil and petroleum products rose from 5.6 billion tonne miles to 9.9 billion tonne miles in the same period. By contrast estimates of the quantity of oil spilled during the period declined. The introduction of crude oil washing coupled with segregated ballast segments for oil tankers helped significantly in reducing operational pollution.

Recent amendments to MARPOL incorporate more stringent requirements including the phasing-in of double hull segments for oil tankers; improvements to pump-room bottom protection on oil tankers exceeding 5000 tonnes deadweight; double bottoms in pump rooms for vessels and more stringent requirements to reduce accidental outflow of oil in the event of grounding or collision. Advances have been made in oily water separating equipment allowing levels of discharge of oil effluent from machinery space bilges to be reduced from 100 to 15 ppm.

Chemicals carried at sea can pose a much greater threat than oil. Some are so dangerous that their carriage in bulk is banned although they may be carried in drums. Harmful chemicals are carried in smaller ships than tankers, and the ship design is quite complex to enable their cargoes to be carried safely. One ship may carry several chemicals, each posing its own problems.

Garbage

Depending on the type of ship, area of operation, and size of crew, ships may be equipped with incinerators, compactors, comminuters or other devices for shipboard garbage processing. Use of such processing equipment makes it possible to discharge certain garbage at sea which otherwise would not be permitted, reducing shipboard space for storing garbage, making it easier to off-load garbage in ports, and enhancing assimilation of garbage discharged into the marine environment. Considering these treatments:

- Incinerators are mainly used intermittently. MARPOL prohibits the incineration of certain items including garbage contaminated with more than traces of heavy metals and refined petroleum products containing halogen compounds. The incineration of sewage sludge and sludge oil is allowed in main or auxiliary power plant or boilers but not in ports, harbours and estuaries.
- Compactors make garbage easier to store, to transfer to port reception facilities and to dispose of at sea when discharge limitations permit.
- Comminuters are used in ships operating beyond 3 nautical miles (NM) from the nearest land, to grind food wastes to a particle size capable of passing through a screen with openings no larger than 25 mm. Such a process is recommended even beyond 12 NM because the particle size hastens assimilation into the marine environment.

The regulations concerning disposal depend upon whether the ship is within a specially designated area or not. Generally within these areas disposal at sea is prohibited except for food waste which can be discharged at 12 NM or more from land. Outside of these areas floating dunnage can be jettisoned if more than 25 NM from land; food, paper, glass and metal if more than 12 NM (3 NM if comminuted or ground). Plastic garbage must be retained onboard for discharge at port reception facilities unless reduced to ash by incineration.

Sewage

Broadly raw sewage may not be discharged at less than 12 NM from land; macerated and disinfected sewage at not less than 4 NM; only discharge from approved sewage treatment plants is permitted at less than 4 NM. Sewage can be heat treated and then burnt. It can be treated by chemicals but the residues have still to be disposed of. The most common system is to use treatment plant in which bacteria are used to break the sewage down. Because the bacteria will die if they are not given enough 'food', action must be taken if the throughput of the system falls below about 25% of capacity, as when, perhaps, the ship is in port. There is usually quite a wide fluctuation in loading over a typical 24 h day. Some ships, typically ferries, prefer to use holding tanks to hold the sewage until it can be discharged in port.

Ballast Water

The IMO has developed the International Convention for the Control and Management of Ships' Ballast Water and Sediments (BWM) to regulate discharges of ballast water and reduce the risk of introducing non-native species. The BWM, expected to come into force in 2012/13, will require all ships to implement a Ballast Water and Sediments Management Plan and carry a Ballast Water Record Book. It will require ballast water treatment to be used in place of ballast water exchange.

The typical ballast water capacity of passenger and general cargo ships is in the range 1500–5000 m^3. Tankers and bulk carriers tend to have capacities greater than 5000 m^3.

Ballast water is needed for safe ship operations, but it may pose serious ecological, economic and health problems due to the multitude of marine species in the ballast water. The transferred species may initiate a reproductive population in the new environment, possibly growing into pest proportions. The effects in some areas have been devastating and the rate of bio-invasions is increasing at an alarming rate with new areas being invaded.

No one ballast water management technique can remove all organisms from ballast tanks and a combination of different methods is needed. Treatment options include:

- A combination of filtration and other treatments. Pre-filtration, using a series of filters of reducing mesh size, is likely to be needed before other treatments are applied. Filters should be back washed after ballast water is taken on board so that any organisms are restricted to their original environment.
- Chemicals such as chlorination or ozone. The chemicals must not themselves pollute the water.
- Mechanical means such as cavitation.
- Magnetic forces which can kill certain invertebrates.
- UV radiation which is used ashore to prevent the spread of disease and is being considered for water treatment plants as a replacement for chlorine.
- Ultrasonic, specific frequencies killing different organisms.

More research is needed and a variety of systems are being developed. It is likely that no one solution will fit all ships.

Noise Pollution

There is growing concern over the effects of increased noise pollution on marine life. This is discussed by Leaper and Renilson (2012). Noise from shipping, in the 10–300 Hz frequency range, can mask the sounds associated with communication, breeding and feeding of marine species leading to reductions in population levels. The main concern is with the general increase in ambient noise and it has been assessed that the noisiest 10% of vessels generate the majority of the noise impact. An IMO correspondence group has been developing guidelines on how noise reductions can be achieved in merchant ships.

Radiated noise is generated by the propeller, machinery and by the hull moving through the water. The energy of this noise is typically a few watts so it is very small compared with the energy used in propulsion, suggesting that small changes in propulsive efficiency could make a large difference to the noise output. Cavitation is the main source of noise and, whilst it cannot be avoided in most ships which require to operate above their cavitation inception speed, its impact can be reduced by improving the flow into the propeller and by good propeller design. This will also improve propulsive efficiency. Fishermen using quiet vessels report larger catches.

A draft BS/ISO standard – 16554 – deals with underwater noise radiated by merchant ships.

Noise transmitted into the air can also be annoying. Carlton and Vlasic (2005) cite a case of a ship's funnel being excited by gas pulses from its diesel generators when in port. The particular peak frequency was annoying to nearby residents.

The Atmosphere

Air pollution is the introduction of chemicals and particulate matter that cause harm or discomfort to humans or damage the atmosphere. The atmosphere is a complex gaseous system, essential to support life on earth. Stratospheric ozone depletion due to air pollution has long been recognised as a threat. Air pollution from ships has a cumulative effect, contributing to air quality problems and affecting the natural environment. IMO's Regulations for the Prevention of Air Pollution from Ships aim to minimise emissions from ships of sulphur oxides (SO_x), nitrous oxides (NO_x), carbon dioxide and VOCs. Sulphur enters the combustion process via the fuel and nitrogen via the air. Deliberate emission of ODS is prohibited.

Greenhouse Gases

The so-called greenhouse gases (GHG) allow sunlight to enter the atmosphere freely some of which is reflected back as infrared radiation. GHG, water vapour and clouds trap some of this reflected heat in the atmosphere. Ideally the amount of energy arriving from the sun should equal the energy radiated back, leaving the temperature of the Earth's surface roughly constant. Many gases exhibit 'greenhouse' properties. Some occur in nature (water vapour, carbon dioxide, methane, and nitrous oxide), while others are exclusively human-made (like gases used for aerosols). If man-made gases affect the balance of heat arriving and being reflected they can lead to global warming. IMO's work in this area is based on a study to establish the amount and relative percentage of GHG emissions from ships. This found that exhaust gases are the primary source of emissions and carbon dioxide is the most important GHG emitted by ships, both in quantity and global warming potential.

Shipping emitted some 1406 million tonnes of CO_2 in 2007. This figure may grow by a factor of 2–3 by 2050 in the absence of corrective action.

A significant potential for the reduction of GHG through technical and operational measures has been identified. If implemented, these measures could increase efficiency and reduce the emissions rate by 25%–75% below the current levels. Many measures appear to be cost-effective, although other factors may discourage their implementation. One approach is the energy efficiency design index (EEDI) for new ships as discussed below. The report assessed the potential reductions of CO_2 emissions to be obtained from applying known technology and practices to the design of new ships and the operation of all ships. For new ships, estimated savings included:

- 5–15% for power and propulsion systems
- 5–15% from the use of low carbon fuels
- 1–10% from using renewable energy.

For ship operations, estimated savings included:

- 1–10% from voyage optimisation
- 1–10% from energy management.

Shipping is, in general, an energy-efficient means of transportation compared to the other modes. In 2007 it was estimated that shipping accounted for about 2.7% to the global emissions of carbon dioxide. ODS include chlorofluorocarbons (CFC) and halons used in older refrigeration and fire-fighting systems and portable equipment and in some insulation foams.

Hydrochlorofluorocarbons (HCFC) were introduced as an intermediate replacement for CFCs but are themselves still classed as ODS. As part of a worldwide movement, the production and use of all these materials is being phased out under the provisions of the Montreal Protocol. The controls in this regulation do not apply to sealed equipment such as small, domestic type, refrigerators and air conditioners.

No new CFC or halon containing systems are permitted to be installed on ships. No HCFC containing systems or equipment are to be installed on ships constructed on or after 1 January 2020. Existing systems and equipment are permitted to continue in service, but the deliberate discharge of ODS to the atmosphere is prohibited.

Nitrogen Oxides

MARPOL's NO_x control requirements apply to diesel engines of over 130 kW output power other than those used solely for emergency purposes.

Sulphur Oxides

SO_x and particulate matter emission controls apply to all fuel oil combustion equipment onboard including main and auxiliary engines, boilers and inert gas generators. The global sulphur cap is being reduced progressively from 3.50% (2012) to 0.50% (2020), subject to a feasibility review. The limits applicable in special emission control areas (ECAs) for SO_x and particulate matter were reduced to 1.00% (2010); being further reduced to 0.10% (2015).

Details of the limits, and of areas designated as ECAs, can be found in IMO's Regulations. The limits are mainly achieved by limiting the maximum sulphur content of the fuel oils used.

Volatile Organic Compounds

VOCs are emitted as gases from certain solids or liquids and include a variety of chemicals, some of which may have short- and long-term adverse health effects. VOCs are emitted by a wide array of products including paints, cleaning materials, pesticides, furnishings and office equipment.

Anti-Fouling Systems

Commonly anti-fouling paints used metallic compounds which slowly leached into the seawater, killing barnacles and other marine life that had attached to the ship. These compounds persisted in the water, killing sea life, harming the environment and possibly entering the food chain. One of the most effective anti-fouling paints, developed in the 1960s, contains the organotin tributylin (TBT), which has been proven to cause deformations in oysters and sex changes in whelks.

The International Convention on the Control of Harmful Anti-fouling Systems on Ships now prohibits the use of harmful organotins compounds which act as biocides in anti-fouling paints used on ships and aims to prevent the future use of other harmful substances.

The Ship Energy Efficiency Design Index (EEDI)

In 2011 IMO introduced a mandatory EEDI for new ships. This is regarded as a cost-effective solution providing an incentive to improve the design efficiency of new ships. Its environmental effect is limited because it applies only to new ships. The EEDI requirements are contained in IMO Interim Guidelines, the intent being to stimulate innovation and technical development of all elements that directly or indirectly influence the energy efficiency of a ship. The Guidelines apply to all commercial ships with special provision being given to ice-navigating ships that need additional power for safety in ice-infested waters. The equation to calculate the energy efficiency of new ships in reducing CO_2 emissions is extremely complex and at this stage the equation may not be applicable to diesel-electric propulsion, turbine propulsion and hybrid propulsion systems.

The main parameters in calculating the EEDI include a non-dimensional conversion factor between fuel consumption and CO_2 emissions; ship speed in calm water; ship capacities; power of main and auxiliary engines; main engine power reduction factor due to innovative mechanical energy efficiency technology; auxiliary engine power reduction due to innovative electrical energy efficiency technology.

Besides the EEDI, IMO has proposed an Energy Efficiency Operational Indicator (EEOI) and a Ship Energy Efficiency Management Plan (SEEMP). These three measures are still being developed.

Of the above measures, the EEDI represents a major technical regulation for marine CO_2 reduction. It is based on a new and complex set of formulations, guidelines and regulations and therefore deserves closer understanding and evaluation. Bazari and McStay (2011) describe the EEDI development trend, explain the regulatory framework and discuss its likely impacts.

Two important questions relate to potential for EEDI reduction and how the industry is to comply with the required reduction rates. The main ways in which the EEDI reduction might be achieved include:

- reducing design speed so reducing installed power. Many container ships have reduced their cruising speeds by about 5 to 18–20 knots. Further reductions could pose increased stresses on their engines;
- energy-efficient technologies, requiring less fuel for the same amount of power;
- innovative/renewable energy technologies, reducing all or a significant portion of CO_2;
- using renewable energy;
- increasing the deadweight for a given lightweight;
- using low carbon fuels.

Ship speed reduction is likely to be the main method of EEDI reduction in the short and medium term. However, the impact of speed reduction on ship economics and safety has yet to be established.

There are many existing and emerging technologies that could reduce EEDI. Some, such as extra smooth foul release paints, are being adopted while others, such as waste heat recovery and hull-propeller improvements, could be used if they prove cost-effective. Generally energy-efficient technologies involve additional costs and complexity.

Renewable energy sources are solar, wind and nuclear power. They are expected to be adopted more in the long term.

Low carbon fuels will directly reduce EEDI via the carbon factor. Possibilities are LNG and biofuels. LNG is likely to be used in the medium to long term as ship design issues are resolved and supply infrastructure mature. Establishing a network of LNG bunkering facilities will be costly and there are related safety concerns. The use of LPG is perhaps easier and some diesel generators are already running on LPG.

Where ships need to operate within IMO's designated Special Emission Control Areas they may use machinery designed to operate on two different fuels. Shipboard arrangements become more complex – separate tanks and handling systems are needed. Quality of fuel supplied must be checked and engineering staff need more training.

Minimising Environmental Impact – the 'Green' Ship

First, the design must be in full compliance with the statutory requirements of MARPOL and associated guidelines. This will reduce the risks of polluting the environment but not necessarily minimise them. Designing the hull to minimise the propulsive power needed is one way of reducing levels of emissions. CFD methods can be used to design the after end of the ship to improve propulsive efficiency. Japan's National Maritime Research Institute (NMRI) is aiming to produce a zero emission ship. Their programme is in three phases.

- First using advances in hydrodynamics will reduce emissions by 50%. Linked to this are:
  - Twin skegs designed to modify the inflow to the propellers producing a swirling flow in the opposite direction to the propeller movement creating something like a contra-rotating propeller. A reduction of fuel consumption of 20% is claimed.
  - Drawing water into the ship and ejecting it just ahead of the main propellers – assisted by small propellers in the tubes close to the outlets.
  - Using boundary layer control.
- Next will be a vessel with hybrid power plant, solar energy and an electrical supply system reducing emissions by 80%.
- Then will be a vessel powered by fuel cells.

The need to reduce pollution can have a significant affect upon the layout of, and equipment fitted in, ships. Sources of waste are grouped in vertical blocks to facilitate collection and treatment. Crude oil washing of the heavy oil deposits in bulk carrier oil tanks and segregated water ballast tanks are becoming common. Steam cleaning of tanks is being discontinued. In warships the average daily arisings from garbage amount to 0.9 kg per person food waste and 1.4 kg per person other garbage. It is dealt with by a combination of incinerators, pulpers, shredders and compactors.

HUMAN FACTORS

General

Human factors are increasingly important in ship design, build and operations. In essence anything affecting, or affected by, human beings on board is embraced by the subject. Here we consider the well-being and comfort of people, and human actions/reactions. The internal environment has a major impact upon the former. If it is too hot or cold, too noisy or the lighting is poor the crew will not perform their duties well and passengers will not relax and enjoy themselves.

Human Factors in Design and Operation

Observing people, what they do and how they react to given situations (particularly emergencies) is important. To understand what is happening, and why, it is necessary to establish the train of thought that leads to any specific action. This, coupled with the need to create the necessary environment, means that there are two major disciplines involved, with the engineer, in the study of human factors. These are:

- Physiologists who study, and recommend, desirable conditions for people to live and work in, including:
  - Air quality, vibration and noise levels, lighting levels, and so on as discussed in the section dealing with the ship's internal environment.
  - What are possible limitations of humans in general? Whilst one can expect crew members to be well trained and able-bodied this may not be true of passengers.
  - Shift patterns for different activities to avoid fatigue.
- Psychologists who examine why humans do what they do in the way they do. For instance:
  - The thought processes followed in analysing a radar display.
  - Can data from ships' sensors be presented in a better, more easily understood way?
  - What constitutes a good display of data from the many machines on board?

Apart from the physical conditions provided by the environment, there is the question of a person's mental state. A good worker needs to feel appreciated and to be well trained for the job to be carried out. These aspects are covered in an IMO Convention on crew training and certification.

Regulations

IMO's Standards of training, certification and watchkeeping (STCW) Convention helps promote the well-being of crews and safety of ships and people by:

- establishing basic requirements on training, certification and watchkeeping for seafarers internationally, prescribing minimum standards to be met.
- encouraging the use of simulators for training and assessment purposes. Simulators are mandatory for training in the use of radar and automatic radar plotting.

- setting out basic principles and mandatory provisions relating to radio and engineering watchkeeping.
- ensuring that those with specific duties related to the cargo and cargo equipment of tankers have completed an appropriate shore-based fire-fighting course.
- specifying training requirements for the launching and recovery of fast rescue boats including launching them in adverse weather conditions.
- laying down training requirements for those on passenger ships, covering crowd management, passenger safety, crisis management and human behaviour training.

THE INTERNAL ENVIRONMENT

Ships must be designed so as to provide a suitable environment for the continuous, efficient and safe working of equipment and crew. The environment should also be one in which crew and passengers will be comfortable. Vibration, noise and shock are all factors in that environment. The vibration levels, for instance, must be kept low for comfort and efficient functioning of machinery. Noise levels must also be kept below certain levels to avoid physical harm and facilitate communications. The vertical accelerations associated with ship motions must be reduced as much as possible at the critical frequencies, to reduce the likelihood of motion sickness. Dobie (2003) reviews the various ways in which the shipboard environment can affect humans and some of his findings are used below.

A complicating factor in assessing habitability standards is the variability of human responses to their environment. Those responses will depend upon age, health, metabolism and sex – what is acceptable to one person may not be to another.

Ship Motions and Motion Sickness

Motion sickness is very unpleasant as are the feelings of nausea that precede it. Besides causing discomfort to everyone on board, motions degrade the performance of the crew, both mentally and physically. Motion sickness can be caused by the absence of expected motion, being caused by 'visual motion'. In essence motion sickness is a response to real or apparent motion to which a person is not adapted.

Seasickness is widespread but also very variable. More than a quarter of inexperienced passengers can be seasick for the first few days in a moderate sea; 90% are seasick in very rough weather. Even in warships half the crew is likely to be sick in severe conditions. Variability arises from the type of stimulus – frequency (around 0.2 Hz is the most provocative), intensity, direction and duration – and the susceptibility of the individual.

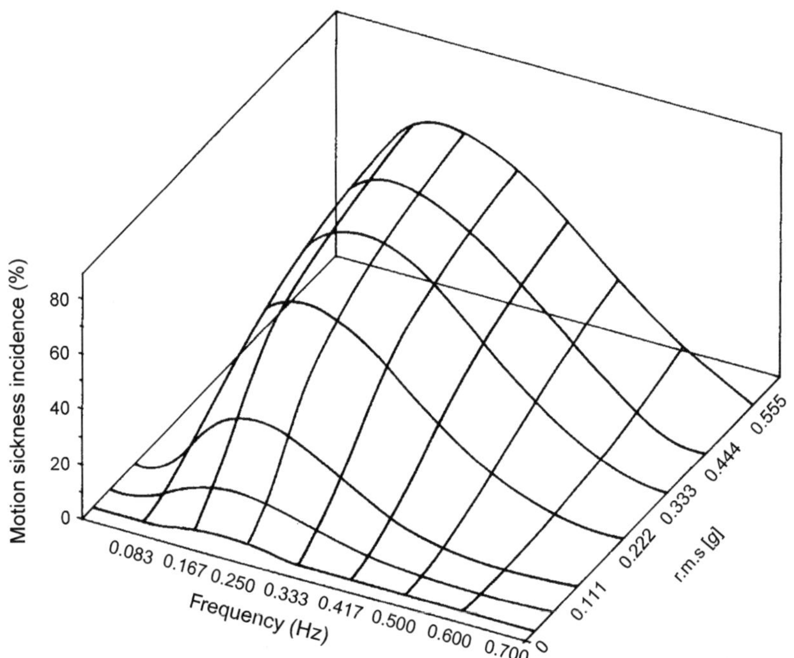

FIGURE 9.10 Motion sickness incidence.

Research indicates that the most important factors, as far as human beings are concerned, is the vertical acceleration they experience, these being mainly due to roll and pitch, and its period. The most critical frequencies are those in the range 0.15–0.20 Hz. Figure 9.10 shows the motion sickness incidence against frequency and r.m.s. acceleration in terms of g.

Motions are very subjective and one measure of comparing the impact of different motions is a *subjective motion (SM) index*. A combination of acceleration and frequency at which people feel the motion to have the same intensity are determined. This level is denoted by SM = 10 and other combinations are rated at SM = 10 n when they are judged to be n times as intense as the base measurement. It has been found that:

$SM = A(f)a^{1.43}$, where a is the acceleration amplitude in g
$A(f) = 30 + 13.53(\log_e f)^2$ with frequency, f, in Hz.

For motion in a given sea SM can be calculated from the statistical variances of the motion displacement, velocity and acceleration experienced at any point, say the bridge. There is no clear-cut limit to acceptable SM but a maximum of 15 has been suggested. An overall figure for personnel spread around the ship can be found by applying weighting factors related to the distribution of people around the ship. Another concept used is that of *motion sickness index (MSI)*. The MSI is the percentage of people likely to be sick when subject to the particular motion for a given time. The greater the motion, or elapsed time, the higher is the MSI.

Reduction in the incidence of seasickness can be achieved by designing the vessel to have lower motions at the critical frequencies, or by training crews to help them adapt to motions – a form of habituation. Providing the individual with an external frame of reference helps.

Whole-Body Motions

When subject to motion, humans experience:

- Motion induced interruptions (MII). MII involves the individual losing balance or slipping so interrupting the task being performed. In ships people are likely to lose balance before slipping. Providing an external reference reduces MII.
- Motion induced fatigue (MIF). Much more muscular effort is involved in maintaining posture while carrying out a task. This is complicated by the loss of sleep. One study showed a 32% reduction in alertness following a reduction in sleep of 1.5 h for one night. The combined effects are important where crews are required to work both day and night.
- Impairment of cognitive performance which appears to be linked to motion sickness.
- Degradation of fine motor skills.

Temperature and Humidity

Heat and odours are important factors in determining a person's reactions to motions as well as general comfort. Thus there is a need to control the air quality in terms of temperature, humidity, purity and smells. Typically about 0.3 m^3 of fresh air is introduced for each person per minute. A person generates about 45 W of sensible heat and 150 W latent heat, depending upon the level of activity. These figures, and the heat from machines, must be allowed for in the design of an air-conditioning system which must cater for a range of ambient conditions. Good insulation helps prevent heat from outside the ship, or from hot spaces within it, getting into general accommodation areas.

As regards moisture in the atmosphere it is the *relative humidity* that is important. This is the ratio of the amount of water present in the air to the maximum amount it can hold at that temperature. The higher the temperature the more water the air can hold. To assess the relative humidity two temperatures are recorded: the *dry bulb* and the *wet bulb*. At 100% relative humidity the two temperatures are the same. The air is then said to be *saturated*. Any lowering of temperature will lead to water condensing out and the temperature at which this occurs is known as the *dew point*. Air-conditioning systems use this fact to control humidity by first cooling and then heating air. At humidity levels below saturation the wet bulb temperature will be lower than the dry bulb, being reduced by evaporation – rather as a human being feels colder when in wet clothing. The degree of cooling will vary with the movement of air.

Thus how comfortable someone will feel depends upon temperature, humidity and air movement. This complicates matters and the concept of *effective temperature* is used. This is the temperature of still, saturated air which would produce the same feelings of comfort.

The problems of atmosphere management are most severe in submarines where the ship remains under water for long periods. Systems are fitted to remove carbon dioxide, add oxygen and remove a wide range of impurities.

Vibration

Vibrations, like motions, are unpleasant and make life on board ship uncomfortable or more difficult. A ship is an elastic structure that vibrates when subject to periodic forces which may arise from within the ship or be due to external factors. Of the former type the unbalanced forces in main and auxiliary machinery can be important. Usually turbines and electric motors produce forces which are of low magnitude and relatively high frequency. Reciprocating machinery on the other hand produces larger magnitude forces of lower frequency. Large main propulsion diesels are likely to pose the most serious problems particularly where, probably for economic reasons, four or five cylinder engines are chosen. They can have large unbalance forces at frequencies equal to the product of the running speed and number of cylinders and of the same order as those of the main hull vibration modes. Vibration forces transmitted to the ship's structure can be much reduced by flexible mounting systems. In critical cases vibration neutralisers can be fitted in the form of sprung and damped weights which absorb energy or active systems can be used generating forces equal but in anti-phase to the disturbing forces. These last are expensive and are not commonly fitted.

Misalignment of shafts and propeller imbalance can cause forces at a frequency equal to the shaft revolutions. With modern production methods the forces involved should be small. A propeller operates in a non-uniform flow and is subject to forces varying at blade rate frequency, which is the product of the shaft revolutions and the number of blades. These are unlikely to be of concern unless there is resonance with the shafting system or ship structure. Even in uniform flow a propulsor induces pressure variations in the surrounding water and on the ship's hull in the vicinity. The variations are more pronounced in non-uniform flow particularly if cavitation occurs. Stable cavitation over a relatively large area is equivalent to an increase in blade thickness and the blade rate pressures increase accordingly. If cavitation is unstable pressure variations may be many times greater. The number of blades directly affects frequency but has little effect on pressure amplitude.

A ship in waves is subject to varying hull pressures as the waves pass. The ship's rigid body responses are dealt with under seakeeping. Some of the wave energy is transferred to the hull causing main hull and local vibrations. The main hull vibrations are usually classified as *springing* or *whipping*. The former is a fairly continuous and steady vibration in the fundamental hull mode due to the general pressure field. The latter is a transient caused by slamming or shipping green seas. Generally vertical vibrations are most important because the vertical components of wave forces are dominant. However, horizontal and torsional vibrations can become large in ships with large deck openings, such as container ships, or in ships of relatively light scantlings.

Acceptance criteria for habitability and comfort of personnel on board ships are set out in ISO 6954. See also BS 6841 which deals with human exposure to whole-body vibration and repeated shock. When issued in 1984 ISO 6954 was intended to apply to crew members of merchant ships but was also often used for passenger comfort standards. It gave peak values of acceleration (1–5 Hz) and velocity (5–100 Hz) defining areas where adverse comments were probable or not. If the acceleration was below 128 mm/s^2 adverse comments were not likely; above 285 mm/s^2 they were. The corresponding velocities were 4 and 9 mm/s. The revised version in 2000 applied to both crew and passengers. Carlton and Vlasic (2005) point out that one limitation of ISO 6954 (1984) was that it used discrete harmonic responses and did not reflect possible combinations of several frequency components. Vibration was measured as a maximum repetitive peak value – obtained by applying a conversion factor to the peak value measured. A factor of 1.8 was used in the absence of a factor determined by measurement. Vibration levels in ISO 6954 (2000) are measured as an overall frequency-weighted r.m.s. acceleration or velocity over the range 1–80 Hz. That is, a single value is used to characterise the total vibration. There is some difficulty in relating these figures to historical data built up around the earlier standard.

ISO 6954 (2000) provides criteria for crew habitability and passenger comfort in terms of overall frequency-weighted r.m.s. values from 1 to 80 Hz for three different ship areas. Table 9.5 gives the figures.

The accelerations, which are what humans mainly respond to, refer to any direction but usually vertical accelerations are most severe together with their duration. Transverse accelerations are rather less important and fore and aft accelerations are generally of little consequence. Humans are also upset by vibration of objects in their field of view.

Continuous exposure to vibration leads to fatigue and decreased efficiency. Acceptable exposure times decrease with increasing r.m.s. acceleration. At 5 Hz, as an example, the acceptable levels reduce from about 3 m/s^2 for 1 min, to about 1.2 m/s^2 for 1 h and about 0.14 m/s^2 for 24 h.

TABLE 9.5 Overall Frequency-Weighted r.m.s Values

Ship Area	Adverse Comment Probable	Acceleration (mm/s^2)	Velocity (mm/s)
Passenger spaces	Yes	143	4
	No	71.5	2
Crew spaces	Yes	214	6
	No	107	3
Work spaces	Yes	288	8
	No	143	4

Noise

Sound levels are subjective because a typical noise contains many components of different frequency and these will affect the human ear differently. In relation to human reactions to noise it is usual to express noise levels in dB(*A*). The *A* weighted dB is a measure of the total sound pressure modified by weighting factors which vary with frequency. The end result reflects more closely a human's subjective appreciation of noise. Humans are more sensitive to high (1000 Hz and over) than low (250 Hz and less) frequencies and this is reflected in the weighting factors.

Noise effects can range from mere annoyance to physical injury. Apart from noise making it hard to hear and be heard (it is very difficult to hold a prolonged conversation over 1 m distance at noise levels of 78 dB(*A*) – critical for communications), crew performance can fall off because prolonged exposure to noise causes fatigue and disorientation. It can annoy and disturb sleep. Intermittent noise is more distracting than continuous and high pitch noise than low. The degree of impairment depends upon the complexity of the task and the duration. High noise levels (about 130–140 dB) will cause pain in the ear and higher levels can cause physical harm to a person's hearing ability. In life ashore the noise level in offices varies from 58 to 65 dB and traffic noise on a busy road is 68 dB.

The International Maritime Organisation (IMO) lay down acceptable noise levels in ships according to a compartment's use (Table 9.6). These levels have been reducing over the years.

Reducing Noise Levels

Generally anything that helps reduce vibration will also reduce noise. Machinery can be isolated but any mounting system must take account of

TABLE 9.6 Acceptable Noise Levels in Ships

Location	Permitted Noise Level (dB(*A*))
Engine room	108 85 if continuously manned
Workshops	80
Bridge	55
Mess room	55
Recreation space	60
Cabins	50
Offices	55

vibration, noise and shock. Because of the different frequencies at which these occur the problem can be difficult. A mount designed to deal with shock waves may actually accentuate the forces transmitted in low-frequency hull whipping. Dual systems may be needed to deal with this problem. Airborne noise can be prevented from spreading by putting noisy items into sound booths or by putting sound absorption material on compartment boundaries. Such treatments must be comprehensive. To leave part of a bulkhead unclad can negate, to a large degree, the advantage of cladding the rest of the bulkhead. Flow noise from pipe systems can be reduced by reducing fluid speeds within them, by avoiding sudden changes of direction or cross section and by fitting resilient mounts. Inclusion of a mounting plate of significant mass in conjunction with the resilient mount can improve performance significantly.

Where mounts are fitted to noisy machinery care is needed to see that they are not 'short circuited' by connecting pipes and cables, and clearances must allow full movement of the machine. In recent years, active noise cancellation techniques have been developed. The principle used is the same as that for active vibration control. The system generates a noise of equivalent frequency content and volume, but in anti-phase to the noise to be cancelled. Thus to cancel the noise of a funnel exhaust a loudspeaker producing a carefully controlled noise output could be placed at the exhaust outlet.

Shock

Warships must be designed specifically to withstand certain levels of shock due to enemy action but all ships may experience shock from wave impacts or collision. Slamming is basically a shock followed by a flexural vibration of the hull.

As with vibration and noise, the effects of shock on sensitive equipment can be mitigated by mounting that equipment on shock mounts. Where alignment of several equipments is important they are mounted on a raft which itself has isolation mounts between it and the hull.

Illumination

The levels of illumination aimed for will depend upon the activity within a compartment. Typically the level in lux will be about 75 in cabins, 100–150 in public rooms, 50 in passageways, 150–200 in machinery spaces. In passenger ships lighting is important not only to provide an adequate level of illumination but also for the moods it can create. The idea of a romantic candle lit dinner is perhaps a cliché but it is nevertheless true that lighting does affect the way people feel.

SUMMARY

There is an interaction between the ship and the ocean environment in which it operates. The greatest impacts of the environment on the ship arise from the wind, waves and temperature. The apparently confused ocean surface can be represented by the summation of a large number of individually small amplitude regular waves and the energy spectrum concept is used to represent the irregular sea surface. Formulations of such spectra include the ITTC two-parameter spectrum. The pollution of the environment is of growing concern and is increasingly the subject of national and international regulation. Ships may pollute the marine environment in a number of ways.

Human factors are important in the design and operation of ships. The internal environment can have a major impact upon the comfort and mood of passengers, and on the efficiency of the crew. The more important factors are temperature, air quality, noise vibrations and ship motions. Many are controlled by international regulations and standards.

This page intentionally left blank

Chapter 10

Seakeeping

INTRODUCTION

In the broadest sense *seakeeping* and *seaworthiness* cover all those features of a vessel which influence its ability to remain at sea in all conditions and carry out its intended mission. They embrace stability, strength, manoeuvrability and endurance as well as the motions of the ship. This chapter concentrates on those aspects of a ship's performance directly attributable to the action of the waves. Seakeeping needs to be considered early in design as most hull form parameters are decided then. Rolling can be reduced by fitting stabilisers but that should be to an already good seakeeping design.

SHIP RESPONSES

The basic dynamic responses to the forces acting on a ship are discussed in Appendix C. Here we consider a ship's motions in waves.

A ship has three principal axes and two degrees of freedom (translation and rotation) for each axis. The terms applied to these basic motions are:

- *surge*, *sway* and *heave* for translations along the fore and aft, transverse and vertical axes respectively;
- *roll*, *pitch* and *yaw* for rotations about the three axes.

Roll, pitch and heave are oscillatory motions and are those of most significance. Applying simple harmonic theory to the motions, the ship following a small disturbance in still water leads to their natural periods. As the ship is flexible other degrees of freedom will be excited but these are dealt with under strength and vibration.

Rolling

If φ is the inclination to the vertical at any instant, and the ship is stable, there will be a moment acting on it tending to return it to the upright (Figure 10.1).

Introduction to Naval Architecture. DOI: http://dx.doi.org/10.1016/B978-0-08-098237-3.00010-2

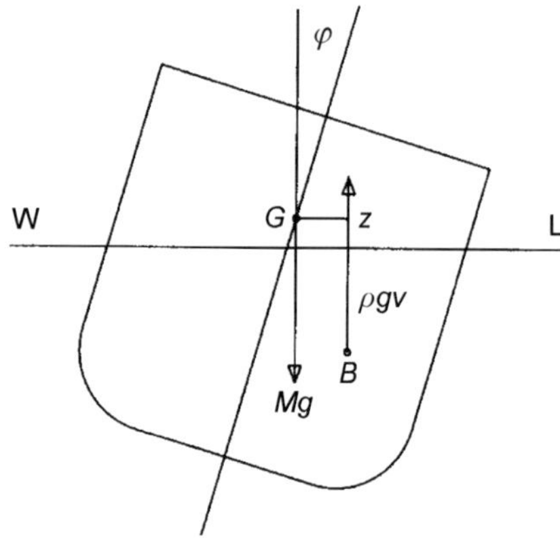

FIGURE 10.1 Rolling.

For small disturbances the value of this moment will be proportional to φ and given by:

$$\text{Displacement} \times GM_{\mathrm{T}} \times \varphi = \Delta \mathrm{GZ}$$

This is the condition for simple harmonic motion with a period T_{φ} defined by:

$$T_{\varphi} = 2\pi\left(\frac{k_x^2}{gGM_{\mathrm{T}}}\right)^{0.5} = \frac{2\pi k_x}{(gGM_{\mathrm{T}})^{0.5}}$$

where k_x is the radius of gyration about a fore and aft axis.

This period is independent of φ and such rolling is said to be *isochronous*. The relationship holds for most ships up to angles of about 10° from the vertical. The greater the GM_{T} the shorter the period and a ship with a short period of roll is said to be *stiff* and one with a long period is termed *tender*. Most people find a longer period motion less unpleasant.

Pitching

This is controlled by a similar equation to that for roll. In this case:

$$T_{\theta} = \frac{2\pi k_y}{(gGM_{\mathrm{L}})^{0.5}} \text{ for small pitch angles } (\theta)$$

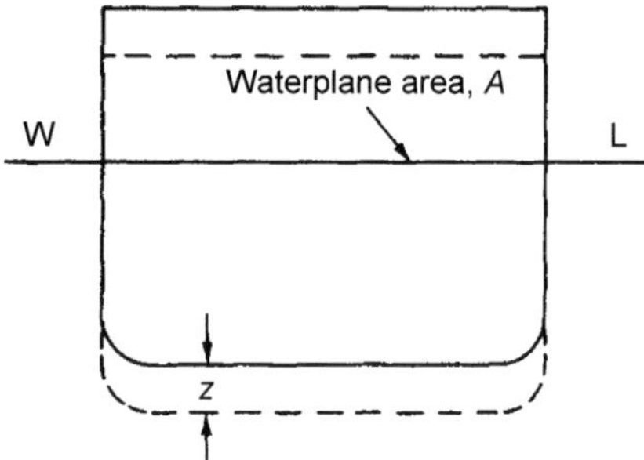

FIGURE 10.2 Heaving.

Heaving

If z is the downward displacement at any instant – Figure 10.2 – there will be a net upward force on the ship tending to reduce the magnitude of z. This force is given by $\rho g A_W z$ and the resulting motion is defined by:

$$\rho \nabla \frac{d^2 z}{dt^2} = -\rho g A_W z$$

where A_W is the waterplane area.

Again the motion is simple harmonic with period:

$$T_z = 2\pi \left(\frac{\nabla}{g A_W} \right)^{0.5}$$

Added Mass and Damping

Added mass and damping will affect these motions and their periods. Added mass values vary with the frequency of motion but, to a first order, this variation can be ignored. Typically the effect for rolling is to increase the radius of gyration by about 5%. In heaving its influence is greater and may amount to as much as an apparent doubling of the mass of the ship. Damping will cause the amplitude of the motion to decrease with time.

SHIP MOTIONS

So far small disturbances have been considered. For larger excursions the proportionality breaks down and the resulting motions become more complex.

Motions in Regular Waves

The apparently random surface of the sea can be represented by the summation of a large number of regular sinusoidal waves, each with its own length, height, direction and phase. Further the response of the ship in such a sea can be taken as the summation of its responses to all the individual wave components. Hence the basic building block for the general study of motions in a seaway is a ship's response to a regular sinusoidal wave.

For simplicity it is assumed that the pressure distribution within the wave is unaffected by the presence of the ship. This common assumption was first made by Froude in his study of rolling and it is known as *Froude's hypothesis*.

Rolling in a Beam Sea

The rolling a ship experiences is most severe in a beam sea. With Froude's hypothesis, the equation for motion will be that for still water with a forcing function added. This force arises from the changes in pressure acting on the hull due to the wave and could be found by integrating the pressure over the whole of the wetted surface.

The resultant force acting on a particle in the surface of a wave must act normal to the surface. If the wavelength is long compared to the beam of the ship, and it is these longer waves which cause more severe rolling, it is reasonable to assume that there is a resultant force acting on the ship normal to an 'effective surface', taking account of all the subsurfaces interacting with the ship. This useful concept was proposed by William Froude, who further assumed that the effective wave slope is that of the subsurface passing through the centre of buoyancy of the ship. With these assumptions and neglecting the added mass and damping, the equation of motion takes the form:

$$\frac{\Delta k_x^2}{g} \times \frac{\mathrm{d}^2\varphi}{\mathrm{d}t^2} + \Delta GM_\mathrm{T}(\varphi - \varphi') = 0$$

where $\varphi' = \alpha \sin \omega t$, α being the maximum slope of the effective wave.

If the subscript 0 relates to unresisted rolling in still water, it can be shown that the solution to the equation for resisted motion takes the form:

$$\varphi = \varphi_0 \sin(\omega_0 t + \beta) + \frac{\omega_0^2 \alpha}{\omega_0^2 - \omega^2}(\sin \omega t)$$

In this expression the first term is the free oscillation in still water and the second is a forced oscillation in the period of the wave. When damping is present the free oscillation dies out in time, leaving the forced oscillation modified somewhat by the damping. In a truly regular wave train the ship would, after a while, roll only in the period of the wave. The highest forced roll amplitudes occur when the period of the wave is close to the natural period of roll when it is said to *resonate*. Thus heavy rolling of a ship at sea is mainly at frequencies close to its natural frequency.

Pitching and Heaving in Regular Waves

For these motions attention is concentrated on head seas as these represent the most severe case. It is not reasonable to assume the wave long in relation to the length of the ship and the wave surface can no longer be represented by a straight line. However, the general approach of a forcing function still applies.

When a ship heads directly into a regular wave train it experiences hydrodynamic forces that can be resolved into a force at the centre of gravity and a moment about that point. As with rolling, the resulting pitch and heave will be greatest when the period of encounter with the waves is close to the natural period of motion. When the two periods are nearly equal resonance occurs and it is only the action of the damping that prevents the amplitudes of motion becoming infinite. The amplitudes in practice may become quite large and in that case the master would normally change speed or course to change the period of encounter to avoid resonance.

The amplitude of the pitching or heaving also depends upon the height of the waves. Simple theories assume that the exciting forces and the resulting motion amplitudes are proportional to the wave height. This applies whilst the motions can be approximated to by a linear equation of motion.

PRESENTATION OF MOTION DATA

The presentation of motion data should be such that it can be applied easily to geometrically similar ships in waves of varying amplitude. This is possible when the motions are linear, the basic assumptions being that:

- translations are proportional to the ratio of linear dimensions in waves whose lengths vary in the same way. For geometric similarity the speed varies so that V^2/L is constant;
- angular motions can be treated in the same way bearing in mind that the maximum wave slope is proportional to wave height;
- all motion amplitudes vary linearly with wave height;
- natural periods of motion vary as the square root of the linear dimension.

These assumptions permit the results of model experiments to be applied to the full-scale ship. In model experiments the motions seem 'rapid' because of the scale effect on period. Thus a 25th scale model will pitch and heave in a period only a fifth that of the full-scale ship. A typical presentation of heave data is as in Figure 10.3. Because wave period is related to wavelength the abscissa can equally be shown as the ratio of wave to ship length. The ordinates of the curve are known as *response amplitude operators* (RAOs) or *transfer functions*.

Motions in Irregular Seas

Usually a designer wishes to compare the seakeeping behaviour of two or more designs. If one design exhibited more acceptable response operators in all waves and at every speed of interest, the decision would be easy. Unfortunately one design will usually be superior under some conditions and inferior in others. The designer needs some way of comparing designs in the generality of wave conditions. The energy spectrum was a useful means of representing the nature of an irregular wave system and is equally valuable in studying a ship's motions in irregular seas. The spectrum needs to be modified to reflect the fact that the ship is moving through the waves, whereas the wave spectra so far discussed are those recorded at a fixed point.

Period of Encounter

As far as ship motions are concerned it is the period of encounter with the waves that is important not the absolute period of the wave. The ship is moving relative to the waves and it will meet successive peaks and troughs in a shorter or longer time interval depending upon whether it is advancing into the waves or is travelling in their direction. The situation can be generalised by considering the ship at an angle to the wave crest line as shown in Figure 10.4.

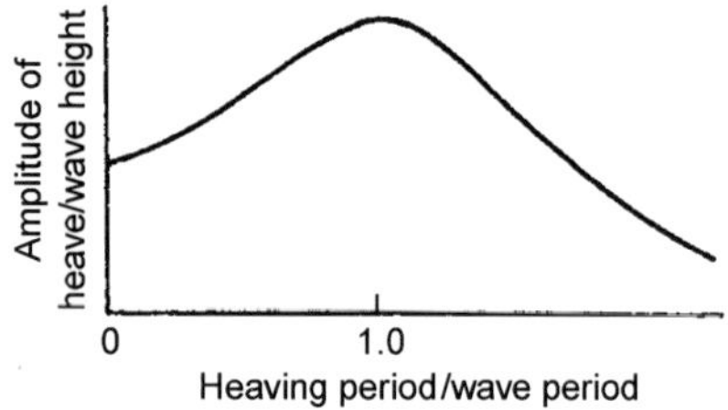

FIGURE 10.3 Response amplitude operators.

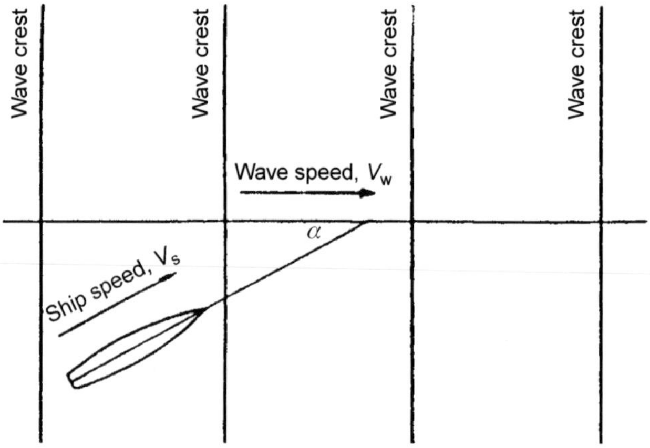

FIGURE 10.4 Period of encounter.

Measured at a fixed point the wave period is $T_w = \lambda/V_w$. If the ship is travelling at V_s at α to the direction of wave advance, in a time T_E the ship will have travelled $T_E V_s \cos\alpha$ in the wave direction and the waves will have travelled $T_E V_w$. If T_E is the period of encounter the difference in the distances is one wavelength λ, and

$$\lambda = T_E(V_w - V_s \cos\alpha)$$

and hence

$$T_E = \frac{\lambda}{V_w - V_s \cos\alpha} = \frac{T_w}{1 - (V_s/V_w)\cos\alpha}$$

Modification of Wave Energy Spectrum

From the expression for T_E it follows that

$$\omega_E = \frac{2\pi}{T_E} = \omega\left(1 - \frac{V_s}{V_w}\cos\alpha\right) = \omega\left(1 - \frac{\omega V_s}{g}\cos\alpha\right)$$

If the abscissae of the ‘absolute’ wave spectra are multiplied by $[1 - (\omega V_s/g) \cos\alpha]$ the abscissae of what is called an *encounter spectrum* are found. Ignoring any influence of the ship's presence on the waves, the area under the spectrum must remain the same, that is

$$S(\omega_E)\mathrm{d}\omega_E = S(\omega)\mathrm{d}\omega \quad \text{and} \quad S(\omega_E) = S(\omega)\mathrm{d}\omega/d\omega_E$$

$$= S(\omega)\left(1 - \frac{2\omega V_s}{g}\cos\alpha\right)^{-1}$$

So the ordinates of the spectrum must be multiplied by $[1 - (2\omega V_s/g) \cos\alpha]^{-1}$. In the case of a ship moving directly into the waves, that is $\alpha = 180^\circ$, the multiplying factors become:

$[1 + \omega V_s/g]$ for the abscissae and $[1 + 2\omega V_s/g]^{-1}$ for the ordinates

Obtaining Motion Energy Spectra

The energy spectrum for any given motion can be obtained by multiplying the ordinate of the wave encounter spectrum by the square of the RAO for the motion concerned at the corresponding encounter frequency. Taking heave, if the response amplitude operator is $Y_z(\omega_E)$ for encounter frequency ω_E, then the energy spectrum for the heave motions is:

$$S_z(\omega_E) = [Y_z(\omega_E)]^2 S_w(\omega_E)$$

where $S_w(\omega_E)$ is the wave energy spectrum.

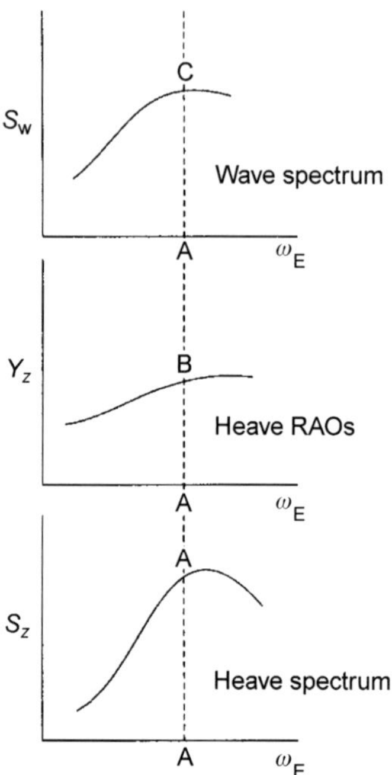

FIGURE 10.5 Motion energy spectrum.

This is illustrated in Figure 10.5 where AA = (AB)^2(AC).

Having created the motion spectrum, its area can be found and various statistical characteristics of the motions deduced as they were for the wave system itself. Thus if m_{ho} is the area under the heave spectrum:

$$\text{average heave amplitude} = 1.25(m_{\mathrm{ho}})^{0.5}$$

$$\text{significant heave amplitude} = 2(m_{\mathrm{ho}})^{0.5}$$

$$\text{average amplitude of } \frac{1}{10} \text{ highest heaves} = 2.55(m_{\mathrm{ho}})^{0.5}$$

Such values as the significant motion amplitude in the given sea can be used to compare the performance of different designs in that sea. There remains the need to consider more than one sea, depending upon the areas of the world in which the design is to operate, and to take into account their probability of occurrence.

A polar plot can be produced showing motions arising when the ship is at different headings to the wave system. Figure 10.6 shows such a plot of root mean square (RMS) roll values for a frigate with and without stabilisation.

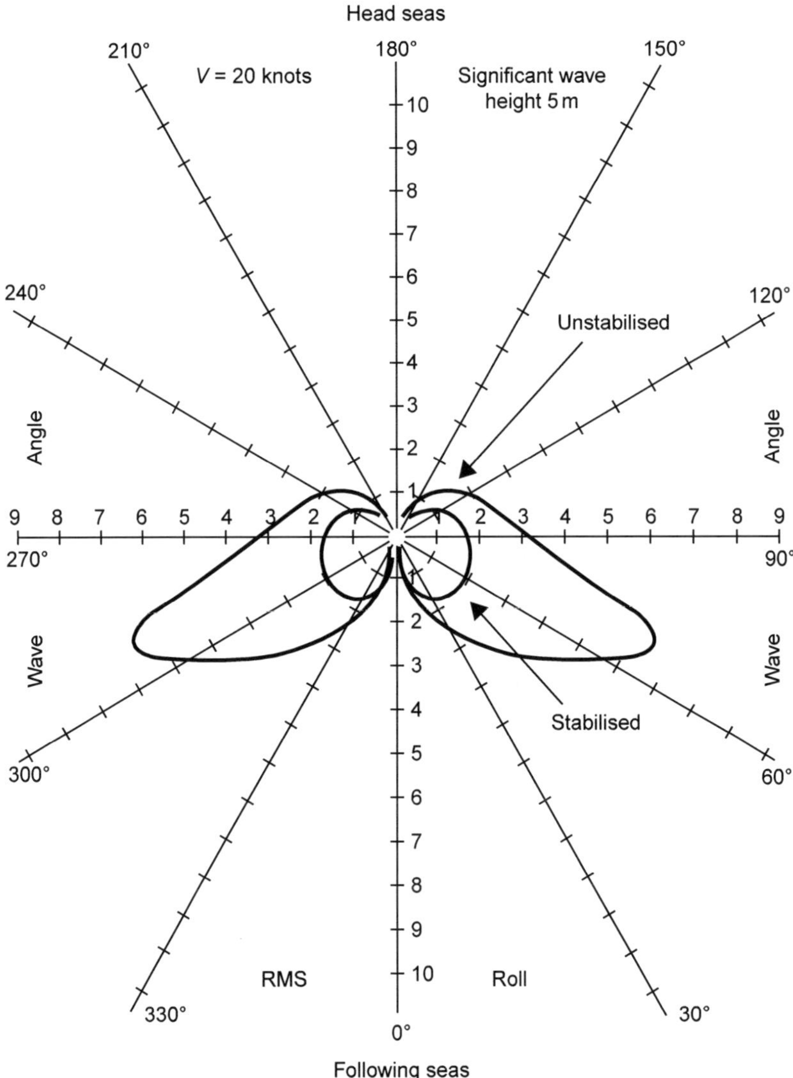

FIGURE 10.6 RMS roll values in irregular seas. *Courtesy of RINA.*

LIMITING FACTORS

Excessive motions are to be avoided if possible. They cause discomfort for passengers and crew, make the crew less efficient and make some tasks difficult, perhaps impossible. Apart from their amplitudes the phasing of motions can have significance. Phasing of pitch and heave generally creates an area of minimum motion about two-thirds of the length from the bow.

This becomes a 'desirable' area and would be used for activities for which motions are significant.

A number of factors, apart from its general strength and stability, may limit a ship's ability to carry out its intended function. Ideally these would be definable and quantifiable but generally this is not possible except in fairly subjective terms. The limits may be imposed by the ship itself, its equipment or the people on board. The seakeeping criteria most frequently used as potentially limiting a ship's abilities are discussed below.

Speed in Waves

A ship experiences a greater resistance in waves and the propulsor is working under less favourable conditions. These combined, possibly, with increased air resistance force a reduction in speed for a given power. The severity of motions, slamming and wetness can usually be alleviated by decreasing speed and a master may additionally reduce speed voluntarily to alleviate their effects. For many ships their schedule is of great importance and ship routeing can be used to avoid the worst sea conditions and so suffer less delay, danger and discomfort with the added advantage of a saving on fuel. Savings of the order of 10–15 h have been made in this way on the Atlantic crossing. Computerised weather routeing systems are now available allowing the master greater control.

Wetness

Wetness is the shipping of heavy spray or green seas over the ship. The bow area is the region most likely to be affected and is assumed in what follows. It may limit a ship's speed and the designer needs a way of assessing the conditions under which it will occur and its severity. To a degree wetness is subjective and it depends upon the wind speed and direction as well as the wave system. It was often studied by running models in waves but it is now usually assessed by calculating the relative motion of the bow and the local sea surface, the assumption being that the probability of deck wetness is the same as that of the relative motion exceeding the local freeboard. The greater the difference, the wetter the ship is likely to be.

Increased freeboard, say by increasing sheer forward, is one means of reducing wetness. At sea the master can reduce wetness by reducing speed and, usually, by changing the ship's heading relative to the predominant waves. Good deck camber helps clear water quickly. A bulwark can be used to increase the effective freeboard but in that case adequate freeing ports are needed to prevent water becoming trapped on the deck. The size of freeing ports to be fitted is laid down in international regulations. The designer would avoid siting other than very robust equipment in the area where green seas are likely. Any vents would face aft and be provided with water traps.

Even so vents do get carried away and water may enter the compartments below. This appears to have contributed to the loss of MV *Derbyshire*. The investigation into its loss highlighted the need to design hatch covers to be capable of withstanding higher green seas loads than had previously been regarded as adequate.

Slamming

Sometimes the local pressures exerted by the water on the ship's hull become very large and slamming occurs. Slamming is a high-frequency transient vibration in response to the impact of waves on the hull, occurring at irregular intervals and is characterised by a sudden change in vertical acceleration followed by a vibration of the ship's girder in its natural frequencies. The region of the outer bottom between 10% and 25% of the length from the bow is the most vulnerable area. The impact may cause physical damage and accelerate fatigue failure in this area. In a big ship those on a bridge well aft may not be aware of the severity of slamming and distant reading indicators are sometimes fitted. Because the duration of the slam is only of the order of a thirtieth of a second, it does not perceptibly modify the rigid body motion of the ship but the ensuing vibration can last for 30 s. A prudent master will reduce speed when slamming badly to reduce its effects. Also a change of direction can help. Lightly loaded cargo ships are particularly liable to slam with their relatively full form and shallow draught forward, and speed reductions may be as high as 40%. Slamming is less likely in high-speed ships because of their finer form.

Slamming is likely when the relative velocity between the hull and water surface is large and when the bow is re-entering the water with a significant length of bottom roughly parallel to the sea surface. It is amplified if the bottom has a low rise of floor. The pressure acting in a slam can be shown to be proportional to the square of the velocity of impact and inversely proportional to the square of the tangent of the deadrise angle.

Propeller Emergence

The probability of the propeller emerging from the water, as the result of ship's motions, can be assessed in a similar way to wetness by calculating the motion of the ship aft relative to the local sea surface. If the propeller does emerge, even partially, it will be less effective in driving the ship and tend to race causing more vibration.

Human Responses

Ship's motions can cause nausea and then sickness. This discomfort can itself make people less efficient and make them less willing to work. Motions can make tasks physically more difficult to accomplish. Thus the movement of weights around the ship, as when replenishing a warship at sea, is made more difficult. Tasks requiring careful alignment of two elements may become impossible without some mechanical aid. Over and above this the motions, and the drugs taken to alleviate the symptoms of motion sickness, may adversely affect a person's alertness and mental dexterity.

In broad terms the effects of motion on human behaviour depend upon the acceleration experienced and its period. The effect is most marked at frequencies between about 0.15 and 0.2 Hz. The designer can help by locating important activities in areas of lower motion, by aligning the operator position with the ship's principal axes, providing an external visual frame of reference and providing good air quality free of odours. See ITTC 7.5-02-05-04.1 (2002).

OVERALL SEAKEEPING PERFORMANCE

The most common cause of large amplitude rolling, as shown by linear theory, is resonance. Large roll angles can also be experienced due to the fact that a ship's effective metacentric height varies as it passes through waves. These are non-linear effects.

An overall assessment of seakeeping performance is complicated by the many different sea conditions a ship may meet and the different responses that may limit the ship's ability to carry out its function. A number of authorities have tried to obtain a single 'figure of merit' but this is difficult. The approach is to take the ship's typical operating pattern over a period long enough to cover all significant activities. From this the following points are deduced:

- probability of meeting various sea conditions, using statistics on wave conditions in different areas of the world;
- likely ship speed and direction in these seas;
- probability of the ship being in various conditions, deep or light load;
- ship responses that are likely to be critical for the ship's operations.

From such considerations the probability of a ship being limited from any cause can be deduced for each set of sea conditions. These combined with the probability of each sea condition being encountered can lead to an overall probability of limitation. The relative merits of different designs can be 'scored' in a number of ways. Amongst those that have been suggested are as follows:

- the percentage of its time a ship, in a given loading condition, can perform its intended function, in a given season at a specified speed,
- a generalisation of this to cover all seasons and/or all speeds,
- the time a ship needs to make a given passage in calm water compared with that expected under typical weather conditions.

It is really a matter for the designer to establish what is important to an owner and assess how this might be affected by wind and waves.

ACQUIRING SEAKEEPING DATA

Computations of performance criteria require good data input, including that for waves, response operators and limitations experienced in ship operations.

Wave Data

The sources of wave data have been discussed elsewhere. The designer must select that data which is applicable to the design under review. The data can then be aggregated depending upon where in the world the ship is to operate and in which seasons of the year.

Response Amplitude Operators

The designer can call upon theory, model testing and full-scale trials. Modern ship motion theories can give good values of responses for most motions. The most difficult are the prediction of large angle rolling, due to non-linear effects. The equations of motion can be written down fairly easily but the problem is in evaluating the various coefficients in the equations. Many modern approaches are based on a method known as *strip theory* or *slender body theory*. The basic assumptions are those of a slender body, linear motion, a rigid and wall-sided hull, negligible viscous effects apart from roll damping and that the presence of the hull has no effect upon the waves. The hull is considered as composed of a number of thin transverse slices or strips – typically 20 or more. The flow about each element is assumed to be two dimensional and the same as would apply if the body were an infinitely long oscillating cylinder of that cross section. In spite of what might appear fairly gross simplification, the theory gives good results in pitch and heave and with adjustment is giving improved predictions of roll. Refinements include allowance for non-linear effects in roll, for dealing with low encounter frequencies and allowing for bilge keels, etc.

Linear three-dimensional theories in which the ship is regarded as a rigid body responding to the encountered waves overcame some of the limitations of strip theory (principally the omission of interactions between the individual strips). With the great increase in size of some ships the assumption of a rigid ship was not acceptable and it became necessary to consider the deformation of the hull under the imposed forces. Hydroelasticity theory was then used to model the ship. Taking a "dry" ship – one without the surrounding fluid – allowed the principal mode shapes and frequencies of the hull responses to be determined. With the fluid causing the external loading taken into account it was possible to produce the generalised, coupled equations of motion to describe the behaviour of a flexible ship in an irregular seaway. This response includes both the rigid body motions and the hull distortions. It is usual to apply finite element methods to represent the ship structure. See Hudson et al. (2010) and Bertram (2012).

To validate new theories, or where theory is judged to be not accurate enough, and for ships of unconventional form, model tests are still required. These are particularly useful for the study of motions which are strongly non-linear.

For many years long narrow ship tanks were used to measure motions in head and following regular waves. Subsequently the wavemakers were modified to create long-crested irregular waves. In the second half of the twentieth century a number of special seakeeping basins were built. In these free running models could be manoeuvred in short- and long-crested wave systems. For motions, the response operators can be measured directly by tests in regular seas but this involves running a large number of tests at different speeds in various wavelengths. Using irregular waves the irregular motions can be analysed to give the regular components to be compared with the component waves. Because the irregular surface does not repeat itself, or only over a very long period, a number of test runs are needed to give statistical accuracy. The number of runs, however, is less than for testing in regular waves. A third type of model test uses the *transient wave* approach. The wavemaker is programmed to generate a sequence of wavelengths which merge at a certain point along the length of the tank to provide the wave profile required. The model is started so as to meet the wave train at the chosen point at the correct time. It then experiences the correct wave spectrum and the resulting motion can be analysed to give the response operators. This method is a special case of the testing in irregular waves. Whilst in theory one run would be adequate several runs are made to check repeatability.

The model can be viewed as an analogue computer in which the functions are determined by the physical characteristics of the model. See ITTC 7.5-02-07-02.1 (2011). To give an accurate reproduction of the ship's motion the model must be ballasted to give the correct displacement, draughts and moments of inertia. It must be run at the correct representative speed. To do all this in a relatively small model is difficult particularly when it has to be self-propelled and to carry all the test equipment. The model cannot be made too large otherwise a long enough run is not achievable in the confines of the tank. Telemetering of data ashore helps. Another approach has been to use a large model in the open sea in an area where reasonably representative conditions pertain.

HAZARDS DUE TO WAVE RESONANCE EFFECTS

Some motions are non-linear, some non-linearity being introduced by the changes in the immersed hull shape as the ship moves relative to the waves. The change can be very significant at large motion amplitudes; the fore end of the ship may leave the water completely before slamming back down into it. Another cause of non-linearity, particularly for rolling, is the differing way in which damping forces vary with velocities of motion. The situation is further complicated by the cross coupling that occurs between motions. These effects become more pronounced at higher motion amplitudes and are important in the study of extreme loadings to which a ship is subject.

An important non-linear effect occurs as a wave travels along the length of the ship and the immersed hull volume and the shape of the waterplane change. This causes a variation in the effective metacentric height. Generally the stability is reduced when the wave crest is amidships and increased when the trough is amidships. With some forms there is a significant loss of stability and large roll angles can build up. Another non-linearity occurs in head or following seas when the wave dominant encounter period is about half the natural roll period. The rolling can build up very rapidly to alarming angles, even in moderate following seas. These types of phenomena were discussed by Perrault et al. (2010) in their stability studies. They can be summarised as follows:

- *Sympathetic rolling* which is usually associated with stern or quartering seas when the ship is travelling at or near the wave group velocity;
- *Asymmetric resonance* which builds up with each wave encountered and produces large-amplitude oscillations in several modes including roll, due to fluctuations in the righting lever with the slow passage of the waves along the ship;
- *Sudden extreme behaviour* due to loss of transverse righting moment when a wave crest is near amidships.
- *Resonant excitation* in a beam sea which excites the ship close to its natural roll period.
- *Parametric rolling* in following seas (or head seas at low forward speed). The periodic variation of righting lever and buoyancy distribution lead to a buildup of excessive roll angles. The most critical situation is when the wave encounter frequency is about half that of the ship's natural roll frequency. Several cruise ships have suffered very large roll angles due to this.
- *Impact excitation* when a steep or breaking wave hits the ship resulting in an extreme roll angle.
- *Broaching* which can occur in following and quartering overtaking seas. A series of passing waves can lead to increased yaw to the point where the ship's head can no longer be controlled and the ship comes broadside to the waves. A ship may "surf" down a large wave with increasing speed with large dynamic side forces generated as it imbeds itself into the proceeding wave. Broaching has been well known as a hazard, particularly

in small ships, for a very long time. The mechanisms concerned have only recently succumbed to analytical treatment because they are highly non-linear.

Modern theoretical methods can cope with many of these problems but they are complex. Fortunately for the designer RAOs derived from model tests in reasonably severe conditions will already include something of these non-linear factors. It does mean, however, that RAOs obtained in different sea spectra may differ somewhat because of differences in the actual non-linear and coupling effects experienced. These variations will be small and can be catered for by averaging data derived from a range of severe spectra. Then the overall motions derived from using the RAOs with some new spectrum would closely align with that a ship would experience in the corresponding seaway.

Wetness and slamming depend upon the actual time history of wave height in relation to the ship. Direct model study of such phenomena can only be made by running the model in a representative wave train over a longish period. However, tests in regular waves can assist in slamming investigations by enabling two designs to be compared or by providing a check on theoretical analyses.

Some full-scale data has been obtained for correlation with theory and model results. These indicate that model tests give good results provided the waves in which they are tested are representative of the actual sea conditions. Direct correlation is difficult because of the need to find sea conditions approximating a long crested sea state during the trial period when the ship is rigged with all the measuring gear. A lot of useful statistical data, however, on the long-term performance can be obtained from statistical recorders of motions and strains during the normal service routine. Such recorders are now fitted in many warships and merchant ships.

DERIVING THE MOTIONS

Based on the above criteria it can be seen that there are three ways of assessing the motions in a given wave system for which the energy spectrum is defined.

- *Statistically.* Once the RAOs have been found (by calculation or model tests) the energy spectra for the various motions can be found. The probabilities of exceeding certain levels of motion can then be assessed.
- *Frequency domain simulation.* Since an irregular sea can be represented by superimposing regular sinusoidal waves in random phase, the responses to those wave components can be found and combined to give the ship response. The actual response at any instant will depend upon the phase relationships at that instant. To give a reasonable representation of a ship's behaviour the simulation must be continued over a long time period.

Otherwise the results may only apply to a relatively quiescent period, or to a particularly severe period.
- *Time domain simulation*. This approach does not use the RAOs. Instead a specific spectrum of waves is assumed and the ship is placed in the resulting sea. The forces and moments acting on the ship are calculated and its changes in attitude deduced. The immersed hull form will be varying with time and cross-coupling effects due to these changes will be built in to the analysis. It will be appreciated that this method is useful in studying wetness and slamming as these depend upon the actual time history of wave height in relation to the ship. Again the simulation must be carried out over a long period to get good results.

These three methods place increasingly heavy demands on computational capacity and skills. The first is very useful in comparing the expected general behaviour of different hull forms. The second is useful for the actual sequence of motions a ship may experience, although not directly related to the sea surface. The last can show how the ship moves relative to the sea surface showing, for instance, how freeboard varies with time. It is the preferred method for studying extreme loading conditions.

Deducing Limiting Criteria

It is not always easy to establish exactly what are limiting criteria for various shipboard operations. They will depend to some extent upon the ability of the people involved. Thus an experienced helicopter pilot will be able to operate from a frigate in conditions which might prove dangerous for a less-skilled pilot. The criteria are usually obtained from careful questioning and observation of the crew. Large motion simulators can be used for scientific study of human performance under controlled conditions. These can throw light upon how people learn to cope with difficult situations. The nature of the usual criteria has already been discussed.

EFFECT OF SHIP FORM

It is difficult to generalise on the effect of ship form changes on seakeeping because changing one parameter, for instance moving the centre of buoyancy, usually changes others. Methodical series data (calculated or from model tests) should be consulted where possible but in very general terms, for a given sea state:

- Increasing displacement and length will reduce motions, noting that for a given angle of pitch a long ship will experience greater heave amplitudes at the bow and stern.
- Increasing displacement at constant length has little effect.

- Increasing length will reduce the likelihood of meeting waves long enough to cause resonance. For a given beam, L/B will be greater and this is generally beneficial.
- Higher freeboard leads to a drier ship.
- Flare forward can reduce wetness but may increase slamming.
- A high length/draught ratio will lead to less pitch and heave in long waves but increase the chances of slamming.
- A bulbous bow can reduce motions in short waves but increase them in long waves.
- High Froude number will decrease the wave encounter period shifting resonance to longer wavelengths.
- Block and prismatic coefficient do not have a major influence.
- Good freeboard is important for reducing wetness.

Because form changes can have opposite effects in different wave conditions, and a typical sea is made up of many waves, the net result is often little change. For conventional forms it has been found (Ewing and Goodrich, 1967) that overall performance in waves is little affected by variations in the main hull parameters. Local changes can be beneficial. For instance fine form forward with good rise of floor can reduce slamming pressures. In producing a balanced design, the naval architect must consider the impact of any form change on all the ship's characteristics.

STABILISATION

A ship's rolling motions can be reduced by fitting a stabilisation system. In principle pitch motions can be improved in the same way but in practice this is very difficult. An exception is the fitting of some form of pitch stabiliser between the two hulls of a catamaran which is relatively shorter than a conventional displacement ship. In this section attention is focused on roll stabilisation. The systems may be *passive* or *active*.

Bilge Keels

Of the passive systems, bilge keels are the most popular and are fitted to the great majority of ships. They are effectively plates projecting from the turn of bilge and extending over the middle half to two-thirds of the ship's length. To avoid damage they do not normally protrude beyond the ship's side or keel lines, but to be effective they need to penetrate the boundary layer around the hull. They cause a body of water to move with the ship and create turbulence thus dampening the motion and causing an increase in period and reduction in amplitude.

Although relatively small in dimension the bilge keels have large levers about the rolling axis and the forces on them produce a large moment opposing the rolling. They can produce a reduction in roll amplitude of more than a third. Their effect is generally enhanced by ahead speed. They are aligned with the flow of water past the hull in still water to reduce their drag in that state. When the ship is rolling the drag will increase and slow the ship a little.

Passive Tanks

These use the movement of water in specially designed tanks to produce a moment opposing the rolling motion. The tank is U-shaped and water moves from one side to the other and then back as the ship inclines first one way and then the other. Because of the throttling effect of the relatively narrow lower limb of the U joining the two sides of the tank, the movement of water can be made to lag behind the ship movements. By adjusting the throttling, that is by 'tuning' the tank, a lag approaching 90° can be achieved. Unfortunately the tank can only be tuned for one frequency of motion. This is chosen to be the ship's natural period of roll as this is the period at which really large motions can occur. The tanks will stabilise the ship at zero speed but the effect of their free surface on stability must be allowed for. They take up valuable space and are not fitted to many ships.

Active Fins

This is the most common of the active systems. One or more pairs of stabilising fins are fitted. They are caused to move by an actuating system in response to signals based on a gyroscopic measurement of roll motions. They are relatively small although projecting out further than the bilge keels. The whole fin may move or one part may be fixed and the after section move. A flap on the trailing edge may be used to enhance the lift force generated. The fins may permanently protrude from the bilge or may, at the expense of some complication, be retractable (Figure 10.7).

The lift force on the fin is proportional to the square of the ship's speed. At low speed they will have little effect although the control system can adjust the amplitude of the fin movement to take account of speed, using larger fin angles at low speed.

Active Tanks

This is similar in principle to the passive tank system but the movement of water is controlled by pumps or by the air pressure above the water surface. The tanks either side of the ship may be connected by a lower limb or two separate tanks can be used. The system can deal with more than one frequency. As with the passive system it can stabilise at zero ship speed. There are no projections outside the hull.

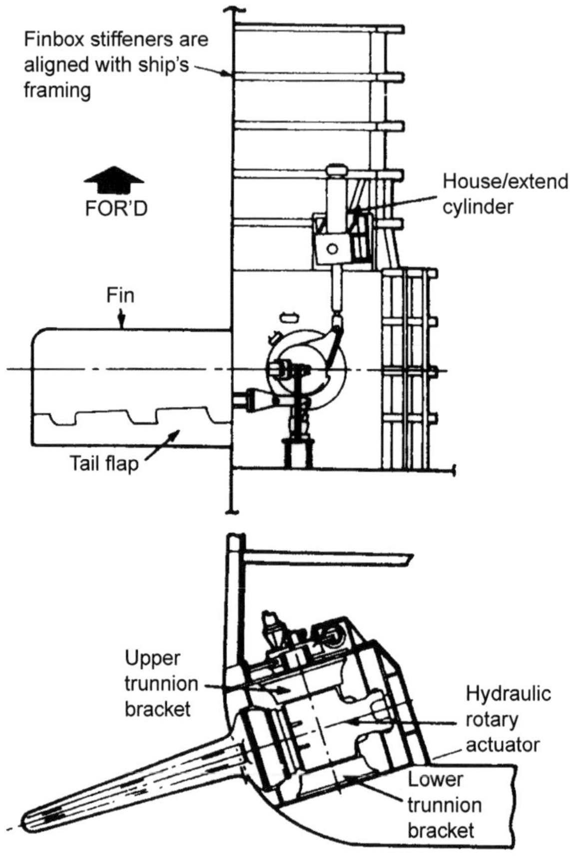

FIGURE 10.7 Stabiliser fin.

Capacity of Stabilising Systems

The capacity of a stabilisation system is usually quoted in terms of the steady heel angle it can produce with the ship underway in still water. This is then checked during still water trials. Alternatively the system may be used to force roll the ship. It is possible to use modern theories to calculate the performance in waves but this would be difficult to check contractually.

SUMMARY

A ship's motions in irregular ocean waves can be synthesised from its motions in regular waves using the energy spectrum concept. Large angle rolling is non-linear and very large amplitudes can build up rapidly under some operating conditions. Factors limiting a ship's seakeeping capabilities include increased resistance, impaired propeller performance, wetness, slamming and degradation of human performance. These factors can be combined to give an overall assessment of the probability that a ship will be able to undertake its intended mission. Motions can be reduced by stabilisation systems, either active or passive.

It has only been possible to deal with the subject in an elementary way. Rigorous treatments are available.

This page intentionally left blank

Chapter 11

Vibration, Noise and Shock

INTRODUCTION

Appendix C outlines the general response of an elastic body to applied forces. The application to the relatively low-frequency ship motions is considered elsewhere and this chapter deals with the higher frequency responses associated with vibration, noise and shock.

Vibration and noise are generated in the same way and generally from the same sources, the difference being the way they are perceived by the observer. Vibration involves frequencies in the range 1−80 Hz. Noise is in the frequency band from 20 to 20,000 Hz which the observer hears as sound. Vibration is felt through part of the body in contact with the vibrating item. The resulting effect on the body may be a 'whole body' vibration or vibration of one part of the body – the hand and/or arm, say.

Shock is rather different as it is a discrete event (perhaps as the result of ship slamming) rather than continuous. The shock 'pulse' can be analysed for its constituent frequencies and the response of the structure, say, assessed by its combined responses to those frequencies.

The main sources of vibration and noise in a ship are:

- The machinery, main and auxiliary. Diesels create greater out-of-balance forces than turbines. Using large, slow running diesels leads to higher exciting forces.
- Propellers. The hull local to the propeller is acted on by periodic forces at frequencies determined by the shaft revolutions and the number of blades. Forces on the propeller blades are bound to vary because the propeller is operating in a non-uniform wake. This variation is greater if the propeller cavitates. Other forces are transmitted along the propeller shaft.
- Trunk and pipe systems such as air-conditioning where movements of air or water can create flow noise.

Vibration and noise can lead to strain, fatigue and increased wear in equipment and structure. They are unpleasant, and can become harmful, to humans. The adoption of lighter hull construction to save weight and the move to higher power for higher speeds mean that vibrations of modern hulls

Introduction to Naval Architecture. DOI: http://dx.doi.org/10.1016/B978-0-08-098237-3.00011-4

can be more pronounced and require careful study. Fortunately better analysis methods and mounting systems are now available.

SHIP VIBRATIONS

Local vibrations involve a smallish part of the structure, perhaps an area of deck. The frequencies are usually higher, and the amplitudes lower, than the main hull vibrations. Because there are so many possibilities and the calculations can be complex they are not usually studied directly during design except where large excitation forces are anticipated. Generally the designer avoids machinery which generates disturbing frequencies close to those of typical ship type structures. Any faults are corrected as a result of trials experience. This is often more economic than carrying out extensive design calculations as the remedy is usually a matter of adding a small amount of additional stiffening.

Main hull vibrations are a different matter. If severe vibrations occur remedial action can be very expensive. They must therefore be looked at in design. The hull may flex vertically and/or horizontally as a beam or twist like a rod about its longitudinal axis. These two modes of vibration are called *flexural and torsional*, respectively. The vertical flexing of the hull is usually of most concern. Except in lightly structured ships, or ones with large deck openings, the torsional mode is usually less important.

Flexural Vibrations

When flexing in the vertical or horizontal planes the structure has an infinite number of degrees of freedom and the mode of vibration is described by the number of *nodes* (points at which there is no movement) which exist in the length. The fundamental mode is the two-node as shown in Figure 11.1A.

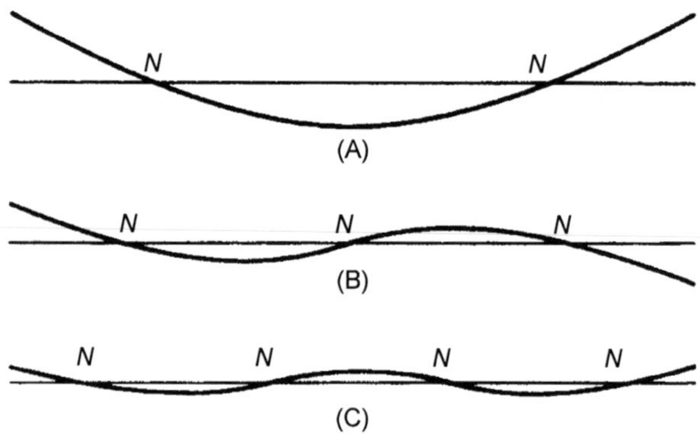

FIGURE 11.1 (A) Two-mode; (B) three-mode; and (C) four-mode.

There is movement at the ends of the ship since there is no rigid support there. This is referred to as a *free-free* mode and differs from that experienced by a structural beam where there would be zero displacement at one end at least. The next two higher modes have three and four nodes, all free-free. Associated with each mode is a natural frequency of free vibration, the frequency being higher for the higher modes. If the ship were of uniform rigidity and uniform mass distribution along its length and was supported at its ends, the frequencies of the higher modes would be simple multiples of the fundamental. In practice ships differ from this although perhaps not as much as might be expected. This is given in Table 11.1 (Dieudonne, 1959) where it will be noted that the greater mass of a loaded ship leads to a reduction in frequency.

Torsional Vibrations

In this case the displacement is angular and a one-node mode of vibration is possible. Figure 11.2 shows the first three modes.

Coupling

It is commonly assumed for analysis purposes that thc various modes of vibration are independent and can be treated separately. In some circumstances, however, vibrations in one mode can generate vibration in another. In this case the motions are said to be *coupled.* In a ship a horizontal vibration will often excite torsional vibration because of the non-uniform distribution of mass in the vertical plane. When exciting forces have components in more than one plane more than one mode of vibration will be excited.

CALCULATIONS

Formulae for Ship Vibration

The formulae for uniform beams suggest that for the ship an approximation will be given by a formula of the type

$$\text{Frequency} = \text{const.}\left(\frac{EI}{ML^3}\right)^{0.5}$$

Suggestions for the value of the constant for different ship types have been made but these can only be very approximate because of the many variables involved in ships. The most important are:

- Mass and stiffness distribution along the length
- Departure from ordinary simple theory due to shear deflection and structural discontinuities
- Added mass
- Rotary inertia.

TABLE 11.1 Typical Ship Vibration Frequencies (cpm)

Ship Type	Length (m)	Condition of Loading	Frequency of Vibration (cpm)						
			Vertical				Horizontal		
			2 Node	3 Node	4 Node	5 Node	2 Node	3 Node	4 Node
Tanker	227	Light	59	121	188	248	103	198	297
		Loaded	52	108	166	220	83	159	238
Passenger ship	136		104	177			155	341	
Cargo ship	85	Light	150	290			230		
		Loaded	135	283			200		
Cargo ship	130	Light	106	210			180	353	
		Loaded	85	168			135	262	
Destroyer	160	Average action	85	180	240		120	200	

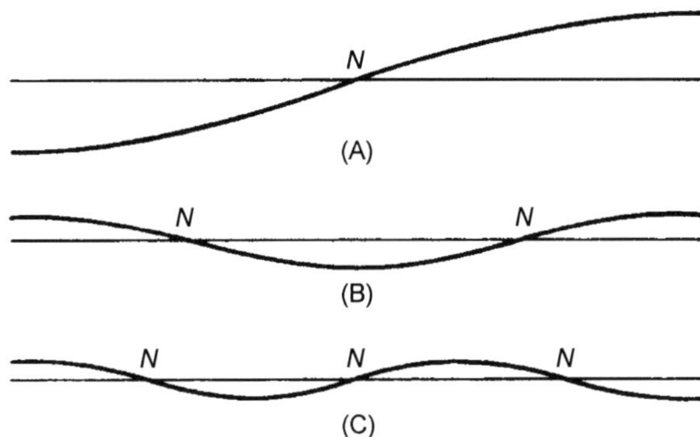

FIGURE 11.2 (A) One-node; (B) two-node; and (C) three-node.

Direct Calculation

Empirical formulae enable a first shot to be made at the frequency of vibration. The accuracy will depend upon the amount of data available from ships on which to base the coefficients. It is desirable to be able to calculate values directly taking account of the specific ship characteristics and a range of loading conditions. These days a full finite element analysis of the hull can be carried out to give the vibration frequencies, including the higher order modes. Such approaches are discussed in some detail by Bertram (2012) who concludes that they need models with a very large number of degrees of freedom and an experienced person to do the work. Even then the results are not that satisfactory. Before such methods became available there were two methods used for calculating the two-node frequency:

- The *deflection method* or *full integral method*
- The *energy method.*

The Deflection Method

In this method the ship is represented as a beam vibrating in simple harmonic motion in which, at any moment, the deflection at any position along the length is $y = f(x)\sin pt$. The function $f(x)$ for non-uniform mass and stiffness distribution is unknown but it can be approximated by the curve for the free-free vibration of a uniform beam.

Differentiating y twice with respect to time gives the acceleration at any point as proportional to y and the square of the frequency. This leads to the dynamic loading. Integrating again gives the shear force and another integration gives the bending moment. A double integration of the bending moment curve gives the deflection curve. At each stage the constants of integration can be evaluated from the end conditions. The deflection curve now obtained can be compared with that originally assumed for $f(x)$. If they differ significantly a second approximation can be obtained by using the derived curve as the new input to the calculation.

In using the deflection profile of a uniform beam it must be remembered that the ship's mass is not uniformly distributed, nor is it generally symmetrically distributed about amidships. This means that in carrying out the integrations for shear force and bending moment the curves produced will not close at the ends of the ship. In practice there can be no force or moment at the ends so corrections are needed. A bodily shift of the baseline for the shear force curve and a tilt of the bending moment curve are used as was the case in strength calculations.

In the calculation the mass per unit length must allow for the mass of the entrained water using one of the methods described for dealing with added virtual mass. The bending theory used ignores shear deflection and rotary inertia effects. Corrections for these are made at the end by applying factors to the calculated frequency.

The Energy Method

This method uses the principle that, in the absence of damping, the total energy of a vibrating system is constant. Damping exists in any real system but for ships it is acceptable to ignore it for the present purpose. Hence the sum of the kinetic and potential energies is constant.

In a vibrating beam the kinetic energy is that of the moving masses and initially this is assumed to be due to linear motion only. Assuming simple harmonic motion and a mass distribution, the kinetic energy is obtained from the accelerations deduced from an assumed deflection profile and frequency. The potential energy is the strain energy of bending.

When the beam is passing through its equilibrium position the velocity will be a maximum and there will be no bending moment at that instant. All the energy is kinetic. Similarly when at its maximum deflection the energy is entirely potential. Since the total energy is constant the kinetic energy in the one case can be equated to the potential energy in the other.

As in the deflection method the initial deflection profile is taken as that of a uniform bar. As before allowance is made for shear deflection and for rotary inertia. Applying this energy method to the case of the simply supported, uniform section, beam with a concentrated mass M at mid-span, and assuming a sinusoidal deflection curve, yields a frequency of

$$\left(\frac{1}{2\pi}\right)\left(\frac{\pi^4 EI}{2ML^3}\right)^{0.5} \text{ compared with } \left(\frac{1}{2\pi}\right)\left(\frac{48EI}{ML^3}\right)^{0.5} \text{ for the exact solution}$$

Since $\pi^4/2$ is 48.7 the two results are in good agreement. This simple example suggests that as long as the correct end conditions are satisfied there is considerable latitude in the choice of the form of the deflection profile.

Calculation of Higher Modes

It might be expected that the frequencies of higher modes could be obtained by the above methods by assuming the appropriate deflection profile to match the mode needed. Unfortunately, instead of the assumed deflection curve converging to the correct one it tends to diverge with successive iterations. This is due to the profile containing a component of the two-node profile which becomes dominant. Whilst ways have been developed to deal with this, one would today choose to carry out a finite element analysis.

Approximate Formulae

It has been seen that the mass and stiffness distributions in the ship are important in deriving vibration frequencies. Such data is not available in the early design stages when the designer needs some idea of the frequencies for the ship. Hence there has always been a need for simple empirical formulae. Schlick (1884) suggested that

$$\text{Frequency} = \text{const.}\left(\frac{EI_a}{ML^3}\right)^{0.5}$$

where I_a is the moment of inertia of the midship section.

This formula has severe limitations and various authorities have proposed modifications to it. Todd (1961) adapted Schlick to allow for added mass, the total virtual displacement being given by

$$\Delta_v = \Delta\left(\frac{B}{3T} + 1.2\right)$$

He concluded that I should allow for superstructures in excess of 40% of the ship length. For ships with and without superstructure the results for the two-node vibration generally obeyed the rule:

$$\text{Frequency} = 238,660\left(\frac{I}{\Delta_v L^3}\right)^{0.5} + 29$$

if I is in m^4, dimensions in m and Δ_v is in MN.

By approximating the value of I, Todd proposed:

$$\text{Frequency} = \text{const.} \times \left(\frac{BD^3}{\Delta_v L^3}\right)^{0.5}$$

Typical values of the constant in SI units were found to be:

Large tankers (full load)	11000
Small tankers (full load)	8150
Cargo ships (60% load)	9200

Many other approximate formulae have been suggested. The simpler forms are acceptable for comparing ships which are closely similar. The designer must use the data available to obtain the best estimate of frequency allowing for the basic parameters which control the physical phenomenon.

VIBRATION LEVELS

Amplitudes

The amplitude of oscillation of a simple mass spring combination depends upon the damping and magnification factor. The situation for a ship is more complex. Allowance must be made for at least the first three or four modes, superimposing the results for each. This can be done by finite element analysis and once the amplitude has been obtained the corresponding hull stress can be evaluated.

The question then arises as to whether the amplitude of vibration is acceptable. Limitations may be imposed by the reactions of humans, equipment or by strength considerations. Sensitive equipment can be protected by placing them on special mounts and this is done quite extensively in warships in particular. Human beings respond mainly to the vertical acceleration they experience. Curves are published (ISO 6954) indicating the combinations of frequency and displacement that are likely to be acceptable. See the discussion on a ship's internal environment.

Checking Levels

It will be appreciated by now that accurate calculation of vibration levels is difficult. It is possible to put a check upon the levels likely to be achieved as the ship nears structural completion by using a vibration exciter. The exciter is simply a device for generating large vibratory forces by rotating an out-of-balance weight. Placed at appropriate positions in the ship it can be activated and the structural response to known forces measured.

Reducing Vibration

Ideally vibration would be eliminated completely but this is not a realistic goal. In practice a designer aims to:

- balance all forces in reciprocating and rotary machinery and in the propeller;
- provide good flow into the propeller and good hull clearance;
- avoid resonance by changing the stiffness of components or varying the exciting frequencies;
- mount vibrating machinery and isolate associated piping, etc;
- use special mounts to shield sensitive equipment from the vibration;
- fit a form of vibration damper, either active or passive;
- maintain equipment to a high standard.

Vibration Testing of Equipment

Most equipment is fitted in a range of ships and in different positions in a ship. Thus their design cannot be tailored to a specific vibration specification. Instead they are designed to standard criteria. Then samples are tested to confirm that the requirements have been met. These tests include endurance testing for several hours in the vibration environment. Table 11.2 gives test conditions for naval equipment to be fitted to a number of warship types.

In Table 11.2 the masthead region is that part of the ship above the main hull and superstructure. The main hull includes the upper deck, internal compartments and the hull.

NOISE

Noise levels are expressed in decibels (dB). In the open, sound intensity falls off inversely as the square of the distance from the source. At half the distance the intensity will be quadrupled. Sound levels are subjective because a typical noise contains many components of different frequency and these affect the human ear differently. To define a noise fully the strength of each component and its frequency must be specified. This is done by presenting a spectral plot of the noise. A noise level expressed in dB(*A*) is used to reflect the human sensitivity to different frequencies.

TABLE 11.2 Vibration Response and Endurance Test Levels for Surface Warships (Time of endurance tests in hours, h)

Ship Type	Region	Standard Test Level Peak Values and Frequency Range	Endurance Tests
Minesweeper size and above	Masthead	1.25 mm, 5–14 Hz	1.25 mm, 14 Hz
		0.3 mm, 14–23 Hz	0.3 mm, 23 Hz
		0.125 mm, 23–33 Hz	0.125 mm, 33 Hz each 1 h
	Main	0.125 mm, 5–33 Hz	0.125 mm, 33 Hz for 3 h
Smaller than minesweeper	Masthead and main	0.2 mm or a velocity of 63 mm/s whichever is less; 7–300 Hz	0.2 mm, 50 Hz for 3 h
	Aftermost ⅛ of ship length	0.4 mm or a velocity of 60 mm/s whichever is less; 7–300 Hz	0.4 mm, 24 Hz for 3 h

Primary sources of noise are the same as those generating vibration, i.e. machinery, propulsors, pumps and fans. Secondary sources are fluids in systems, electrical transformers, and the sea and waves interacting with the ship. Noise from a source may be transmitted through the air surrounding the source or through the structure to which it is attached. The structure on which a machine is mounted can have a marked influence on the amounts of noise transmitted, but it is difficult to predict the transmission losses in typical structures; airborne noise may excite structure on which it impacts and directly excited structure will radiate noise to the air. Much of the noise from a propulsor will be transmitted into the water. That represented by pressure fluctuations on the adjacent hull will cause the structure to vibrate transmitting noise both into the ship and back into the water. Other transmission paths will be through the shaft and its bearings.

SHOCK

All ships are liable to collisions and in wartime they are liable to enemy attack. The most serious threats to a ship's survival are probably collision or an underwater explosion. Collisions are dealt with under damage stability. The detonation of an explosive device leads to the creation of a pulsating bubble of gas containing about half the energy of the explosion (Figure 11.3). This bubble migrates towards the sea surface and towards the hull of any ship nearby. It causes pressure waves which strike the hull. The frequency of the pressure waves is close to the fundamental hull frequencies

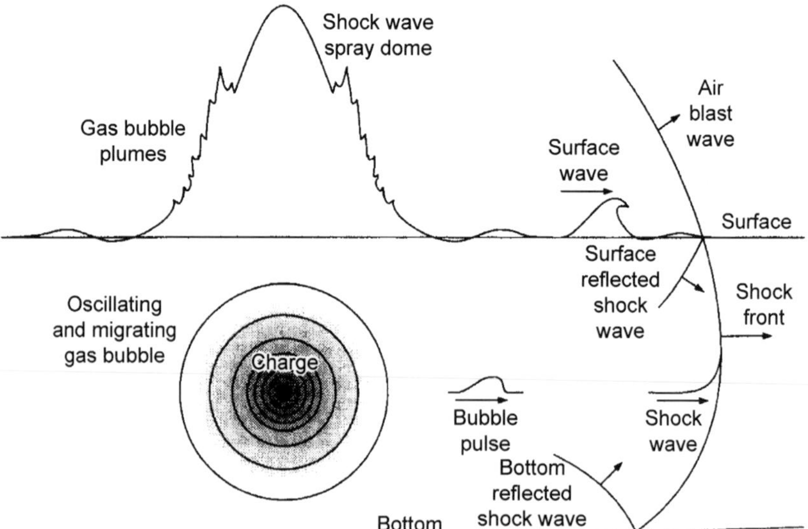

FIGURE 11.3 Underwater explosion. *Courtesy of RINA.*

of small ships, such as frigates and destroyers, and can cause considerable movement and damage. A particularly severe vibration, termed *whipping*, occurs when the explosion is set off a little distance below the keel. The pressure waves act on a large area of the hull and the ship *whips*. This whipping motion can lead to buckling, and even breaking, of the hull girder.

Another major feature of any underwater explosion is the shock wave containing about a third of the total energy of the explosion. This shock wave is transmitted through the water, and so into and through the ship's structure. It causes shock and may lead to hull rupture. The intensity of shock experienced depends upon the size, distance and orientation of the explosion relative to the ship.

Generally an individual equipment is fitted to more than one design and in different positions in any one ship so they must be able to cope with a range of shock conditions. The approach is to design to generalised shock grade curves. The overall design can be made more robust by providing shock isolation mounts for sensitive items and by siting system elements in positions where the structure offers more shock attenuation. This has the advantages that the item itself does not have to be so strong and the mounts can assist in attenuating any noise the equipment produces, reducing its contribution to the underwater noise signature.

In warships essential equipment is designed to remain operable up to a level of shock at which the ship is likely to be lost by hull rupture. The first of class of each new design of warship is subjected to a *shock trial* in which its resistance to underwater shock is tested by exploding charges, up to 500 kg, fairly close to the hull.

SUMMARY

Cyclic forces acting on a ship's hull will generate vibration and noise. Levels of both must meet acceptable standards. Some simplified formulae can assist in preliminary design assessments provided data from similar ships are available. Considerable advances have been made in recent years in methods of analysis (computational fluid dynamics and finite element methods) available to tackle vibration, but these are beyond the scope of this book. Equipments which generate exciting forces can be isolated from the structure by mounting them, so reducing transmissions of vibration and noise. Also sensitive equipment can be mounted (individually or in groups on a raft) to protect them from the effects of vibration and/or shock. Mounting systems are usually passive but active systems are available. Having calculated, during design, the vibration amplitudes expected, these can be checked as the build nears completion, by running a vibration generator on board. Finally the ship's acceptance trials are the final demonstration of how successful a designer has been in reducing vibration levels to acceptable limits.

This page intentionally left blank

Chapter 12

Manoeuvring

INTRODUCTION

All ships need to control their speed and follow an intended course when in transit. When entering congested waterways or harbours, they must be able to follow accurately a particular track relative to land. Vessels used for oil drilling or extraction often need to hold a particular position relative to the seabed with great precision.

Thus a ship must have the means of producing ahead and astern thrust, turning moments and lateral thrust. The last two are provided by rudders of various types assisted, in some cases, by lateral thrust units at the bow and/or stern. Ahead and astern thrust is usually provided by the main propulsion system as dealt with in Chapter 8. Because rudders are usually positioned close behind the propulsors there is an interaction between the two. Where more than one shaft is fitted, a turning moment can be produced by going ahead on one shaft and astern on the other.

The ease with which a vessel can maintain a straight course, or be made to turn, depends upon its *directional stability*. A number of measures are used to define the manoeuvring characteristics of a ship. These are defined and measured in still water conditions. The influence of wind, waves and current must be allowed for in applying the data to practical seagoing conditions. Wind effects can be very important especially in ships with large superstructures such as cruise liners and ferries. Strong winds may prevent a ship entering a port or turning into the wind if it has large windage areas aft. When operating close aboard another ship, close to a bank, or in shoaling water, the ship experiences additional forces that may throw it off the intended course.

A submarine operates in three dimensions with a need to control its position and attitude in depth as well as azimuth. Submarines are dealt with in one section and the rest of the chapter is devoted to surface vessels.

A comprehensive treatment of rudders and control surfaces, including stabilisers and hydroplanes, is to be found in Molland and Turnock (2007).

Introduction to Naval Architecture. DOI: http://dx.doi.org/10.1016/B978-0-08-098237-3.00012-6

DIRECTIONAL STABILITY AND CONTROL

When a ship, at rest in still water, is disturbed in the horizontal plane there are no hydrostatic forces to return it to its original position or to increase the movement. The ship is in neutral equilibrium. When a moving ship is disturbed in yaw it is acted upon by hydrodynamic forces which may be stabilising or destabilising. If stabilising, the ship will take up a new steady line of advance once the disturbance is removed. Unless some corrective action is applied, by using the rudder for example, this will not be the original line of advance. A ship is said to be *directionally stable* if, after being disturbed in yaw, it takes up a new straight line path. Clearly this stability differs from that discussed in considering inclinations from the vertical.

An arrow is an example of a very directionally stable body. If gravity is ignored the flight of an arrow is a straight line. If it is disturbed, say by a gust of wind, causing it to take up an angle of attack relative to its line of motion, the aerodynamic forces on its tail feathers will be much greater than those on the shank. The disturbing force will push the arrow sideways and the moment from the force on the tail will reduce the angle of attack. The arrow will oscillate a little and then settle on a new straight line path. The arrow, and likewise a weathercock, has a high degree of directional stability.

For a ship form it is not clear from looking at the lines whether it will be directionally stable or not. Good stability requires that the resultant hydrodynamic moment following a disturbance should tend to reduce yaw. The disturbing force is said to act at the hull's *centre of lateral resistance*. For stability this must be aft of the centre of gravity and it is to be expected that a cut away bow, a large skeg aft and trim by the stern would all tend to improve stability. That is about as much as one can deduce from the general hull shape at this stage. A degree of directional stability is desirable otherwise excessive rudder movements will be needed to maintain a straight course. Too much directional stability makes a ship difficult to turn.

Consider a small disturbing force in the horizontal plane. In general this will be the net effect of external forces (wind, say) and ship generated forces (propeller forces and rudder movements, say). The fore and aft component of this force will merely cause the ship to slow or speed up a little. The transverse component will lead to a sideways velocity and acceleration, and angular velocity and acceleration in yaw. As the ship responds, the force varies and additional hydrodynamic forces will be brought into play. When the disturbing force is removed these hydrodynamic forces will persist for a while. They will either tend to increase the deviations in course already experienced or decrease them. In these cases the ship is said to be directionally unstable or directionally stable, respectively.

Ignoring the fore and aft components, a small applied transverse force can be regarded as a transverse force at the centre of gravity and a moment about that point. Following a short period of imbalance the ship will settle down with a steady transverse velocity and yaw velocity, at which the hydrodynamic forces induced on the hull balance the applied force and moment. There will be a point along the length of the ship at which an applied force leads only to a transverse velocity with no yaw velocity. That is, the ship's head will remain pointing in the same direction. This point is commonly called the *neutral point* and is usually about a third of the length from the bow.

If the sideways force is applied aft of the neutral point and to starboard the ship will turn to port. If it is applied forward of the neutral point the ship turns in the direction of the force. The greater the distance the force is from the neutral point the greater the turning moment on the ship. Thus rudders placed aft are more effective than rudders at the bow would be, by a factor of about five for typical hull forms. Also they can benefit from the propeller race aft and are less vulnerable in a collision.

MANOEUVRING

Turning a Ship

For a ship to move in a circle requires a force to act on it, directed towards the centre of the circle. That force is not provided by the rudder directly. The rudder exerts a moment on the ship which produces an angle of attack between the ship's heading and its direction of advance. This angle of attack causes relatively large forces to act on the hull and it is the component of these, directed towards the centre of the circle, which turns the ship. The fore and aft components will slow the ship down which is a noticeable feature of a ship's behaviour in turning. The force on the rudder itself actually reduces the resultant inward force.

Evaluating Manoeuvrability

What is desired of a ship in manoeuvring can be discussed in general terms, but it is not easily quantified so as to set performance standards. Large oceangoing ships spend most of their transit time in the open seas, steering a steady course. They can use tugs to assist with manoeuvring in confined waters, so the emphasis will probably be on good directional stability. Poor inherent directional stability can be compensated for by fitting an auto pilot, but the rudder movements would be excessive and the steering gear would need more maintenance. For ships such as short haul ferries the designer would aim for rapid rudder response to help the ships avoid collision and to assist berthing. It is possible to set out the equations of motion and, by calculation or model tests, obtain the coefficients for a theoretical assessment of manoeuvrability in different operational scenarios. A simulator can be used to demonstrate the performance achieved and/or for training purposes. For comparing different designs it is useful to study a number of standard manoeuvres. Those most commonly used are described below.

Turning Circle

The path of a ship turning in a circle is shown in Figure 12.1.

As the rudder is put over there is a force which pushes the ship sideways in the opposite direction to which it wishes to turn. As the hydrodynamic forces build up on the hull the ship slows down and starts to turn in a steadily tightening circle until a steady-state speed and radius of turn is reached. A number of parameters are used to define the turning performance. They are:

- The *drift angle*, which at any point is the angle between the ship's head and its direction of motion. This varies along the length, increasing with distance aft. Unless otherwise specified the drift angle at the ship's centre of gravity is understood.
- The *advance*, which is the distance travelled by the ship's centre of gravity, in the original direction of motion, from the instant the rudder is put over. Usually the advance quoted is that for a 90° change of heading although this is not the maximum value.

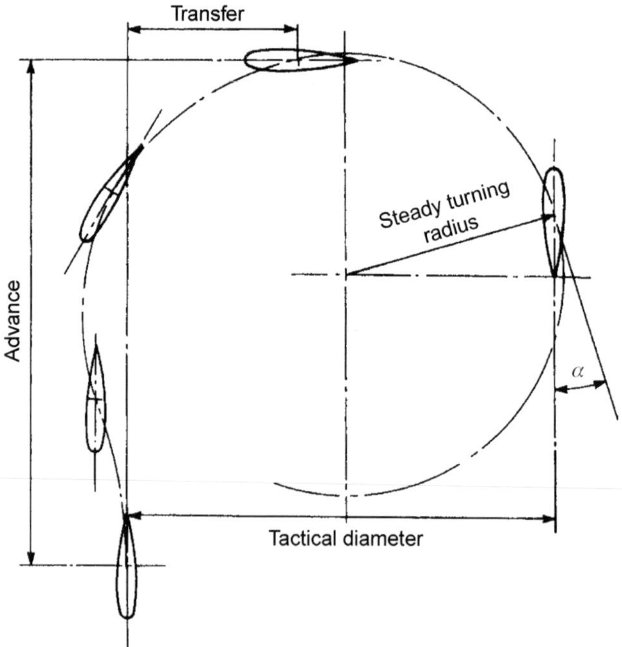

FIGURE 12.1 Turning circle.

- The *transfer*, which is the lateral displacement of the ship's centre of gravity from the original path. Usually transfer is quoted for 90° change of heading.
- The *tactical diameter*, which is the value of the transfer for 180° change of heading although this is not the maximum transfer. It is usual to quote a *tactical diameter to length ratio, TD/L*. Modern frigates at high-speed and full rudder turn with a *TD/L* of about 3. In merchant vessels it has been found that *TD/L* decreases with increasing block coefficient. Smaller turning circles may be required for special service craft and lateral thrust units or azimuthing propellers would be used. A value of 4.5 would be regarded as good for most merchant ships but a value greater than 7 as very poor.
- The *diameter of the steady turning circle*. The steady state is typically reached at some point between 90° and 180° change of heading.
- The *steady speed on turn*. Due to the fore and aft component of the hydrodynamic forces the ship slows down during the turn. Unless engine power is increased it may be only 60% of the approach speed. The steady speed is reached as the diameter steadies. If a ship does need to reverse direction, as might be the case of a frigate hunting a submarine, the time to turn through 180° is likely to be more important than a really small diameter of turn. Because of the loss of speed on turn such ships would choose a lesser rudder angle to get round quickly and reducing the acceleration needed after the turn.
- The *turning rate*. The quickest turn might not be the tightest. A frigate would turn at about 3° per second. Half this rate would be good for merchant ships and values of 0.5–1 would be more typical.
- The *pivoting point*. This is the foot of the perpendicular from the centre of the turning circle to the middle line of the ship, extended if necessary. This is the point at which the drift angle will be zero and it is typically about 1/6 of the length from the bow.
- The *angle of heel during the turn*. A ship typically heels in to the turn as the rudder is initially applied. On the steady turn it heels outwards, the heeling moment being due to the couple produced by the athwartships components of the net rudder and hull hydrodynamic forces and the acceleration force acting at the centre of gravity which is caused by the turning of the ship. The heeling moment is countered by the ship's stability righting moment.

If the steady radius of turn is R (Figure 12.2) and the steady heel is φ and the transverse components of the forces on the hull and rudder are F_{h} and F_{r}, acting at KH and KR above the keel then

$$F_{\mathrm{h}} - F_{\mathrm{r}} = \frac{\Delta V^2}{Rg}$$

and the heeling moment is

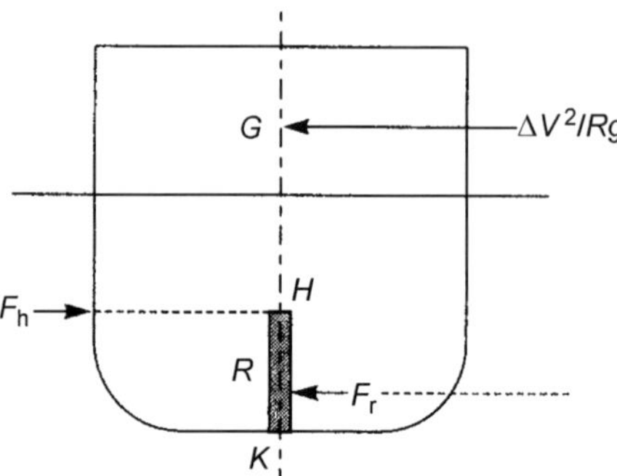

FIGURE 12.2 Ship heeling in turn.

$$\frac{\Delta V^2}{Rg}KG + F_r(KR) - F_h(KH) = (F_h - F_r)(KG - KH) - F_r(KH - KR)$$

For most ships, $(KH - KR)$ will be small and the heeling moment becomes $(F_h - F_r)GH$. This leads to an angle of heel such that

$$\Delta GM \sin\varphi = (F_h - F_r)GH = \frac{\Delta V^2}{Rg}GH, \quad \text{giving } \sin\varphi = \frac{GH}{GM} \times \frac{V^2}{Rg}$$

This is only an approximation to the angle as it is difficult to estimate the centre of lateral resistance for a heeled hull. In some high-speed turns the heel can be quite pronounced. It is important in passenger carrying ships and will be a factor in the choice of metacentric height.

Zig-Zag Manoeuvre

A ship does not often turn through large angles so the turning circle is not realistic for a ship in service. It is also difficult to measure the initial reaction to the rudder accurately in this manoeuvre. On the other hand a ship does often need to turn through angles of 10–30°. The initial response of the ship to the rudder being put over can be vital in trying to avoid a collision. This initial response is studied in the *zig-zag manoeuvre*. In it the ship proceeds on a straight course at a steady speed, a rudder angle of 20° is applied and held until the ship's head has changed by 20° and then the rudder is reversed to 20° the other way and held until the ship's head has changed 20° in the opposite direction. The manoeuvre is repeated for different speeds, rudder angles and heading changes.

The important measurements from the manoeuvre (Figure 12.3) are:

- The *overshoot* angle. This is the increase in the heading after the rudder is reversed. Large angles would represent a ship in which the helmsman would have difficulty in deciding when to take action to check a turn. Values of 5.5° and 8.5° would be reasonable aims for ships at 8 and 16 knots,

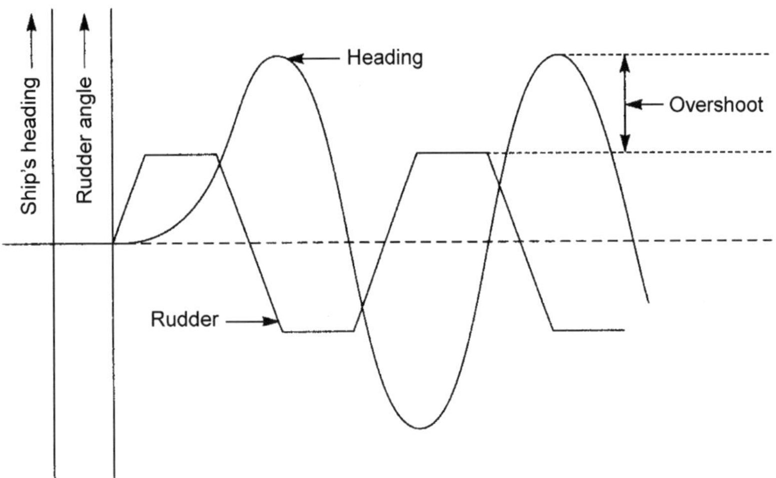

FIGURE 12.3 Zig-zag manoeuvre.

respectively, varying roughly with speed. The angle does not depend upon ship length.

- The times to the first rudder reversal and the first maximum heading change. It has been suggested (Burcher, 1991) that for reasonable designs, times to change heading by 20° would be of the order of 80–30 s for a 150 m ship over the range 6–20 knots. The time would be roughly proportional to length.
- The steady overshoot angle and the period of the cycle once a steady condition is reached.

Spiral Manoeuvre

This is a manoeuvre aimed at giving a feel for a ship's directional stability. From an initial straight course and steady speed the rudder is put over say 15° to starboard. After a while the ship settles to a steady rate of turn and this is noted. The rudder angle is then reduced to 10° starboard and the new steady turn rate noted. This is repeated for angles of 5°S, 5°P, 10°P, 15°P, 10°P and so on. The resulting steady rates of turn are plotted against rudder angle (Figure 12.4).

If the ship is stable there will be a unique rate of turn for each rudder angle. If the ship is unstable the plot has two 'arms' for the smaller rudder angles, depending upon whether the rudder angle is approached from above or below the value. Within the rudder angles for which there is no unique response it is impossible to predict which way the ship will turn, let alone the turn rate, as this will depend upon other disturbing factors present in the ocean. The manoeuvre does not give a direct measure of the degree of

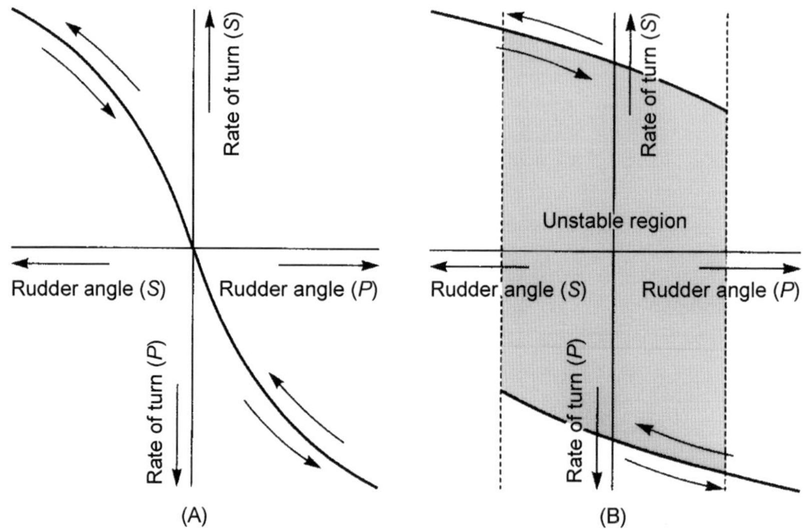

FIGURE 12.4 Spiral manoeuvre.

stability, although the range of rudder angles over which response is indeterminate is a rough guide. To know the minimum rudder angle needed to ensure the ship turns in the desired direction is very useful.

Pull-Out Manoeuvre

This manoeuvre (Burcher, 1991) is also related to the directional stability of the ship. The rudder is put over to a certain angle and held until the ship is turning at a steady rate. The rudder is returned to amidships and the change in the turn rate with time is noted. For a stable ship the turn rate reduces to zero and the ship takes up a new steady straight line course. A plot of the log of the rate of turn against time is a straight line after a short transition period. If the ship is unstable the turn rate will not reduce to zero but there will remain some steady rate of turn. The area under the plot of turn rate against time gives the total heading change after the rudder angle is taken off. The smaller this is the more stable the ship.

If the ship is conducting turning trials it will be in a state of steady turning at the end of the run. If the rudder is centred the pull-out manoeuvre can be carried out immediately for that speed and rudder angle.

Crash Stopping

In stopping a large ship can travel several miles before coming to rest and move laterally from its original course. Although turning under control is likely to be the better manoeuvre to avoid a collision, stopping trials are carried out to determine in a crash stop the head reach (distance travelled in the original direction), sideways translation to port or starboard and the stopping time. The action taken to stop depends upon the type of propeller; a fixed pitch propeller will be stopped and then put into reverse; a controllable pitch propeller can have its pitch reversed more quickly. Diesel ships have a higher percentage astern power than steam turbine ships and have a better stopping performance The rudder is kept amidships for trials but in normal service may be used in an attempt to control heading.

Barrass (2004) gives data for a range of ships and some approximate formulae.

MANOEUVRING DEVICES

This section reviews briefly some of the more common manoeuvring devices.

Rudders

These have a streamlined section to give a good lift to drag ratio and are of double-plate construction. They can be categorised according to the degree of balance – how close the centre of pressure is to the rudder axis. A balanced rudder requires less torque to turn. Rudders are termed *balanced*, *semi-balanced* or *unbalanced*. The other method of categorisation is the arrangement for suspending the rudder from the hull. Some have a pintle at the bottom of the rudder, others one at about mid-depth and others have no lower pintle. The last are termed *spade rudders* and it is this type which is most commonly fitted in warships. To prevent air-drawing, and loss of rudder effectiveness, the hull form aft should project aft of the rudder trailing edge and provide good immersion.

Different rudder types are shown in Figures 12.5–12.7.

Special Rudders

A number of special rudders have been developed, the aim being to improve the lift to drag ratio. A *flap rudder* (Figure 12.8) uses a flap at the trailing edge to improve the lift by modifying aerofoil shape. Typically as the rudder turns, the flap goes to twice the angle of the main rudder but in some rudders the flaps can be moved independently. A variant is the *Flettner rudder* which uses two narrow flaps at the trailing edge. The flaps move so as to assist the main rudder movement reducing the torque required of the steering gear.

In semi-balanced and unbalanced rudders the hull structure ahead of the rudder can be shaped to augment the lateral force at the rudder.

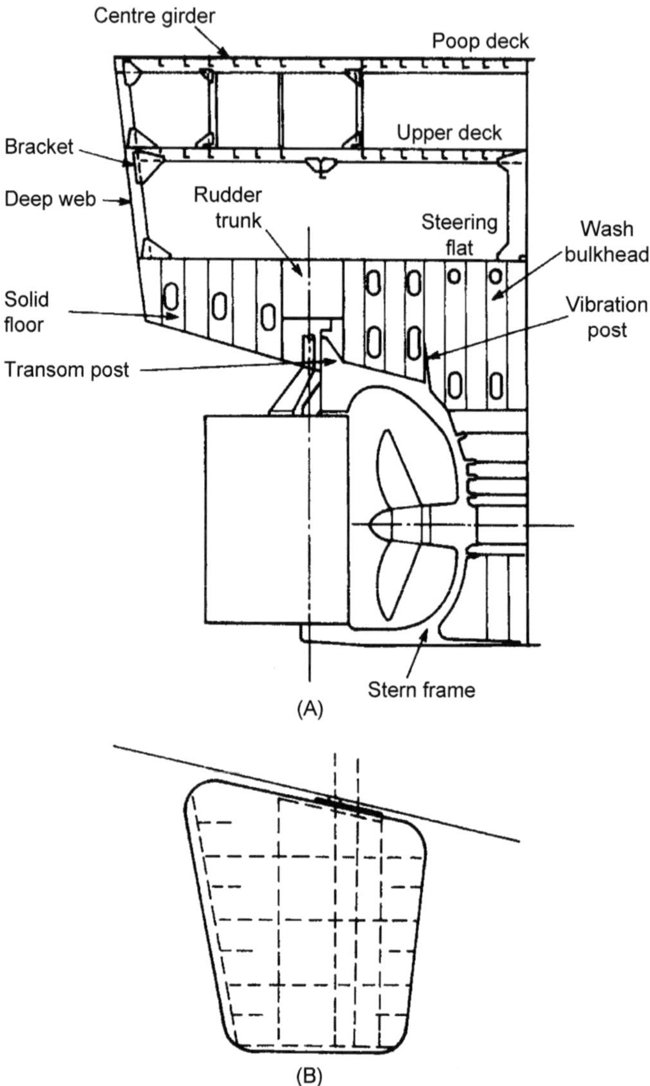

FIGURE 12.5 Balanced rudders: (A) simplex and (B) spade.

Active Rudders

These are usually spade-type rudders incorporating a faired housing with a small electric motor driving a small propeller. This provides a 'rudder' force even when the ship is at rest when the hydrodynamic forces on the rudder are zero. They are used in ships requiring good manoeuvrability at very low speeds.

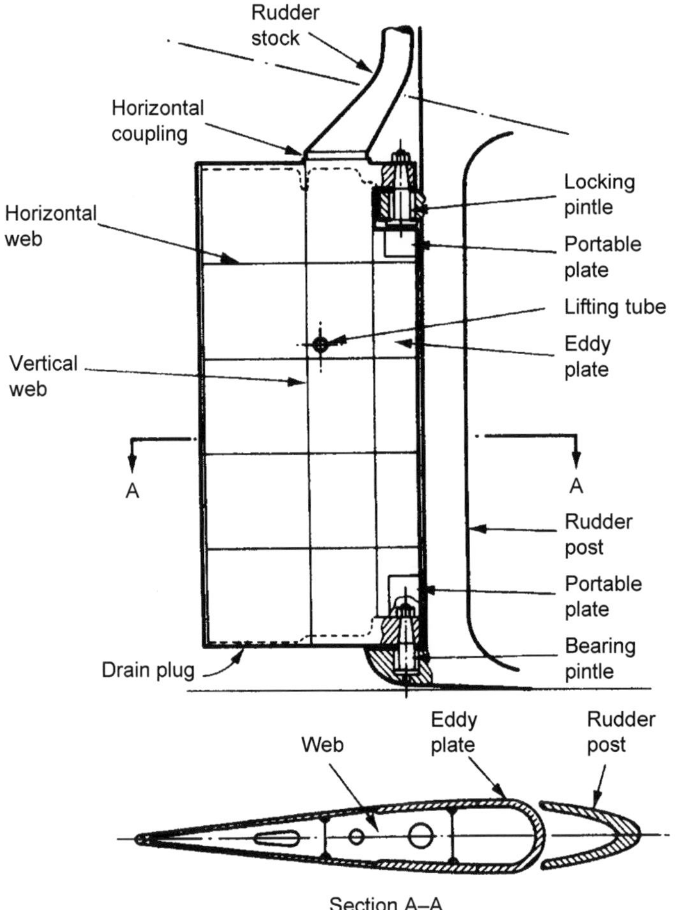

FIGURE 12.6 Unbalanced rudder.

Kitchen Rudder

This rudder (Figure 12.9) is a two-part tube shrouding the propeller and turning about a vertical axis. For ahead propulsion the two halves of the tube are opened to fore and aft flow. For turning the two halves can be moved together to deflect the propeller race. For stopping the two halves can be moved to block the propeller race and reverse its flow.

Other Devices

Vertical Axis Propeller

This type of propeller is essentially a horizontal disc carrying a number of aerofoil-shaped vertical blades. As the disc turns the blades are caused to

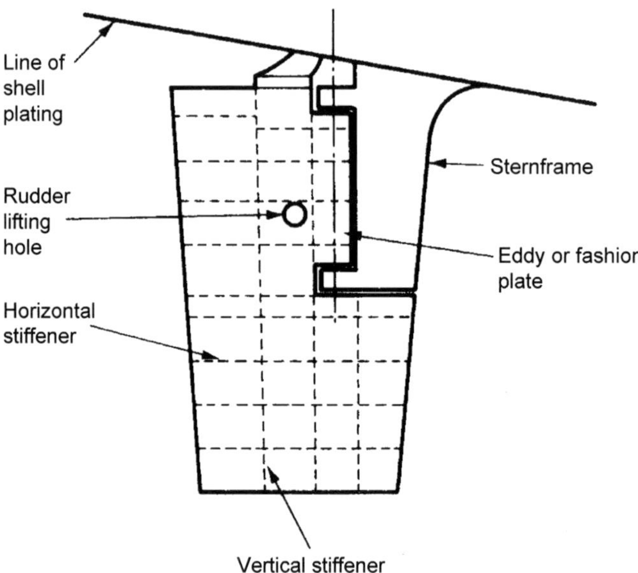

FIGURE 12.7 Semi-balanced rudder.

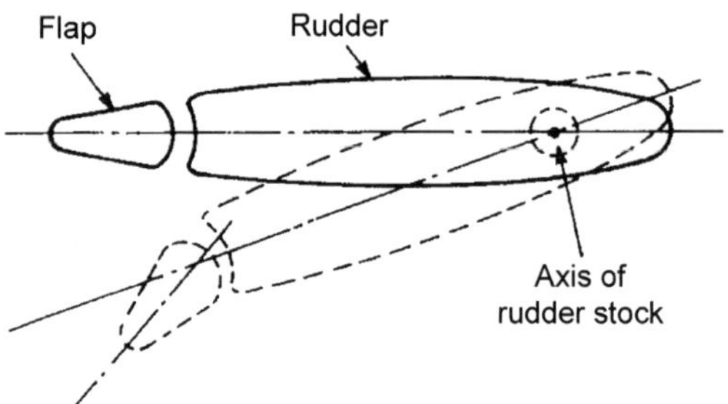

FIGURE 12.8 Flap rudder.

turn about their vertical axes so that they create a thrust. For normal propulsion the blades are set so that the thrust is fore and aft. When the ship wishes to turn the blades are adjusted so that the thrust is at an angle. They can produce lateral thrust even at low ship speed.

Cycloidal Rudder

This rudder was developed by Voith Schneider as a derivation of their cycloidal, or vertical axis, propeller. Like the propeller, the rudder has a rotor casing with a vertical axis of rotation. The cycloidal rudder has two blades

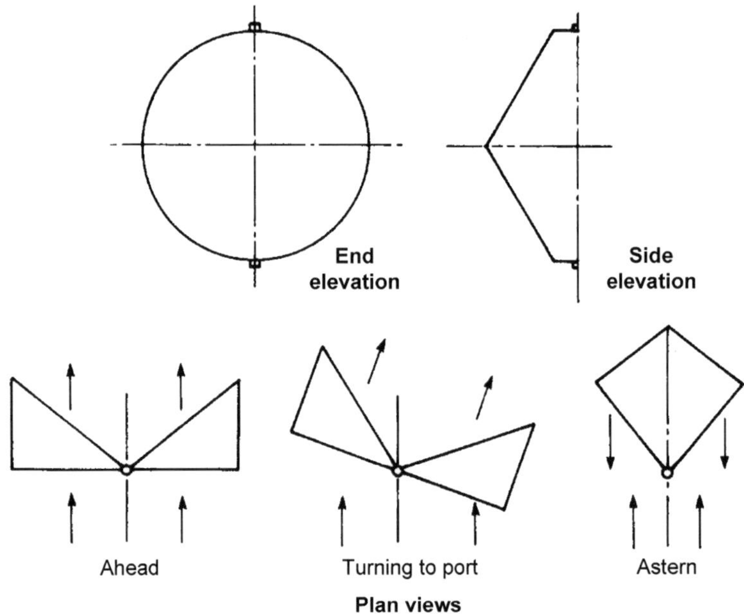

FIGURE 12.9 Kitchen rudder.

only, each much deeper than the blades of a vertical axis propeller. It has two modes of operation:

- *Passive*: In this mode the rotor does not rotate continuously but turns through limited angles so that the locked rudder blades develop steering forces as does a conventional rudder.
- *Active*: In this mode the rudder is operated like a vertical axis propeller to develop a thrust controllable in magnitude and direction (0–360°).

The cycloidal rudder has been shown, by trials, to have good shock resistance and low water-borne noise levels.

Lateral Thrust Units

It is sometimes necessary to control a ship's head and course independently. This situation can arise in vessels which need to follow a precise path relative to the ground in conditions of wind and tide. This requires the ability to produce lateral thrusts at the bow and/or the stern. Bow rudders are relatively ineffective and the alternative is to use a thrust unit – usually a contra-rotating propeller, in a transverse tube. Such devices are called *lateral thrust units* or *bow thrust units* when fitted forward. Their efficiency is seriously reduced by a ship's forward speed, the thrust being roughly halved at about 2 knots. Some offshore rigs, and ships needing to hold a precise position over the seabed, have dynamic positional control provided by a number of computer-controlled lateral thrust units.

Rotating Propulsion Pods

Many ships, including some large cruise ships, are using rotating, or azimuthing, pods to carry their main propulsion propellers. Such vessels can be steered by rotating these units.

RUDDER AREA, FORCES AND TORQUES

Area

To a first approximation the rudder area (A in m^2) (the total if two rudders are fitted) required can be scaled from a similar ship in relation to the ship length (L) multiplied by the draught (T). Molland and Turnock (2007) quote a formula:

$$A/LT = 0.01[1 + 50C_B^2(B/L)^2]$$

where B is the beam.

Barrass (2004) suggests

$$A = K \times L_{BP} \times T$$

where T is summer loaded waterline draught.

The K (%) values listed include (higher figures in the range applying to faster vessels):

container ships and passenger liners K (%) =	1.2–1.7;
general cargo ship	1.5;
oil tankers and bulk carriers	1.7;
Ro–Ro ships	2.0–3.0.

Forces

Rudders are streamlined to produce high lift with minimum drag. Only conventional rudders are considered here. These are symmetrical to produce the same lift characteristics whichever way they are turned. The force on the rudder, F, depends upon the cross-sectional shape area, A, the velocity, V, through the water and the angle of attack, α:

$$F = \text{const. } \rho A V^2 f(\alpha)$$

The constant depends upon the cross section and the rudder profile, in particular the ratio of the rudder depth to its chord length and the degree of rounding off on the lower corners. The lift is also sensitive to the clearance between the upper rudder surface and the hull. If this is very small the lift is augmented by the mirror image of the rudder in the hull. $f(\alpha)$ increases roughly linearly with α up to the stall angle which is typically about 35°. $f(\alpha)$ will then decrease.

Various approximate formulae have been proposed for calculating F. An early one was

$$F = 577AV^2 \sin \alpha \text{ N}$$

In this an allowance was made for the effect of the propeller race by multiplying V by 1.3 for a rudder immediately behind a propeller and by 1.2 for a centreline rudder behind twin screws. Other formulations based on the true speed of the ship are:

$F = 21.1AV^2\alpha$ N for ahead motion
$F = 19.1AV^2\alpha$ N for astern motion
$F = 18.0AV^2\alpha$ N

The first two were proposed for twin rudders behind twin screws and the third for a centreline rudder behind a single screw. If wind or water tunnel data is available for the rudder cross section this should be used to calculate the lift and the centre of pressure position.

Torques

The torque needed to turn a rudder depends upon the position on the rudder at which the rudder force acts – the *centre of pressure*. For a rectangular flat plate of breadth B at angle of attack α, this can be taken as $(0.195 + 0.305 \sin \alpha)B$ aft of the leading edge. For a typical rudder section it has been suggested that the centre of pressure for a rectangular rudder can be taken at $K \times$ (chord length) aft of the leading edge, where

$K = 0.35$ for a rudder aft of a fin or skeg, the ship going ahead
$= 0.31$ for a rudder in open water

The open water figure is used for both configurations for a ship going astern.

For a non-rectangular rudder an approximation to the centre of pressure position can be obtained by dividing the rudder into a number of rectangular sections and integrating the individual forces and moments over the total area. This method can also be used to estimate the vertical location of the centre of pressure, which dictates the bending moment on the rudder stock or forces on the supporting pintles.

Worked Example 12.1

A rudder with an area of 20 m^2 when turned to 35° has the centre of pressure 1.2 m from the stock centreline. If the ship speed is 15 knots and the rudder is located aft of the single propeller, calculate the diameter of the stock able to take this torque, assuming an allowable stress of 70 MN/m^2.

Solution

Using the simple formula from above to calculate the rudder force and a factor of 1.3 to allow for the screw race:

$$\begin{aligned} F &= AV^2 \sin\alpha \\ &= 577 \times 20 \times (15 \times 1.3 \times 0.5144)^2 \times \sin 35^\circ \\ &= 0.666 \text{ MN} \end{aligned}$$

Torque on rudder stock = 0.666 × 1.2 = 0.799 MNm.

This can be equated to qJ/r, where r is the stock radius, q is the allowable stress, and J is the second moment of area about a polar axis equal to $\pi r^4/2$. Hence

$$r^3 = 2T/\pi q = \frac{0.799 \times 2}{70\pi} = 0.00727$$

$$r = 0.194 \text{ m and diameter of stock} = 0.388 \text{ m}$$

In practice it would be necessary to take into account the shear force and bending moment on the stock in checking that the strength was adequate. The bending moment and shear forces will depend upon the way the rudder is supported. If astern speeds are high enough the greatest torque can arise then as the rudder is less well balanced for movements astern.

SHIP HANDLING

Several additional aspects of ship handling need to be discussed.

Handling at Low Speed

At low speed any hydrodynamic forces on the hull and rudders are small since they vary as the square of the speed. The master must use other means to manoeuvre the ship, including:

- Using one shaft, in a twin shaft ship, to go ahead while the other goes astern.
- When leaving, or arriving at, the dockside a stern or head rope can be used as a pivot while going ahead or astern on the propeller.
- Using the so-called *paddle wheel effect* which is a lateral force arising from the non-axial flow through the propeller. The force acts so as to cause the stern to swing in the direction it would move had the propeller been a wheel running on a hard surface. In twin screws the effects generally balance out when both shafts are acting to provide ahead or astern thrust. In coming

alongside a jetty a short burst astern on one shaft can 'kick' the stern in towards the jetty or away from it depending which shaft is used.

- Using one of the special devices described above.

Broaching

The conditions under which a ship may broach are discussed under seakeeping.

Shallow and Confined Water

In shallow water the water flow under the ship is restricted. The flow speeds up causing a reduction in pressure leading to a vertical force and trimming moment resulting in a bodily sinkage and trim of the ship by the stern. This can cause a ship to ground in water which is nominally significantly deeper than the draught (Dand, 1981).

The sinkage is known as *squat* and has become more important with the increasing size of tankers and bulk carriers. Squat is present even in deep water due to the different pressure field around the ship at speed. It is accentuated, as well as being more significant, in shallow water. In a confined waterway a *blockage* effect occurs once the ship's sectional area exceeds a certain percentage of the waterway's cross section. This is due to the increased speed of the water which is trying to move past the ship.

For narrow channels a *blockage factor* (Barrass, 2004) has been defined as

$$\text{Blockage factor, } S, = \frac{\text{ship cross-sectional area } (A_S)}{(\text{cross-sectional area of river or canal } (A_C)}$$

The vessel can be regarded as effectively in open water if the breadth of the river is greater than the 'width of influence' defined by 7.04(ship beam)/$(C_B)^{0.85}$, where C_B is the block coefficient.

The blockage factor in a channel of width B and depth H is then

$$S = (b \times t)/(B \times H)$$

where b and t are the ship's breadth and draught, respectively.

A formula for estimating the maximum squat at speed V in open or confined water is

$$\text{Maximum squat} = \frac{[C_B \times S^{0.81} \times V^{2.08}]}{20}\ \text{m}$$

where V is the ship speed relative to the water.

Two simplified formulae for maximum squat (Barrass, 2004) are

$$\text{Maximum squat} = \frac{C_B \times V^2}{100}\ \text{m}$$

for open water with water depth/draught = 1.1–1.4.

$$\text{Maximum squat} = \frac{C_B \times V^2}{50}\ \text{m}$$

Another approximate approach (Dand, 1977) is to take squat as 10% of the draught or as 0.3 m for every 5 knots of speed.

The general effect of shallow water is to degrade manoeuvrability by increasing the turning circle diameter and time and reducing the ability of the rudder to check any yawing. It increases the time and distance travelled in a crash stop.

Interaction Between Ships

In a ship's pressure field there is a marked increase in pressure near the bow and stern with a suction over the central portion of the ship. This pressure field acts for quite an area around the ship. Anything entering and disturbing the pressure field will cause a change in the forces on the ship and suffer forces on itself. If one ship passes close to another in overtaking it, the ships initially repel each other. This repulsion force reduces to zero as the bow of the overtaking ship reaches the other ship's amidships and an attraction force builds up. This is at a maximum soon after the ships are abreast after which it reduces and becomes a repelling force as the two ships part company. When running abreast the ships experience bow outward moments. Such forces are very important for ships when they are replenishing at sea.

Similar considerations apply when a ship approaches a fixed object. For a vertical canal bank or jetty the ship experiences a lateral force and yaw moment. Open structure jetties will have much less effect than solid ones.

DYNAMIC STABILITY AND CONTROL OF SUBMARINES

Submarines can travel at high speed, but sometimes their mission requires them to move very slowly. These two speed regimes pose quite different situations as regards their *dynamic stability* and control in the vertical plane. The submarine's static stability dominates the low-speed performance but has negligible influence at high speed. For motions in the horizontal plane the submarine's problems are similar to those of a surface ship except that the submarine, when deep, experiences no free surface effects. At periscope depth the free surface becomes important as it affects the forces and moments the submarine experiences, but again mainly in the vertical plane.

A submarine must avoid hitting the seabed or exceeding its safe diving depth and, to remain covert, must not break surface. It has a layer of water in which to manoeuvre which is only about two or three ship lengths deep. At high speed there is little time to take corrective action should anything go wrong. By convention submarines use the term pitch angle for inclinations about a transverse horizontal axis (the trim for surface ships) and the term trim is used to denote the state of equilibrium when submerged. To trim a submarine it is brought to neutral buoyancy with the centres of gravity and buoyancy in line. A submarine is trimmed each time it dives.

The approach to the problem is like that used for the directional stability of surface ships, but bearing in mind that:

- the submarine is positively stable in pitch angle. So if it is disturbed in pitch while at rest it will return to its original trim angle;
- the submarine is unstable for depth changes due to the compressibility of the hull;
- it is not possible to maintain a precise balance between weight and buoyancy as fuel and stores are used up.

The last two considerations mean that the control surfaces must be able to provide vertical forces to counter any out-of-balance force and moment in the vertical plane. To control depth and pitch separately requires two sets of control surfaces, the *hydroplanes*, one forward and one aft.

A mathematical treatment of motions in the vertical plane leads to a number of interesting relationships as shown by Nonweiler (1961). If M_w and Z_w represent the rates at which the hydrodynamic pitching moment (M) and vertical force (Z) on the submarine vary with the vertical velocity, and V is the speed, then:

- The steady path in the vertical plane cannot be a circle unless BG is zero.
- There is a point along the length at which an applied vertical force causes a depth change but no change in pitch angle. This point is called the *neutral point* and is the equivalent of the neutral point for horizontal motions, already referred to. The neutral point is M_w/Z_w ahead of the centre of gravity (Figure 12.10).
- A second point, known as the *critical point*, is distant $mgBG/VZ_w$ aft of the neutral point. A vertical force applied at the critical point will cause

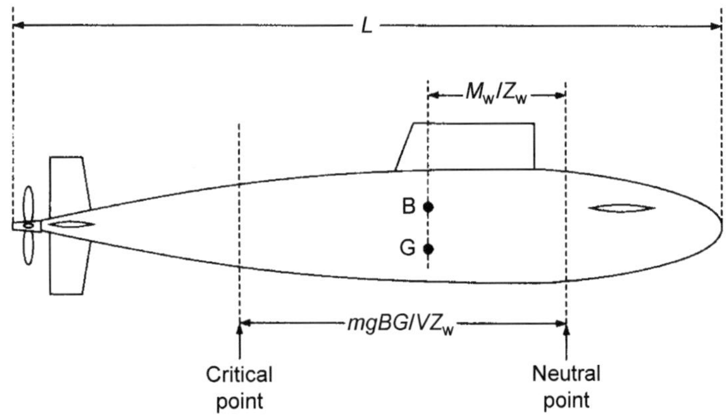

FIGURE 12.10 Neutral and critical points.

no change of depth but will change the pitch angle. A downward force forward of the critical point will increase depth, a downward force aft of the critical point will reduce depth. Thus at this point there is a reversal of the expected result of applying a vertical force.

- As speed drops the critical point moves aft. At some speed, perhaps 2 or 3 knots, the critical point will fall on the after-hydroplane position. The speed at which this happens is termed the *critical speed*.

UNDERWATER VEHICLES, GENERAL

Apart from military submarines which are usually quite large there are now many other underwater vehicles, including:

- Streamlined vehicles, of moderate size and limited depth capability, for taking tourists underwater to see the marine life. Generally their handling is similar to larger submarines but their structural strength need not be so high.
- Small manned vehicles capable of operating at great depth. In this case, strength is paramount and the capsule carrying the crew may well be spherical for efficiency. They are usually provided with lights, cameras and manipulators to assist in survey of the seabed and wrecks and the retrieval of small items. The finding of the *Titanic* at the bottom of the Atlantic helped to bring these vehicles to the public notice.
- Small unmanned vehicles operated remotely from a mother ship. Usually the mother ship is a surface ship but it can be another, larger, submersible. These vehicles have the advantage that they can venture into areas that would be regarded as too hazardous for manned vehicles. They are also fitted with cameras and lights and they may have their own manipulators.

The small submersibles generally operate at low speed and they are not streamlined. Indeed they are often simply a framework on which cameras and other equipment are fixed. Control is by means of propellers, some giving vertical, and some horizontal, thrust. If an umbilical is fitted for power and/or control, the drag on the cable is high when the craft is at depth. This will restrict speed and the vehicle may experience trouble with holding its position against underwater currents.

PREDICTIONS OF MANOEUVRABILTY

Models have been used for many years to predict a new design's manoeuvrability. Model tests are of two types:

- Free running models performing the manoeuvres described previously. This requires a large basin which is usually provided with wave makers so that performance in waves can be measured directly. The results can be compared to full-scale data to obtain correlation factors. Models are

run at the corresponding Froude number so the wave patterns they create will be similar to those of the ship. There are scaling effects due, in part, to viscosity and large models are run to improve correlation. Scaling effects are not very significant in turning but generally a ship is more directionally stable than the model. See ITTC 7.5-02-06-01.
- Captive model tests to obtain the coefficients to be used in the solution of the mathematical equations of motion. The coefficients are measured either by a combination of straight line running and rotating arm tests or by a planar motion mechanism (PMM).

The PMM is a horizontal oscillator used in a long towing tank and capable of forcing the model to oscillate sinusoidally in pure sway or yaw or a combination of the two. The manoeuvring forces and moments are assumed to be low-frequency values and independent of frequency. See ITTC 7.5-02-06-02.

One theoretical approach is to use slender body theory with the coefficients in the equations of motion provided from models. Whilst computational fluid dynamics methods are improving they are not yet at the stage where they can be relied upon for accurate manoeuvring predictions. Better results are possible if modifying factors are available from previous designs.

MODIFYING A SHIP'S MANOEUVRING PERFORMANCE

As with other aspects of ship performance it is difficult to generalise on the effect of design changes on a ship's manoeuvring qualities, because so many factors interact. What is true for one form may not be true for another. Broadly, however, it can be expected that:

- Stern trim improves directional stability and increases turning diameter.
- A larger rudder can improve directional stability and give better turning.
- Decrease in draught can increase turning rate and improve directional stability. This is perhaps due to the rudder becoming more dominant relative to the immersed hull.
- Higher length to beam ratios lead to greater directional stability.
- Quite marked changes in metacentric height, whilst affecting the heel during a turn, have little effect on turning rate or directional stability.
- For surface ships at a given rudder angle the turning circle increases in diameter with increasing speed but rate of turn can increase. For submarines turning diameters are little affected by speed.
- A large skeg aft will increase directional stability and turning circle diameter.
- Cutting away the below-water profile forward can increase directional stability.

By and large the hull design of both a surface ship and a submarine is dictated by considerations other than manoeuvring. If model tests show a need to change the handling performance, this would normally be achieved by modifying the areas and positions of the control surfaces and skegs, which usually suffices.

SUMMARY

A ship must be able to move in a controlled way, following a desired path in both open and confined waters. Whilst, usually, specific standards are not laid down, various manoeuvres can be used to characterise a ship's directional stability and turning performance. These include the spiral, zig-zag and pull-out manoeuvres, and the turning circle. A variety of rudders and other devices such as bow thrusters are used to control a ship's heading and position. These can hold a vessel in a fixed spot, relative to the seabed as may be required for a drilling ship. Shallow water generally degrades a ship's manoeuvring performance. Submarine and submersibles must be controlled in three dimensions using hydroplanes to control pitch angle and depth.

Chapter 13

Structures

INTRODUCTION

At sea in rough weather a ship is heavily strained. It moves violently and the structure becomes distorted. The waves give an impression of utter confusion. This illustrates the two fundamental difficulties facing a naval architect – identifying the loading on the structure and calculating its response. The stresses generated in the material of the ship and the resulting deformations must be kept within acceptable limits and each element of the structure must play its part. There is no opportunity to build a prototype and getting things wrong can have serious consequences.

This chapter deals with the overall strength of the hull, the strength of structural elements, some of the complicating factors and how structures may deteriorate and fail. From the overall strength and loading of the hull it is possible to consider the adequacy of the strength of its constituent parts, the plating and grillages. Global calculations indicate stresses or strains acting on the complete hull and which must also be included in the design of local details. Many local strength problems in a ship can be solved by methods employed in general mechanical engineering if the load and boundary conditions are adequately defined.

Strength in still water and in waves is considered. Even in still water the ship is subject to the forces of hydrostatic pressure and the weight of the ship and all it carries. Care is necessary when loading ships in port to ensure that the structure is not overloaded. In 1994 the OBO carrier *Trade Daring*, a ship of 145,000 dwt, broke in half while loading iron and manganese ore.

A ship's ability to withstand very high occasional loading in waves is ensured by designing to stress levels which are likely to be met perhaps only once in the life of the ship. Failure is much more likely to be due to a combination of fatigue and corrosion. These cumulative failure mechanisms are increasingly determining the ship's structural design and its likely useful life span.

No ship can be completely safe but the probability of structural failure of the hull can be assessed in statistical terms and set at a low level judged to be acceptable. Using computations or model tests, the loads a ship is likely to experience can be assessed by combining its responses to different waves

Introduction to Naval Architecture. DOI: http://dx.doi.org/10.1016/B978-0-08-098237-3.00013-8

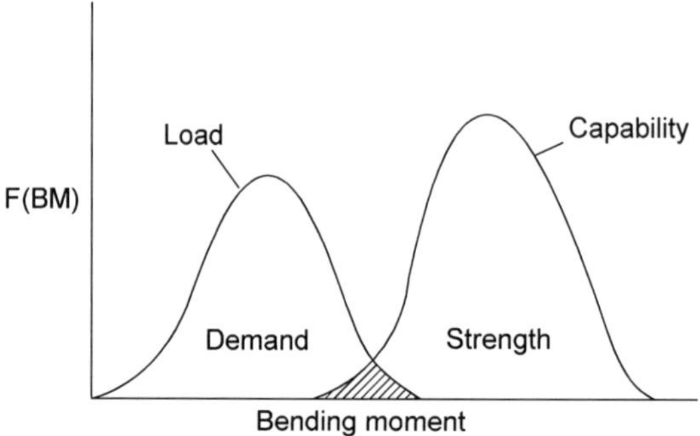

FIGURE 13.1 Load and capability curves.

with the sea conditions it is likely to meet during its life. Allowing for the fact that actual conditions met may differ from those assumed leads to a lifetime 'load' or 'demand' curve. The strength of the ship as calculated leads to single figures for the bending moment and shear stress it can safely endure but in practice this strength is variable and will be a distribution showing the probability of the structure being able to sustain a given load – its 'strength' or 'capability'. Figure 13.1 shows a plot of the load and strength curves in terms of bending moment. Damage is likely in the shaded area because the load is greater than the capability of the structure to take that load. This concept is developed under 'Overall structural safety' after consideration of the many factors involved.

NATURE OF A SHIP'S STRUCTURE

There is an internationally accepted way of referring to the main elements of a ship's hull structure, aiding clarity of communication. *International Association of Classification Societies* (IACS) REC 082 (2003) gives a glossary of structural terms and illustrates typical hull details.

Some ships are made from glass-reinforced plastics (GRP) but the vast majority are of steel with possibly some aluminium in the superstructure areas. High-speed ships may have aluminium hulls for lightness. The following remarks relate to metal ships. Although GRP ships obey the same principles the material behaves rather differently. The complete structure is composed of panels of plating, most of which will be roughly rectangular and supported on the four edges. They are subject to normal and in-plane loads. Together with their supporting stiffeners in the two directions, a group of plating elements constitutes a grillage which may be nominally in one plane or curved in one or two directions. Grillages are combined to create the hull, decks and bulkheads, all mutually supportive. Additional support is provided by pillars and strong frameworks such as hatch coamings.

Since the major forces the hull must withstand are those due to longitudinal bending, it is primarily longitudinally structured, whilst taking account of transverse strength needs. It is usual in dry cargo ships to find the decks and bottom longitudinally stiffened and the side structure transversely stiffened. The principal longitudinal elements are the decks, shell plating, inner bottom, all in the form of grillages, with additional longitudinal strengthening. The plating itself is relatively thin and the spacing of the stiffeners must be such as to prevent buckling. Sided longitudinals in the double bottom are *intercostal*. That is, they are cut at each floor and welded to them. The resulting 'egg box'-type construction of the double bottom is very strong and capable of taking large loads.

If decks, stiffened by transverse beams, were supported only at the sides of the ship, they would need to be very strong to carry the loads. Their dimensions (*scantlings*), would become large. Introducing some support at intermediate positions reduces the span of the beams leading to a better structure in terms of strength to weight ratio. This could be done by pillars but these restrict access in the holds. Usually heavy longitudinal members are used supported in turn by a few pillars and heavy transverse members at the hatches. The hatch end beams are themselves supported by longitudinal centreline bulkheads clear of the hatch opening. In areas which are predominantly longitudinally stiffened, deep transverse members are used for support.

Most structural elements contribute to the overall strength of the ship girder and have some local strength function as well. For instance, the bottom and side shell must withstand water pressures normal to their surfaces, acting as struts with end and lateral loading. Side structure must withstand the loads due to coming alongside a jetty. Decks and bulkheads must withstand the weight of, and dynamic forces from, equipment mounted on them.

MODES OF FAILURE

A ship's structure might fail in various ways with different possible consequences. The ship is essentially an elastic beam floating on the water surface and subject to a range of fluctuating and quasi-steady loads. Those loads will generate bending moments and shear forces which may act over the ship as a whole or be localised. The global loads include the action of the sea. The local ones include the forces on items composed of gravity and dynamic forces due to the ship's motions. Then there is the thrust due to the main propulsion forces.

Failure can be said to occur when the structure can no longer carry out its intended function. If, in failing, one element sheds its load on to another which can withstand it there is no great safety problem, although remedial action may be necessary. If, however, there is a 'domino' effect and the surrounding structural elements fail in their turn the result can be loss of the ship. Failure may be due to the structure:

- becoming distorted due to being strained past the yield point, leading to permanent set. The distortion may mean systems cannot function. For instance, the shafts may be unable to turn because the bearings are out of alignment.
- fracture/cracking. This occurs when the material can no longer sustain the load applied and it breaks. Locally the stresses acting may be augmented by *built-in stresses* and/or *stress concentrations*. The loading may exceed the ultimate strength of the material or, more likely, failure will be due to fatigue leading to cracking and then fracture. Even where the crack does not lead to complete fracture it can cause leakage and thus cause the structure to fail in one of its intended functions. Some steels are prone to brittle fracture at low temperatures.
- buckling or instability. Shear stresses can lead to buckling of individual structural elements forming the hull. Very large deflections can occur under relatively light loads in supporting structure which can then behave like a crippled strut.

The general approach to a study of a ship's structural strength is to assess the overall loading of the hull, determine the likely stresses and strains this engenders and the ability of the main hull girder to withstand them. Then local forces can be superimposed on the overall effects to ensure that individual elements of the structure are adequate and will play their part.

Stresses

Stress can be used to judge some aspects of failure but not all. However, first we consider the stresses in a hull and how realistic calculations of them might be.

The loading patterns and the ship structure are complex. Whilst modern research and computer methods provide an ability to deal with greater complexity, some simplification of the load and structure is still needed. Finite element analysis (FEA), discussed in outline later, is a very powerful tool but the finer the mesh used in way of a discontinuity, say the tip of a crack, the higher the stress obtained by calculation. In the limit it becomes infinite. Clearly some yielding will take place but the naval architect is left with the task of deciding what is acceptable. This can be determined by comparing theory with model or full-scale experiments. Thus model tests are important in improving our understanding of structural strength and modes of failure and for validating numerical computations.

Traditionally the naval architect has treated the problem of overall hull strength as a comparable static one, making fairly gross simplifications and relying upon a comparison with the results of similar calculations for previously successful ships. Although the stresses derived were nominal, and might bear no relation to the actual stresses, the new ship was likely to be satisfactory in service provided it resembled the ships with which it was compared. A drawback of the method was that it was a 'play safe' one. It could not show whether the new ship was grossly overdesigned or close to the acceptable limit. The growing importance of ensuring structural weight is kept to a minimum has driven the naval architect to adopt more realistic design methods as they have become available. In any case novel hull forms cannot be compared with past forms and using new materials complicates the situation.

It is now possible to use computational methods to find the fluid loading on the hull as it moves through, and responds to, waves. Finite element methods (FEMs) can then be used to determine the strains (deflections) the ship experiences as a result of those loads. That is, the ship is allowed to flex rather than being treated as a rigid body. The distortions of the hull will in turn affect the fluid loading. Such analyses provide the bending moments and shear forces acting on the ship structure and the corresponding global hull stresses. They also provide the accelerations at any point in the ship and hence the dynamic forces acting on equipment. These are part of the overall loading but are also vital to the calculation of local stresses – in machinery seatings, say. These hydroelastic computational methods represent the actual situation much more closely than the old style quasi-static calculations but they must still be used with some caution because they cannot yet take account of every factor including some of the non-linearities involved.

FORCES ON A SHIP

The traditional quasi-static method of assessing longitudinal strength is described here as it illustrates the principles concerned.

Forces in Still Water

The buoyancy forces acting on a ship must equal in total the sum of the weight of the ship but over any given unit length the weight and buoyancy forces will not balance. If the mass per unit length at some point is m and the immersed cross-sectional area is a, then at that point:

$$\text{buoyancy per unit length} = \rho g a$$

where ρ is the water density

$$\text{the weight per unit length} = mg$$

and the net force per unit length $= \rho g a - mg$

Integrating along the length there will be, at any point, a force tending to shear the structure such that:

$$\text{Shear force, } S = \int (\rho g a - mg) dx$$

integration being from one end to the point concerned.

Integrating a second time gives the longitudinal bending moment. That is:

$$\text{Longitudinal bending moment, } M = \int S \, \mathrm{d}x = \iint (\rho g a - mg) dx \, \mathrm{d}x$$

The integrals are evaluated by dividing the ship into a number of sections, say 40, calculating the mean buoyancy and weight per unit length in each section, and evaluating the shearing force and bending moment by approximate integration.

For any given loading of the ship the draughts at which it floats can be calculated. Knowing the weight distribution, and finding the buoyancy distribution from the Bonjean curves, gives the net load per unit length. Certain approximations are needed to deal with distributed loads such as shell plating. Also the point at which the net force acts may not be in the centre of the length increment used. Typically the weight distribution at any point is assumed to have the same slope as the curve of buoyancy plotted against length. These approximations are not usually of great significance and certain checks can be placed upon the results. First the shear force and bending moment must be zero at the ends of the ship. If after integration there is a residual force or moment this is usually corrected arbitrarily by assuming the difference can be spread uniformly along the ship length. Besides causing stresses in the structure the forces cause a deflection of the ship longitudinally. Simple beam theory shows that the deflection y at any point is given by:

$$EI \times \frac{\mathrm{d}^2 y}{\mathrm{d}x^2} = \text{bending moment.}$$

When the ship is distorted so as to be concave up it is said to *sag* and the deck is in compression with the keel in tension. When the ship is convex up it is said to *hog*. The deck is then in tension and the keel in compression.

High still water forces and moments, besides being bad in their own right, mean higher values in waves as the values at sea are the sum of the still water values and those due to a superimposed wave. The still water values indicate which are likely to be stressful ship loading conditions.

The static forces of weight and buoyancy also act upon a transverse section of the ship as shown in Figure 13.2.

The result is a transverse distortion of the structure which the structure must be strong enough to resist. The hydrostatic loads tend to dish plating between the supporting frames and longitudinals. The deck grillages must

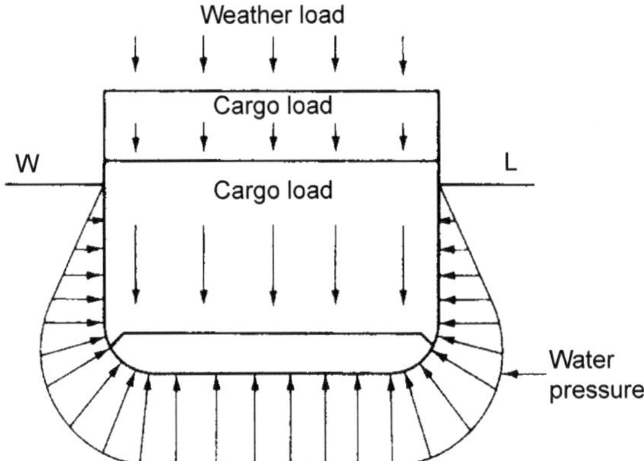

FIGURE 13.2 Loads on ship section.

support the loads of equipment and cargo. Thus structure contributes to the longitudinal, transverse and local strength. Longitudinal strength in a seaway is considered first.

LONGITUDINAL STRENGTH IN WAVES

The mass distribution is the same in waves as in still water assuming the same loading condition. The differences in the forces acting are the buoyancy forces and the inertia forces on the masses arising from the motion accelerations, mainly those due to pitch and heave. For the present the inertia forces are ignored and the problem is treated as a quasi-static one by considering the ship balanced on a wave. The buoyancy forces vary from those in still water by virtue of the different draughts at each point along the length due to the wave profile and the pressure changes with depth due to the orbital motion of the wave particles. This latter, the *Smith effect*, is usually ignored in the standard calculation. Ignoring the dynamic forces and the Smith effect is acceptable as the results are used for comparison with figures from previous, successful, ships calculated in the same way with the same assumptions.

The Static Longitudinal Strength Approach

In this approach the ship is assumed to be poised, in a state of equilibrium, on a trochoidal wave of length equal to that of the ship. To a first order it can be assumed that bending moments will be proportional to wave height. Two heights have been commonly used – $L/20$ and $0.607\ (L)^{0.5}$ where L is in metres. The latter is more generally used because it represents more closely the wave proportions likely to be met in deep oceans. Steeper waves

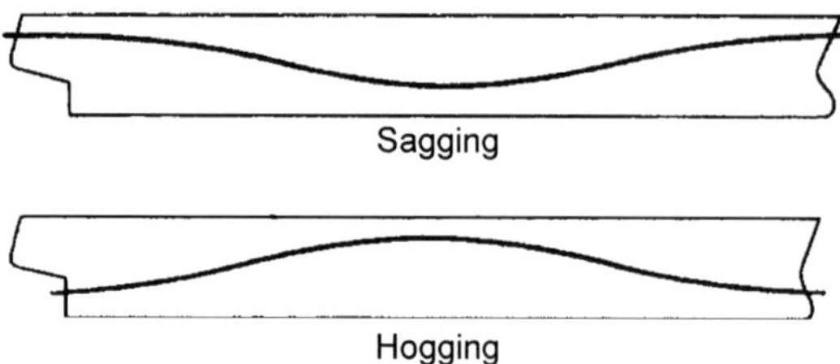

FIGURE 13.3 Ship on wave.

have been used for smaller vessels operating in areas such as the North Sea. It is a matter for the naval architect to decide in the light of the intended service areas of the ship.

Two conditions are considered (Figure 13.3) with a wave crest amidships and with wave crests at the ends of the ship. By moving the ship to various positions in relation to the wave crest the cycle of bending moment experienced by the ship can be computed. The bending moments obtained include the still water moments. It is useful to separate the two as, whilst the still water bending moment depends upon the mass distribution besides the buoyancy distribution, the bending moment due to the waves themselves depends only on the geometry of the ship and wave.

Balancing the ship on the wave is not easy and involves a number of successive approximations to the ship's attitude before equilibrium is achieved. One method of facilitating the process was proposed by Muckle (1954). Once a balance has been obtained the buoyancy and mass distribution curves follow and the SF and BM calculated as for the vessel in still water.

Shearing Force and Bending Moment Curves

Typical curves are shown in Figure 13.4. Both shearing force and bending moment must be zero at the ends of the ship. The shearing force rises to a maximum value at points about a quarter of the length from the ends and is zero near amidships. The bending moment curve rises to a maximum at the point where the shearing force is zero, and has points of inflexion where the shearing force has a maximum or minimum value.

The influence of the still water bending moment on the total moment is shown in Figure 13.5. If the still water moment is changed by varying the mass distribution the total moment alters by the same amount. Whether the greater bending moment occurs in sagging or hogging depends on the type of ship and, inter alia, upon its block coefficient. At low block coefficients the sagging bending moment is likely to be greater than the hogging, the difference reducing as block coefficient increases.

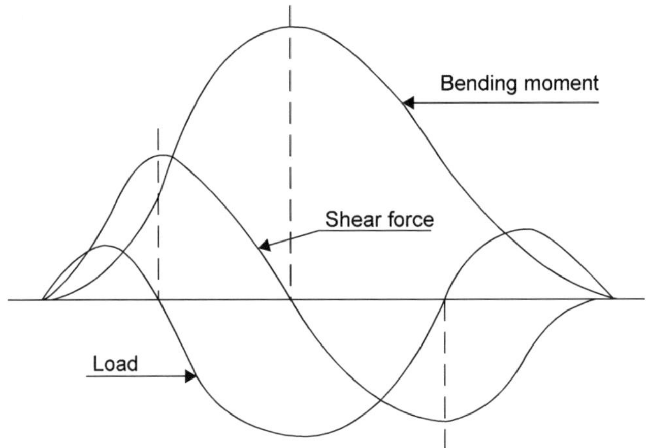

FIGURE 13.4 Shearing force and bending moment.

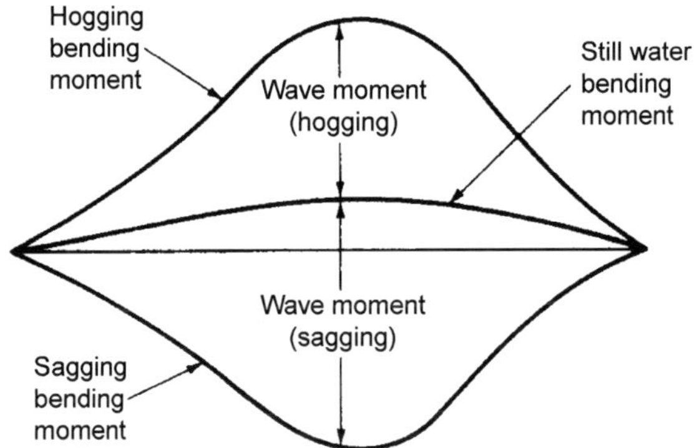

FIGURE 13.5 Still water and wave bending moments.

Influence Lines

The ship will not often be in the condition assumed in the standard calculation. For small weight changes, influence lines can be used to show the effect on the maximum bending moment due to a unit weight added at any point along the length. Lines are drawn for the hogging and sagging conditions. The lines are found by taking a unit weight at some point along the length and calculating the parallel sinkage and trim this causes to the ship balanced on the wave. It can be shown that if the maximum bending moment occurs at k and the centre of flotation is at f aft of amidships respectively, the increase in maximum bending moment per unit weight can be represented by two straight lines as in Figure 13.6 in which $E = I_a/I$ and $F = M_a/A$, and

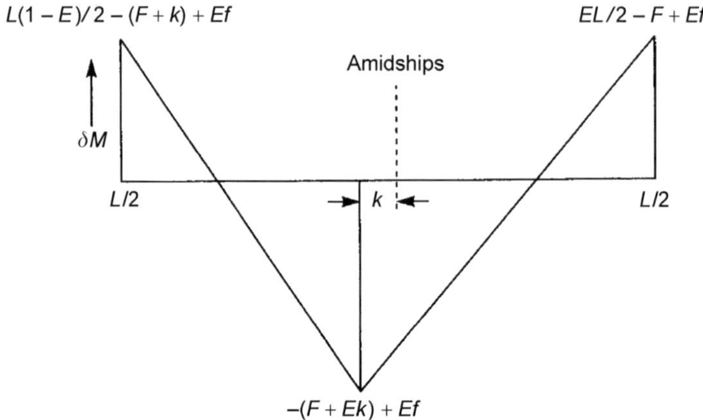

FIGURE 13.6 Influence lines.

M_a and I_a are the first and second moments of the area of the waterplane about an axis at the point of maximum bending moment. *A* and *I* are the area and least second moment of area of the complete waterplane.

In this approximation the centre of flotation and the point of maximum bending moment are assumed to be close, and *k* and *f* are positive if aft of amidships. The ordinate at any point represents the increase in the maximum bending moment if a small unit weight is added at that point.

The Dynamic Situation

Considering a ship balanced on the crest, or in the trough, of a wave is an artificial approach although it has served the naval architect well over many years. In reality the ship in waves will be subject to constantly changing forces. Also the accelerations of the motions will cause dynamic forces on the masses comprising the ship and its contents. These factors must be taken into account in a dynamic analysis of longitudinal strength. The naval architect can extend the programs for predicting ship motion to give the forces acting on the ship. Such calculations have been compared with data from model experiments and full-scale trials and found to correlate quite well.

The *strip theory* for calculating ship motions was outlined in the chapter on seakeeping. The ship is divided into a number of transverse sections, or strips, and the wave, buoyancy and inertia forces acting on each section are assessed allowing for added mass and damping. From the equations so derived the motions of the ship, as a rigid body, can be determined. The same process can be extended to deduce the bending moments and shear forces acting on the ship at any point along its length. This provides the starting point for modern treatments of longitudinal strength using CFD methods to enable non-linearities to be dealt with and to avoid the simplifying assumption of a rigid body. The ship is treated as an elastic body which distorts under the imposed forces and those distortions modify the flow around, and the forces on, the hull.

As with the motions, the bending moments and shear forces in an irregular sea can be regarded as the sum of the bending moments and shear forces due to each of the regular components making up that irregular sea. The bending moments and shear forces can be represented by *response amplitude operators (RAOs)* and energy spectra derived in ways analogous to those used for the motion responses. From these the root mean square, and other statistical properties, of the bending moments and shear forces can be obtained. By assessing the various sea conditions the ship is likely to meet on a voyage, or over its lifetime, the history of its loading can be deduced.

The RAOs can also be obtained from model experiments using a segmented model which is run in waves. The bending moments and shear forces are derived from measurements taken on balances joining the sections. The dynamic properties of the model must replicate those of the ship (See ITTC 7.5-02-07-02.1). Except in extreme conditions the forces acting on the model in regular waves are found to be proportional to wave height. This confirms the general validity of the linear superposition approach to forces in irregular seas.

Results can be obtained for a range of ship speeds, the tests being done in regular waves of various lengths or in irregular waves. The merits of different testing methods were discussed under seakeeping, as was obtaining the encounter spectrum for the seaway.

The process by which the pattern of bending moments the ship is likely to experience, is illustrated in Worked Example 13.1. The RAOs may have been calculated or derived from experiment.

Worked Example 13.1

Bending moment response operators for a range of encounter frequencies are:

RAO MN	0	103	120	106	95	77	64
ω_e rad/s	0	0.4	0.8	1.2	1.6	2.0	2.4

A sea spectrum, adjusted to represent the average conditions over the ship life, is defined by:

ω_e	0	0.4	0.8	1.2	1.6	2.0	2.4
Spectrum ordinate, m^2/s	0	0.106	0.325	0.300	0.145	0.060	0

The bending moments are the sum of the hogging and sagging moments, the hogging moment represented by 60% of the total. The ship spends 300 days at sea each year and has a life of 25 years. The average period of encounter during its life is 6 s. Calculate the value of the bending moment that is only likely to be exceeded once in the life of the ship.

Solution

The bending moment spectrum can be found by multiplying the wave spectrum ordinate by the square of the appropriate RAO. For the overall response the area under the spectrum is needed. This is best done in tabular form using Simpson's First Rule.

In Table 13.1 $E(\omega_e)$ is the ordinate of the bending moment spectrum. The total area under the spectrum is given by:

$$\text{Area} = \frac{0.4}{3} 31,381.6 = 4184.2\ \text{MN}^2\text{m}^2/s^2$$

The total number of stress cycles during the ship's life:

$$= \frac{3600 \times 24 \times 300 \times 25}{6} = 1.08 \times 10^8$$

Assuming the bending moment follows a Rayleigh distribution, the probability that it will exceed some value M_e is given by:

$$\exp - \frac{M_e^2}{2a}$$

where $2a$ is the area under the spectrum.

Table 13.1 Bending Moment Calculation

ω_e	$S(\omega_e)$	*RAO*	$(RAO)^2$	$E(\omega_e)$	Simpson's Multiplier	Product
0	0	0	0	0	1	0
0.4	0.106	103	10,609	1124.6	4	4498.4
0.8	0.325	120	14,400	4680.0	2	9360.0
1.2	0.300	106	11,236	3370.8	4	13,483.2
1.6	0.145	95	9025	1308.6	2	2617.2
2.0	0.060	77	5929	355.7	4	1422.8
2.4	0	64	4096	0	1	0
					Summation	31,381.6

In this case it is desired to find the value of bending moment that is only likely to be exceeded once in 1.08×10^8 cycles, that is its probability is $(1/1.08) \times 10^{-8} = 0.926 \times 10^{-8}$. Thus M_e is given by:

$$0.926 \times 10^{-8} = \exp \frac{M_e^2}{4184.2}$$

Taking natural logarithms both sides of the equation:

$$-18.5 = \frac{-M_e^2}{4184.2} \text{ giving } M_e = 278 \text{ MN m}$$

The hogging moment will be the greater component at 60%. Hence the hogging moment that is only likely to be exceeded once in the ship's life is 167 MN m.

Response of the Structure

When using more advanced approaches the response of the structure will be obtained using FEMs. Here we consider a simple beam approach to illustrate the principles involved. Having determined the shear forces and bending moments it is necessary to find the stresses in the structure and the overall deflection of the hull. For a beam in which the bending moment at some point x from one end is M, the stress f at z from the neutral axis of the section is given by:

$$f = \frac{Mz}{I}$$

where I is the second moment of area about the neutral axis of the section at x. If *Z is the maximum value of* z, I/Z is known as the *section modulus*.

The maximum stresses will occur when z is a maximum, that is at the top and bottom of the section. This relationship was derived for beams subject to pure bending and in which plane sections remained plane. Although a ship's structure is much more complex, applying the simple formula has been found to give reasonable results.

SECTION INERTIA AND MODULUS

A value for the section inertia, and hence modulus, at any point along the length of the ship, is needed to convert bending moments into stresses. The first section to be considered is that for amidships as it is in that area that the maximum bending moments are likely. Two cases have to be considered. The first is when all the material is the same. The second is when different materials are present in the section.

To contribute effectively to the section inertia, material must be continuous for a reasonable length fore and aft. Typically the members concerned are the side and bottom plating, keel, deck plating, longitudinal deck and shell stiffeners and any longitudinal bulkheads. The structure must not be such that it is likely to buckle under load and fail to take its fair share of the load. Because wide panels of thin plating are liable to shirk their load it is usual to limit the width of the plating contributing to 70 times its thickness if the stiffener spacing is greater than this. Having decided which structural elements will contribute, the section inertia/modulus is calculated in a tabular form.

An *assumed neutral axis* (ANA) is chosen at a convenient height above the keel. The area of each element above and below the ANA, the first and second moments about the ANA and the second moments about each element's own centroid are calculated. The differences of the first moments divided by the total area gives the distance of the true NA from the ANA. The second moments of area give the moment of inertia about the ANA and this can be corrected for the position of the true NA. This is illustrated in Table 13.2 for the diagrammatic cross section shown in Figure 13.7, noting that the second moments of thin horizontal members about their own centroids will be negligible.

It will be noted that material in the centre of the top two decks is not included. This is to compensate for the large hatch openings in these decks. Because of the ship's symmetry about its middle line, it is adequate to carry out the calculation for one side of the ship and then double the resulting answer. In this example the ANA has been taken at the keel.

TABLE 13.2 Calculation of Properties of Simplified Section in Figure 13.7

Item	Scantlings	Area (m^2)	Lever about Keel (m)	Moment about Keel (m^3)	Second Moment (m^4)	Second Moment about Own Centroid (m^4)
Upper deck	6×0.022	0.132	13	1.716	22.308	0
Second deck	6×0.016	0.096	10	0.960	9.600	0
Side shell	13×0.014	0.182	6.5	1.183	7.690	2.563
Tank top	10×0.018	0.180	1.5	0.270	0.405	0
Bottom shell	10×0.020	0.200	0	0	0	0
Centre girder	1.5×0.006	0.009	0.75	0.007	0.005	0.002
Summations		0.799		4.136	40.008	2.565

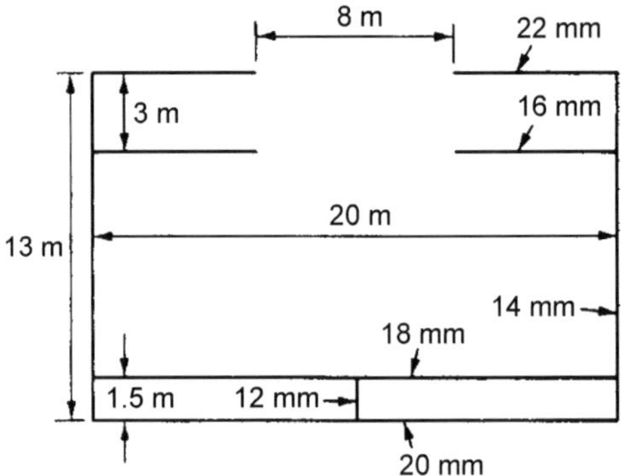

FIGURE 13.7 Diagrammatic ship cross section.

From Table 13.2, the height of the NA above the keel = 4.136/0.799 = 5.18 m
Second moment of area of half section about keel = 40.008 + 2.565 = 42.573 m^4
and about the actual NA = 42.573 − 0.799 (5.18)^2 = 21.163 m^4

For the whole section the section modulus, Z, values are:

$$Z_{\text{deck}} = \frac{42.326}{7.82} = 5.41 \ \text{m}^3$$

$$Z_{\text{keel}} = \frac{42.326}{5.18} = 8.17 \ \text{m}^3$$

If the bending moment for the structure is known, the stresses in the deck and keel can be found. If it was desired to increase the section modulus to reduce the stresses, the best place to add material would be in the keel or upper deck, whichever had the higher stress. That is, to add material as far from the NA as possible. However, the consequential change in the lower of the two original stresses must be watched.

Sections with Two Materials

Some ships have a strength cross section composed of two different materials; the hull may be steel and the superstructure aluminium. Other materials might be wood or reinforced plastic. In such cases it is convenient to think in terms of an effective modulus in just one of the materials. Usually this would be in terms of steel.

The stress, σ, in a beam at a point z from the NA is Ez/R, where R is the radius of curvature. Provided transverse sections of the beam or ship remain plane, this relationship will hold as the extension or strain at any given z will be the same. For equilibrium of the section, the net force across it must be zero. Hence using subscripts s and a for steel and aluminium:

$$\sum(\sigma_s A_s + \sigma_a A_a) = 0 \quad \text{and} \quad \sum\left(\frac{E_s A_s z_s}{R}\right) + \left(\frac{E_a A_a z_a}{R}\right) = 0$$

that is:

$$\sum\left(A_s z_s + \frac{E_a}{E_s}\right)A_a z_a = 0$$

The corresponding bending moment is:

$$\begin{aligned} M &= \sum(\sigma_s A_s z_s + \sigma_a A_a z_a) \\ &= \frac{E_s}{R}\sum\left(A_s z_s^2 + \frac{E_a}{E_s} A_a z_a^2\right) \\ &= \frac{E_s I_E}{R} \end{aligned}$$

where I_E is the effective second moment of area.

The composite cross section can therefore be considered made up of material, s, if an effective area of material a is used in place of the actual area. The effective area is the actual area multiplied by the ratio E_a/E_s. For different steels the ratio is effectively unity, for aluminium alloy/steel it is about $\frac{1}{3}$ and for GRP/steel it is about $\frac{1}{10}$.

Changes to Section Modulus

It is often desirable to change the section modulus in the early design stages. Consider the addition of an element of structure above the neutral axis, but below the upper deck, as in Figure 13.8. Assume the element is of area a and that the original section had area A, and radius of gyration k. With the addition of a the NA is raised by $az/(A + a) = \delta z$.

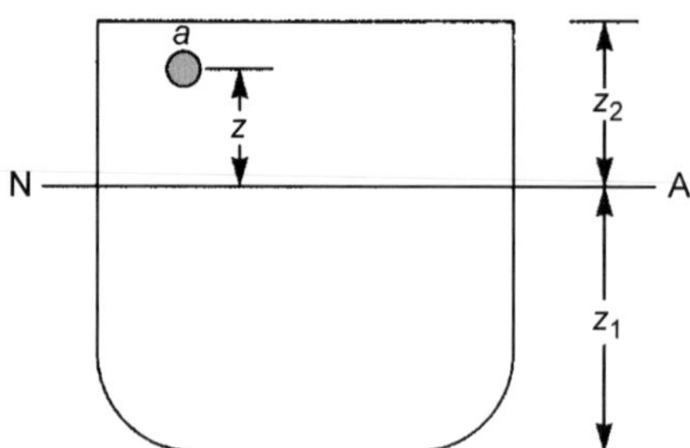

FIGURE 13.8 Changing section modulus.

The new second moment of inertia is:

$$I + \delta I = I + az^2 - (A + a)\left(\frac{az}{A+a}\right)^2 \quad \text{and} \quad \delta I = \frac{Aaz^2}{A + a}$$

For a given bending moment, the stresses will not increase if the section modulus is not reduced. This condition is that:

$$\frac{I + \delta I}{z + \delta z} - \frac{I}{z} \text{ is greater than zero, that is } \frac{\delta I}{I} \text{ greater than } \frac{\delta z}{z}$$

At the deck, as depicted, δI is positive and $\delta z/z_2$ is negative so $\delta I/I$ is always greater than $\delta z/z$ provided the material is added within the section. At the keel the condition becomes:

$$\frac{Aaz^2}{Ak^2(A + a)} \text{ must be greater than } az/z_1(A + a) \text{ or } z \text{ must be greater than } k^2/z_1$$

Thus to achieve a reduction in keel stress the material must be added at a height greater than k^2/z_1 from the neutral axis.

Corresponding relationships can be worked out for material added below the neutral axis. If the new material is added above the main deck then the maximum stress will occur in it rather than in the main deck.

SUPERSTRUCTURES

Superstructures and deckhouses can contribute to the longitudinal strength but will not be fully efficient in so doing. They should not be ignored since, whilst 'playing safe' in calculating the main hull strength, the superstructure itself might not be strong enough to take the loads imposed on it at sea. Superstructures are potential sources of stress concentrations, particularly at their ends, which should be above a transverse bulkhead and not close to highly stressed areas.

A superstructure is joined to the main hull at its lower boundary. As the ship sags or hogs this boundary becomes compressed and extended respectively. Thus the superstructure tends to be arched in the opposite sense to the main hull. If the two structures are not to separate, there will be shear forces due to the stretch or compression and normal forces trying to keep the two in contact.

The ability of the superstructure to accept these forces, and contribute to the section modulus for longitudinal bending, is regarded as an efficiency. It is expressed as:

$$\text{Superstructure efficiency} = \frac{\sigma_0 - \sigma_a}{\sigma_0 - \sigma}$$

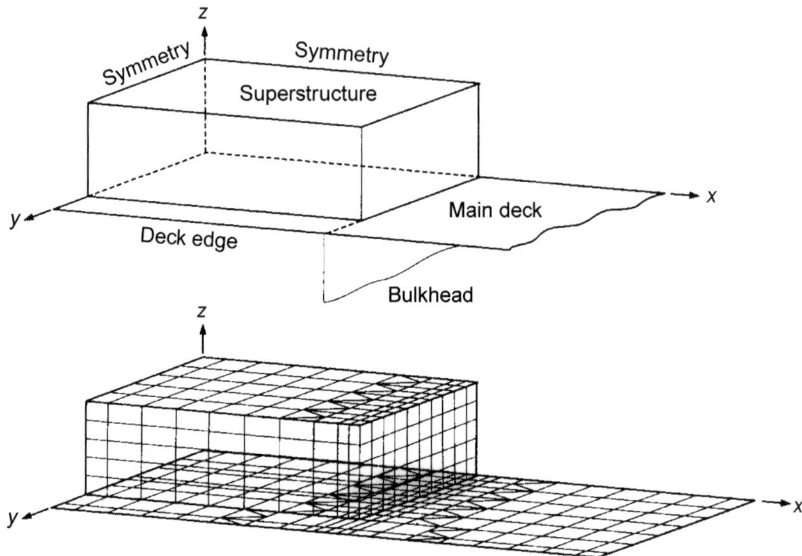

FIGURE 13.9 Superstructure mesh. *Courtesy of RINA.*

where σ_0, σ_a and σ are the upper deck stresses if no superstructure were present, the stress calculated and that for a fully effective superstructure respectively.

The efficiency of superstructures can be increased by making them long, extending them the full width of the hull, keeping their section reasonably constant and paying careful attention to the securings to the main hull. Using a low modulus material for the superstructure such as GRP, can ease the interaction problems. With a Young's modulus of the order of $\frac{1}{10}$ of that of steel, a GRP superstructure makes little contribution to the longitudinal strength. In the past expansion joints have been used at points along the length of the superstructure to stop it taking load. Unfortunately this introduces a source of potential stress concentration and is now avoided.

An FEA can be carried out to ensure the stresses are acceptable where the ends joined the main hull. A typical mesh is shown in Figure 13.9.

Worked Example 13.2

The midship section of a steel ship has the following particulars:

Cross-sectional area of longitudinal material = 2.3 m^2

Distance from neutral axis to upper deck = 7.6 m

Second moment of area about the neutral axis = 58 m^4

A superstructure deck is to be added 2.6 m above the upper deck. This deck is 13 m wide, 12 mm thick and is constructed of aluminium alloy. If the ship must withstand a sagging bending moment of 450 MNm, calculate the superstructure efficiency if, with the superstructure deck fitted, the stress in the upper deck is measured as 55 MN/m^2.

Solution:

Since this is a composite structure, the second moment of an equivalent steel section must be found first. The stress in the steel sections can then be found and, after the use of the modular ratio, the stress in the aluminium.

Taking the Young's modulus of aluminium as 0.322 that of steel, the effective steel area of the new section is:

$$2.3 + (13 \times 0.012)0.322 = 2.35 \text{ m}^2$$

The movement upwards of the neutral axis due to adding the deck:

$$(13 \times 0.012)(7.6 + 2.6)\frac{0.322}{2.35} = 0.218 \text{ m}$$

The second movement of the new section about the old NA is:

$$58 + 0.322(13 \times 0.012)(7.6 + 2.6)^2 = 63.23 \text{ m}^4$$

The second moment about the new NA is $= 63.23 - 2.35(0.218)^2 = 63.12 \text{ m}^4$
The distance to the new deck from the new NA $= 7.6 + 2.6 - 0.218 = 9.98$ m
Stress in new deck (as effective steel) $= 450 \times 9.98/63.12 = 71.15 \text{ MN/m}^2$
Stress in deck as aluminium $= 0.322 \times 71.15 = 22.91 \text{ MN/m}^2$

The superstructure efficiency relates to the effect of the superstructure on the stress in the upper deck of the main hull. The new stress in that deck, with the superstructure in place, is given as 55 MN/m^2. If the superstructure had been fully effective it would have been:

$$\frac{450(7.6 - 0.218)}{63.12} = 52.63 \text{ MN/m}^2$$

$$\text{With no superstructure the stress was} = \frac{450 \times 7.6}{58} = 58.97 \text{ MN/m}^2$$

$$\text{Hence the superstructure efficiency} = \frac{58.97 - 55}{58.97 - 52.63} = 62.6\%$$

STANDARD CALCULATION RESULTS

Bending Stresses

Due to the arbitrary nature of the standard strength calculation any stresses derived are not the stresses one would expect to measure on a ship at sea. By comparison with previously successful designs, certain values of the derived stresses have been established as acceptable. Because the comparison is made with other ships, the stress levels are often expressed in terms of the ship's principal dimensions. Two formulae which although superficially quite different yield similar stresses are:

Acceptable stress $= 77.2(L/304.8 + 1)$ MN/m^2 with L in metres
Acceptable stress $= 23(L)^{1/3}$ MN/m^2

Until 1960 the classification societies used tables of dimensions to define the structure of merchant ships, so controlling indirectly their longitudinal strength. Vessels falling outside the rules could use formulae such as the above in conjunction with the standard calculation but would need approval for this. The societies then changed to defining the applied load and structural resistance by formulae. Although stress levels as such are not defined they are implied. In the 1990s the major societies agreed, under the IACS, a common standard for longitudinal strength (Unified Requirement Section [URS] S11). This is based on the principle that there is a very remote probability that the load will exceed the strength over the ship's lifetime.

The still water loading, shear force and bending moment are calculated by the simple methods already described. To these are added the wave induced shear force and bending moments represented by the formulae:

$$\text{Hogging BM} = 0.19MCL^2BC_b \;\; \text{kNm}$$

$$\text{Sagging BM} = -0.11MCL^2B(C_b + 0.7) \;\; \text{kNm}$$

where dimensions are in metres and:

$$C_b \geq 0.6$$

$$\text{and } C = 10.75 - \left(\frac{300-L}{100}\right)^{1.5} \quad \text{for} \quad 90 \leq L \leq 300 \text{ m}$$

$$= 10.75 \quad \text{for} \quad 300 < L < 350 \text{ m}$$

$$= 10.75 - \left(\frac{L-350}{150}\right)^{1.5} \quad \text{for} \quad 350 \leq L \leq 500 \text{ m}$$

M is a distribution factor along the length. It is taken as unity between $0.4L$ and $0.65L$ from the stern; as $2.5x/L$ at x metres from the stern to $0.4L$ and as $1.0 - (x - 0.65L)/0.35L$ at x metres from the stern between $0.65L$ and L.

The IACS propose taking the wave induced shear force as:

$$\text{Positive SF} = 0.3F_1CLB\,(C_b + 0.7) \text{ kN}$$

$$\text{Negative SF} = -0.3F_2CLB(C_b + 0.7)\text{kN}$$

F_1 and F_2 vary along the length of the ship. If $A = 190C_b/[110\ (C_b + 0.7)]$, then moving from the stern forward in accordance with:

Length	Distance from stern				
	0	**0.2 − 0.3**	**0.4 − 0.6**	**0.7 − 0.85**	**1**
F_1	0	0.92A	0.7	1.0	0
F_2	0	0.92	0.7	A	0

Between the values quoted the variation is linear.

The formulae apply to a wide range of ships but special steps are needed when a new vessel falls outside this range or has unusual design features that might affect longitudinal strength.

The situation is kept under constant review and as more advanced computer analyses become available, as outlined later, they are adopted by the classification societies. Because they cooperate through IACS the classification societies' rules and their application are similar although they do vary in detail and should be consulted for the latest requirements when a design is being produced.

STRENGTH OF INDIVIDUAL STRUCTURAL ELEMENTS

In deciding which structure to include in the section modulus care is necessary to ensure that the elements chosen can in fact contribute and will not 'shirk' their share of the load. The basic structural element is a plate with some form of edge support. Combining the plates and their supporting members leads to grillages. Bulkheads, decks and shell are built up from grillages. Most of the key elements are subject to cyclic loading, at times being in tension and at others in compression. Whilst a structure may be more than adequate to take the direct stresses involved, premature failure can occur through buckling in compression due to instability. This may be aggravated by initial deformations and by lateral pressure on the plating as occurs in the shell and boundaries of tanks containing liquids.

Buckling

A structure subject to axial compression will be able to withstand loading up to a *critical load* below which buckling will not occur. Above this load a lateral deflection occurs and collapse will eventually follow. Euler showed that for an ideally straight column, with ends free to rotate, the critical load is:

$$P_{cr} = \frac{\pi^2 EI}{l^2}$$

where

l = column length
I = second moment of area of the cross section

This formula assumes the ends of the column are pin jointed. The critical stress follows as:

$$p_{cr} = \frac{\pi^2 EI}{Al^2} = \frac{\pi^2 E}{(l/k)^2}$$

where k is the radius of gyration.

If the ends of the strut are prevented from rotating, the critical load and stress are increased fourfold. The ratio l/k is called the *slenderness ratio*. For a strip of plating between supporting members, k will be proportional to the plate thickness. Thus the slenderness ratio can be expressed as the ratio of the plate span to thickness.

When a panel of plating is supported on its four edges, the support along the edges parallel to the load application has a marked influence on the buckling stress. For a long, longitudinally stiffened panel, breadth b and thickness t, the buckling stress is approximately:

$$\frac{\pi^2 Et^2}{3(1 - \nu^2)b^2}$$

where ν is the Poisson's ratio for the material.

For a broad panel, length S, with transverse stiffening, the buckling stress is:

$$\frac{\pi^2 Et^2 \left[1 + \left(\frac{S}{b}\right)^2\right]^2}{12(1 - \nu^2)S^2}$$

The ratio of the buckling stresses in the two cases, for plates of equal thickness and the same stiffener spacing is: $4[1 + (S/b)^2]^{-2}$

Assuming the transversely stiffened panel has a breadth five times its length, this ratio becomes 3.69. Thus the critical buckling stress in a longitudinally stiffened panel is almost four times that of the transversely stiffened panel, demonstrating the advantage of longitudinal stiffening.

The above formulae assume initially straight members, axially loaded. In practice there is likely to be some initial curvature. Whilst not affecting the elastic buckling stress this increases the stress in the member due to the bending moment imposed. The total stress on the concave side may reach yield before instability occurs. On unloading there will be a permanent set. Practical formulae attempt to allow for this and one is the Rankine–Gordon formula. This gives the buckling load on a column as:

$$\frac{f_c A}{1 + C(l/k)^2}$$

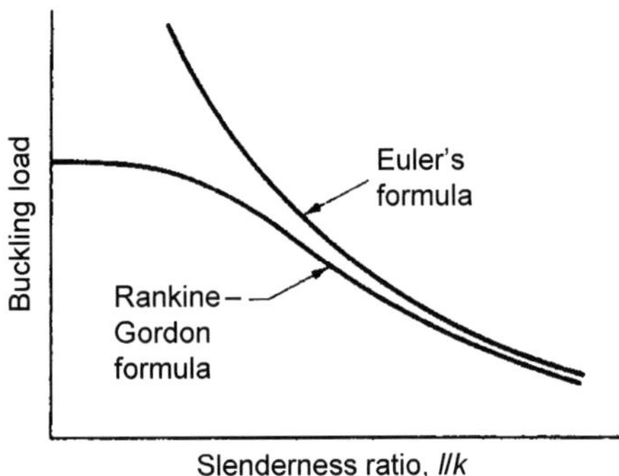

FIGURE 13.10 Comparison of strut formulae.

where

f_c and C are constants depending on the material; C depends upon the fixing conditions
A is the cross-sectional area
l/k is the slenderness ratio

The Euler and Rankine–Gordon formulae are compared in Figure 13.10. At high slenderness ratio the two give similar results. At low slenderness ratios failure due to yielding in compression occurs first.

In considering the buckling strength of grillages the strength of the stiffening members must be taken into account besides that of the plating. The stiffening members must also be designed so that they do not trip. Tripping is the torsional collapse of the member when under lateral load. Tripping is most likely in asymmetrical sections where the free flange is in compression. Small brackets can be fitted to support the free flange and so reduce the risk.

Worked Example 13.3

In Worked Example 13.2 on the aluminium superstructure determine whether a transverse beam spacing of 730 mm would be adequate to resist buckling.

Solution

Treating the new transversely stiffened deck as a broad panel and applying Euler's equation for a strut, its buckling stress is given by the formula:

$$\frac{\pi^2 Et^2[1+(S/b)^2]^2}{[12(1-\nu^2)]S^2}$$

Taking Poisson's ratio, v, as 0.33 the critical stress is:

$$\pi^2 \times 67\,000 \times (0.012)^2 \frac{[1+(0.73/13)^2]^2}{12[1-(0.33)^2](0.73)^2} = 17.79 \text{ MN/m}^2$$

Since the stress in the aluminium deck is 22.91 MN/m^2 this deck would fail by buckling. The transverse beam spacing would have to be reduced to about 620 mm to prevent this.

These relationships indicate the key physical parameters involved in buckling but do not go very far in providing solutions to ship-type problems.

Shear Stresses

Turning now to the shear stresses generated in the hull, the simple formula for shear stress in a beam at a point distant y from the neutral axis is:

$$\text{Shear stress} = FA\bar{y}/It$$

where

F = shear force
A = cross-sectional area above y from the NA of bending
$\bar{y}$ = distance of centroid of A from the NA
I = second moment of complete section about the NA
t = thickness of section at y

The distribution of shear stress over the depth of an I-beam section is illustrated in Figure 13.11. The stress is greatest at the neutral axis and zero at the top and bottom of the section. The vertical web takes by far the greatest load, typically in this type of section over 90%. The flanges, which take most of the bending load, carry very little shear stress.

In a ship in waves the maximum shear forces occur at about a quarter of the length from the two ends. In still water large shear forces can occur at other positions depending upon the way the ship is loaded. As with the I-beam it will be the vertical elements of the ship's structure that will take the majority of the shear load. The distribution between the various elements,

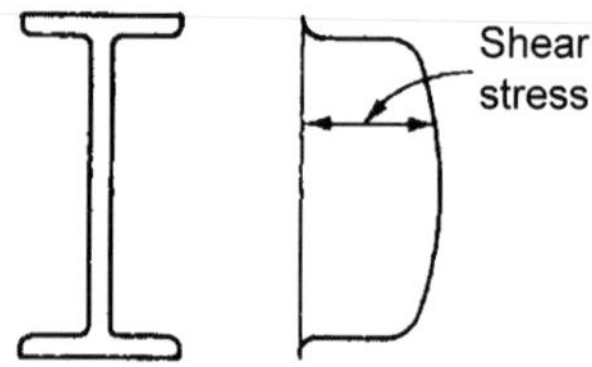

FIGURE 13.11 Shear stress.

the shell and longitudinal bulkheads say, is not so easy to assess. The overall effects of the shear loading are to:

- distort the sections so that plane sections no longer remain plane. This will affect the distribution of bending stresses across the section. Generally the effect is to increase the bending stress at the corners of the deck and at the turn of bilge with reductions at the centre of the deck and bottom structures. The effect is greatest when the hull length is relatively small compared to hull depth.
- increase the deflection of the structure above that which would be experienced under bending alone. This effect can be significant in vibration.

Hull Deflection

Consider first the deflection resulting from the bending of the hull. From beam theory:

$$\frac{M}{I} = \frac{E}{R}$$

where R is the radius of curvature.

If y is the deflection of the ship at any point x along the length, measured from a line joining the two ends of the hull, it can be shown that:

$$R = \frac{-[1+(dy/dx)^2]^{1.5}}{\mathrm{d}^2y/dx^2}$$

For the ship only relatively small deflections are involved and $(\mathrm{d}y/\mathrm{d}x)^2$ will be small and can be ignored in this expression. Thus:

$$-\frac{\mathrm{d}^2y}{\mathrm{d}x^2} = \frac{1}{R} \quad \text{and} \quad M = -EI\frac{\mathrm{d}^2y}{\mathrm{d}x^2}$$

The deflection can be written as:

$$y = \iint \frac{M}{EI}\mathrm{d}x\,\mathrm{d}x + \mathrm{A}x + \mathrm{B}$$

where A and B are constants.

In a simple approach a designer can calculate the value of I at various positions along the length and evaluate the double integral by approximate integration methods.

Since the deflection is, by definition, zero at both ends B must be zero. Then:

$$\mathrm{A} = \frac{1}{L}\iint_L \frac{M}{EI}\mathrm{d}x\,\mathrm{d}x \quad \text{and} \quad y = -\iint \frac{M}{EI}\mathrm{d}x\,\mathrm{d}x + \frac{x}{L}\iint_L \frac{M}{EI}\mathrm{d}x\,\mathrm{d}x$$

The shear deflection is more difficult to calculate. An approximation can be obtained by assuming the shear stress uniformly distributed over the 'web' of the section. If, then, the area of the web is A_w, then:

$$\text{Shear stress} = \frac{F}{A_w}$$

If the shear deflection over a short length, dx, is:

$$dy = \frac{F}{A_w C} dx$$

where C is the shear modulus.

The shear deflection can be obtained by integration. The ratio of the shear to bending deflections varies as the square of the ship's depth to length ratio and would be typically between 0.1 and 0.2.

TRANSVERSE STRENGTH

The loads on a transverse section of the ship in waves are those calculated from the motions of the ship including the inertia and gravity forces although in this case it is their transverse distribution that is of interest. Also there may be forces generated by the movement of liquids within tanks, *sloshing* as it is termed. In addition to the dynamic loading in a seaway the section must be able to withstand the loads at the waterline due to berthing and the racking strains imposed during docking.

The best approach is to analyse the three dimensional section of the ship between main transverse bulkheads as a whole, having ascertained the boundary conditions from a global FEA of the complete hull.

For berthing loads it may be adequate to isolate a grillage in way of the waterline and assess the stresses in it due to the loads on fenders in coming alongside. However, the difficulty of assessing the end fixities of the various members due to the presence of the others, and the influence of longitudinal stiffening, make it unreasonable to deal with side frames, decks and double bottom separately. These influences are likely to be critical. For instance, a uniformly loaded beam, simply supported at its ends, has a maximum bending moment at its centre with zero moments at its ends. If the ends are fixed the maximum bending moment reduces by a third and is at the ends.

The usual approximation is to take a slice through the ship comprising a deck beam, side frame and elements of plating and double bottom structure. This section is loaded and analysed as a framework. The transverse strength of a superstructure is usually analysed by the same technique. The frameworks the naval architect is concerned with are portals in the superstructure, ship-shape rings in the main hull and circular rings in submarines. Transverse bulkheads provide great strength against racking of the framework. Some of this support will be transmitted to frames remote from the bulkhead by longitudinal

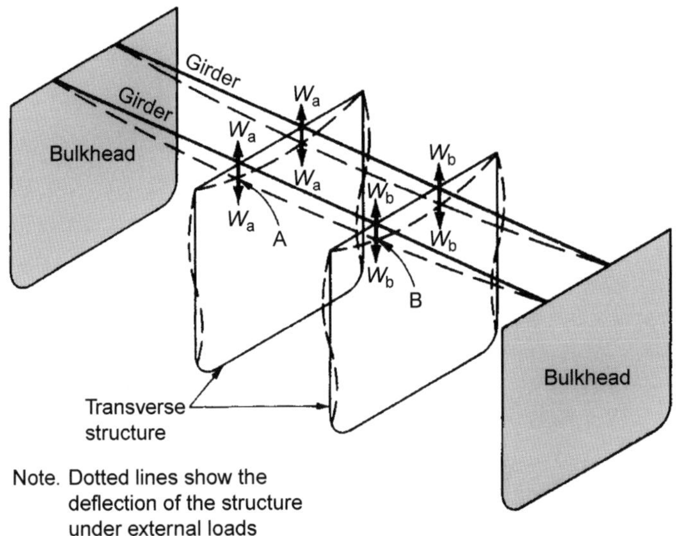

FIGURE 13.12 Transverse strain.

members although these will themselves deflect under the loading as illustrated in Figure 13.12. Ignoring this support means results are likely to be conservative and should really be used as a guide to distributing structure and for comparison with similar successful designs, rather than to obtain absolute values of stress or deflection.

It is not appropriate in this book to deal with the analysis of frameworks in detail. There are many textbooks available to which reference should be made for detailed explanations and for an understanding of all the underlying assumptions. Very briefly, however, the methods of analysis available are:

- *Energy methods* which are based on the theorem of Castigliano which postulates that the partial derivative of the total strain energy with respect to each applied load is equal to the displacement of the structure at the point of application in the direction of the load.
- *Moment distribution methods* which are iterative processes. All members of the framework are initially considered fixed rigidly and the bending moments at the joints calculated. Then one joint is relaxed by allowing it to rotate. The bending moment acting is distributed between the members forming the joint according to their inertias and lengths. Half the distributed moment is transmitted to a member's far end which is still held rigid. Joints are relaxed in turn and the process repeated until the moments are in balance.
- *Slope-deflection methods*. If M is the bending moment at some point along a beam the area under the curve of M/EI between two points on the

beam gives the change in slope between those points. Further, if the moment of the curve between the points is taken about the first point, the moment gives the perpendicular distance of the first point from the tangent at the second point. By expressing the changes in deflection at the ends of portal members in terms of the applied loads and the (unknown) moments at their ends, a series of equations are produced which can be solved to give the unknown moments.

SOME OTHER STRUCTURAL CONSIDERATIONS

Stress Concentrations

There are several reasons why local stresses may exceed considerably those in the general vicinity. The design may introduce points at which the loads in a large structural element are led into a relatively small member. It is useful in looking at a structure to consider where the load in a member can go next. If there is no natural, even, 'flow' then a concentration of stress can occur. Some such details are bound to arise at times, in way of large deck openings for instance, or where the superstructure ends. In such cases the designer must try to minimise the stress concentration. Well rounded corners to hatch openings are essential and added thickness of plating abreast the hatches reduces the stress for a given load. The magnitude of this effect can be illustrated by the case of an elliptical hole in an infinitely wide plate subject to uniform tensile stress across the width. If the long axis of the ellipse is $2a$ and the minor axis is $2b$, then with the long axis across the plate the stresses at the ends of the long axis will be augmented by a factor $[1 + (2a/b)]$. If the hole is circular this concentration factor becomes 3. There will be a compressive stress at the ends of the minor axis equal in magnitude to the tensile stress in the plate. In practice there is little advantage in giving a hatch corner a radius of more than about 15% of the hatch width. The side of the hatch should be aligned with the direction of stress otherwise there could be a further stress penalty of about 25%.

Apart from design features, stress concentrations can be introduced as the ship is built. Structural members may not be accurately aligned either side of a bulkhead or floor. This is why important members are made continuous and less important members are made *intercostal*, i.e. they are cut and secured either side of a continuous member. Other concentrations are occasioned by defects in the welding and other forming processes. Provided the size of these defects is not large, local redistribution of stresses can occur due to yielding of the material. However, large defects, found perhaps as a result of radiographic inspection, should be repaired. Important structure should not have stress concentrations increased by cutting holes in them or by welding attachments to them apart from those absolutely necessary.

Built-In Stresses

Plates and sections will already have been subject to strain before construction starts. They may have been rolled and unevenly cooled. Then in the shipyard they will be shaped and welded. As a result they will already have *residual stresses and strains* before the ship itself is subject to any load. These built-in stresses can be quite large and even exceed the yield stress locally. Built-in stresses are difficult to estimate but in frigates (Somerville et al., 1977) it was found that welding the longitudinals introduced a compressive stress of 50 MPa in the hull plating, balanced by regions local to the weld where the tensile stresses reached yield. In general the remaining strains are an unknown quantity but add to the probability of failure under extra applied loads, particularly in fatigue. Modern welding methods and greater accuracy of build geometry reduce the levels of built-in strain but they remain.

Fatigue

Fatigue is the most common mechanism leading to failure in general engineering structures (Nishida, 1994). It is very important in ships which are expected to remain in service for 20 years or more. Even when there is no initial defect present, repeated stressing of a member causes a crack to form on the surface after a certain number of cycles. This crack will propagate with continued stress repetitions. Any initial crack-like defect will propagate with stress cycling. *Crack initiation* and *crack propagation* are different in nature and need to be considered separately.

Characteristically a fatigue failure, which can occur at stress levels lower than yield, is smooth and usually stepped. If the applied stressing is of constant amplitude the fracture can be expected to occur after a defined number of cycles. Plotting the stress amplitude against the number of reversals to failure gives the traditional S–N curve for the material under test. As the number of reversals increases the acceptable applied stress decreases.

In their *Fatigue Assessment of Ship Structures (2010)*, Det Norske Veritas base fatigue design on use of S–N curves derived from fatigue tests and associated with a 97.6% probability of survival.

The S–N curves are applicable for normal and high-strength steels used in hull structures. The S–N curves for welded joints include the effect of the local weld notch. They are also defined as hot spot S–N curves. If a butt weld is machined or ground flush without weld overfill a better S–N curve can be used.

The basic design S–N curve is given as:

$$\log \mathrm{N} = \log a - \mathrm{m} \log \Delta\sigma$$

where

N = predicted number of cycles to failure for stress range ($\Delta\sigma$)
m = negative inverse slope of S–N curve
log a = intercept of log N-axis by S–N curve

Log a and m values are given below:

a. Parameters for air or with cathodic protection. (For unprotected joints fatigue life is reduced by a factor of 2.)
Welded joint: Log $a = 12.164$ and $m = 3.0$ for $N \leq 10^7$
Log $a = 15.606$ and $m = 5.0$ for $N \geq 10^7$
Base material: Log $a = 15.117$ and $m = 4.0$ for $N \leq 10^7$
Log $a = 17.146$ and $m = 5.0$ for $N \geq 10^7$

b. Parameters for base material in corrosive environment.
Base material: Log $a = 12.436$ and $m = 3.0$

Thus for the base material in a corrosive environment the S–N curve is a straight line. Plots of S–N curves for commonly occurring structural configurations are given in British Standards. The standard data refer to constant range of stressing. Under these conditions the results are not too sensitive to the mean stress level provided it is less than the elastic limit. At sea, however, a ship is subject to varying conditions. This can be treated as a spectrum for loading in the same way as motions are treated. A transfer function can be used to relate the stress range under spectrum loading to that under constant amplitude loading. Based on welded joint tests, it has been suggested (Petershagen, 1986) that the permissible stress levels, assuming 20 million cycles as typical for a merchant ship's life, can be taken as four times that from the constant amplitude tests. This should be associated with a safety factor of four thirds.

Unfortunately using high-tensile steels does not, in practical shipbuilding structures, lead to longer fatigue life. Fatigue life of a steel structure, then, is seen to be largely independent of the steel's ultimate strength but will depend upon the stress level, structural continuity, weld geometry and imperfections. IACS REC 056 (1999) covers the fatigue assessment of ships.

Cracking and Brittle Fracture

In any practical structure cracks are bound to occur. Indeed the build process makes it almost inevitable that there will be a range of crack-like defects present before the ship first goes to sea. This is not in itself a problem but significant cracks must be looked for and corrected before they can cause a failure. They can extend due to fatigue or brittle fracture mechanisms. Even in rough weather fatigue cracks generally grow only slowly. However, under certain conditions, a brittle fracture can propagate at about 500 m/s. The *MV Kurdistan* broke in two in 1979 (Corlett et al., 1988) due to brittle fracture.

The MV *Tyne Bridge* suffered a 4-m crack (Department of Transport, 1988). At one time it was thought that thin plating did not suffer brittle fracture but this has been disproved and it is vital to avoid the possibility of brittle fracture. The only way of ensuring this is to use steels which are not subject to this type of failure under service conditions encountered (Sumpter et al., 1989) and temperature is very important.

The factors governing brittle fracture are the stress level, crack length and material *toughness*. Toughness depends upon the material composition, temperature and strain rate. In structural steels failure at low temperature is by cleavage. Once a crack is initiated the energy required to cause it to propagate is so low that it can be supplied from the release of elastic energy stored in the structure and failure is very rapid. At higher temperatures fracture initiation is by growth and coalescence of voids and subsequent extension occurs only by increased load or displacement (Sumpter, 1986). The temperature for transition from one fracture mode to the other is called the *transition temperature*. It is a function of loading rate, structural thickness, notch acuity and material microstructure. The lower the transition temperature the tougher is the steel.

Unfortunately there is no simple physical test to which a material can be subjected that will determine whether it is likely to be satisfactory in terms of brittle fracture. This is because the behaviour of the structure depends upon its geometry and method of loading. The choice is between a simple test like the *Charpy test* and a more elaborate and expensive test under more representative conditions such as a *Crack Tip Opening (Displacement) test.* The Charpy test is still widely used for quality control and IACS specify 27J at −20°C for Grade D steel and 27J at −40°C for Grade E steel.

Since cracks will occur, it is necessary to use steels which have good crack arrest properties. It is recommended (Sumpter et al., 1989) that one with a crack arrest toughness of 150–200 MPa(m)$^{0.5}$ is used. To provide a high level of assurance that brittle fracture will not occur, a Charpy crystallinity of less than 70% at 0°C should be chosen. For a good crack arrest capability and virtually guaranteed fracture initiation avoidance the Charpy crystallinity at 0°C should be less than 50%. Special crack arrest strakes are provided in some designs. The steel for these should show a completely fibrous Charpy fracture at 0°C. It is not only the toughness of the steel that is important; weld deposits should at least match the toughness of the parent metal.

Statistical Recording at Sea

Many ships have been fitted with *statistical strain gauges* to record strain, most using electrical resistance gauges. They usually record the number of times the strain lies in a certain range during recording periods of 20 or 30 min. From these data histograms can be produced, curves fitted to them

TABLE 13.3 Sea Conditions

Weather Group	Beaufort Number	Sea Conditions
I	0–3	Calm or slight
II	4–5	Moderate
III	6–7	Rough
IV	8–9	Very rough
V	10–12	Extremely rough

and cumulative probability curves produced to show the likelihood that certain strain levels will be exceeded.

Strain levels are usually converted to stress values based on a knowledge of the scantlings of the structure. These are approximate, involving assumptions about the section modulus. However, if the guidelines followed are those used in designing the structure the data are valid for comparisons with predictions. Comparisons are based on statistical probabilities. It is necessary to record the sea conditions applying during the recording period. The sea conditions are recorded on a basis of visual observation related to Beaufort numbers in five groups as in Table 13.3.

For a general picture of a ship's structural loading during its life the recording periods must be random or the results may be biased. If the records are taken when the master feels the conditions are leading to significant strain the results will not reflect the many periods of relative calm a ship experiences. If they are taken at fixed time intervals during a voyage they will reflect the conditions in certain geographic areas if the ship follows the same route each time.

The data give:

- the ship's behaviour during each recording period. The values of strain, or the derived stress, are likely to follow a Rayleigh probability distribution.
- the frequency with which the ship encounters different weather conditions.
- the variation of responses in different recording periods within the same weather group.

The last two are likely to follow a Gaussian, or normal, probability distribution.

The data recorded in a ship are factual. To use them to project ahead for the same ship the data need to be interpreted in the light of the weather conditions the ship is likely to meet, based on *Ocean Wave Statistics*. For a new ship the different responses of that ship to the waves in the various weather groups are also needed. These can be derived from theory or model experiment.

TABLE 13.4 Percentage of Time Spent at Sea in Each Weather Group

	Weather Group				
	I	II	III	IV	V
General routes	51	31	14	3.5	0.5
Tanker routes	71	23	5.5	0.4	0.1

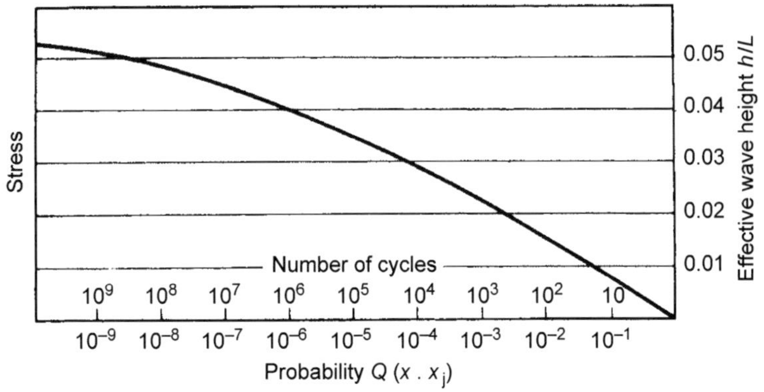

FIGURE 13.13 Probability curve.

Typically ships spend the majority of their time in relatively calm conditions. This is illustrated by Table 13.4. When the probabilities of meeting various weather conditions and of exceeding certain bending moments or shear forces in those various conditions are combined the results can be presented in a curve such as Figure 13.13. This shows the probability that the variable x (stress, shear force or bending moment) will exceed some value x in a given number of stress cycles.

The problem faced by a designer is to decide upon the level of bending moment or stress any new ship should be able to withstand. An overly strong structure will be heavier than it need be and the ship will carry less payload. If the structure is too weak the ship is likely to suffer damage. Repairs cost money and lose the ship time at sea. Ultimately the ship may be lost.

If a ship life of 25 years is assumed, and the ship is expected to spend on average 300 days at sea per year, it will spend 180,000 h at sea during its life. If its stress cycle time is t seconds it will experience:

$$180,000 \times 3600/t \text{ stress cycles.}$$

Taking a typical stress cycle time of 6 s leads to just over 10^8 cycles. If, in Figure 13.13 an ordinate is erected at this number of cycles, a stress is obtained which is likely to be exceeded once during the life of the ship. That is there is a probability of 10^{-8} that the stress will be exceeded and this probability is now commonly accepted as a reasonable design probability.

Effective Wave Height

This probabilistic approach to strength is more realistic than the standard calculation in which the ship is assumed balanced on a wave. It would be interesting though, to see how the two might roughly compare. This can be done by balancing the ship, represented by the data in Figure 13.13, on waves of varying height to length ratio, the length being equal to the ship length. The stresses so obtained can be compared with those on the curve and an ordinate scale produced of the *effective wave height*. That is, the wave height that would have to be used in the standard calculation to produce that stress. Whilst it is dangerous to generalise, the stress level corresponding to the standard $L/20$ wave is usually high enough to give a very low probability that it would be exceeded suggesting the standard calculation is conservative.

HORIZONTAL FLEXURE AND TORSION

Generally the forces which cause vertical bending also produce forces and moments causing the ship to bend in the horizontal plane and to twist about a fore and aft axis. The motions of rolling, yawing and swaying will introduce horizontal accelerations but the last two are modes in which the ship is neutrally stable. It is necessary therefore to carry out a detailed analysis of the motions and derive the bending moments and torques acting on the hull. Since these flexures will be occurring at the same time as the ship experiences vertical bending, the stresses produced can be additive. For instance the maximum vertical and horizontal bending stresses will be felt at the upper deck edges. However, the two loadings are not necessarily in phase and this must be taken into account in deriving the composite stresses.

Fortunately the horizontal bending moment maxima are typically only some 40% of the vertical ones. Due to the different section moduli for the two types of bending the horizontal stresses are only about 35% of the vertical values for typical ship forms. The differing phase relationships mean that superimposing the two only increases the deck edge stresses by about 20% over the vertical bending stresses. These figures are quoted to give some idea of the magnitude of the problem but should be regarded as very approximate.

Horizontal flexure and torsion assume greater significance for ships with large hatch openings such as container ships. It is not possible to deal with them in any simple way although their effects will be included in statistical data recorded at sea if the recorders are sited carefully.

LOAD-SHORTENING CURVES

Theoretical and experimental studies by Smith et al. (1992) show that the stiffness and strength of rectangular plate elements of an orthogonally stiffened shell are strongly influenced by imperfections and residual stresses in the structure arising from the fabrication process and initial deformations of plate and stiffener. These studies were the culmination of a large research programme involving longitudinally loaded plates with stringers (stiffeners) b apart, between transverse frames a apart. The plate thickness was t, the radius of gyration of a stringer with a width b of plating was r and the stringer area was A_s. The stress was σ and strain ε with subscript o denoting yield. Stringers used were tee bars and flat plate. The following parameters were used:

$$\text{Plate slenderness, } \beta = \frac{b}{t}\left(\frac{\sigma_o}{E}\right)^{0.5}$$

$$\text{Stringer slenderness, } \lambda = \frac{a}{r\pi}\left(\frac{\sigma_o}{E}\right)^{0.5}$$

$$\text{Stiffener area ratio} = \frac{A_s}{A} \text{ where } A = A_s + bt$$

The outcome of the research was a series of *load-shortening curves* as shown in Figure 13.14. These are for a range of stringer and plate slenderness with average imperfections. Average imperfections were defined as a residual stress 15% of yield and a maximum initial plate deflection of $0.1\beta^2$.

The results are sensitive to stiffener area ratio, particularly for low λ and high β, Figure 13.14, in which σ'_u is the ratio of the average compressive stress at failure over the plate and stiffener cross section to the yield stress. Peak stresses in Figure 13.14 correspond to the strengths indicated in Figure 13.15B. Figure 13.16 shows the influence of lateral pressure on compressive strength for the conditions of Figure 13.14. The effect is most marked for high λ and increases with β. Q is the corresponding head of sea water.

The importance of the load-shortening curves is that they allow a designer to establish how elements of the structure will behave both before and after collapse and hence the behaviour of the ship section as a whole. Even after collapse elements can still take some stress. However, from Figure 13.14 for λ equal to or greater than 0.6 the curves show a drastic reduction in strength post collapse. For that reason it is recommended that designs be based on λ values of 0.4 or less and β values of 1.5 or less.

Using such approaches leads to a much more efficient structure than would be the case if the designer did not allow the yield stress to be exceeded.

FINITE ELEMENT ANALYSIS

FEA techniques, or FEMs, are the basis of modern computer-based analysis methods in structures. These are very powerful techniques using the

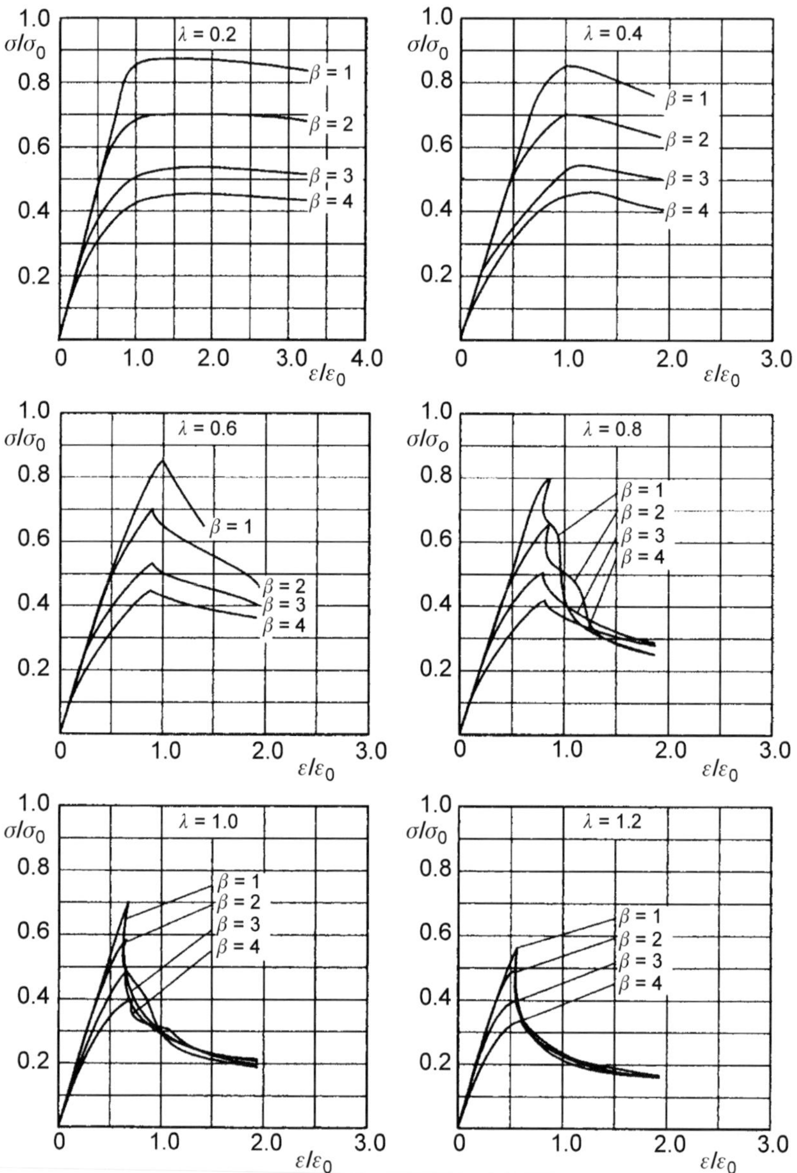

FIGURE 13.14 Load-shortening curves. *Courtesy of RINA.*

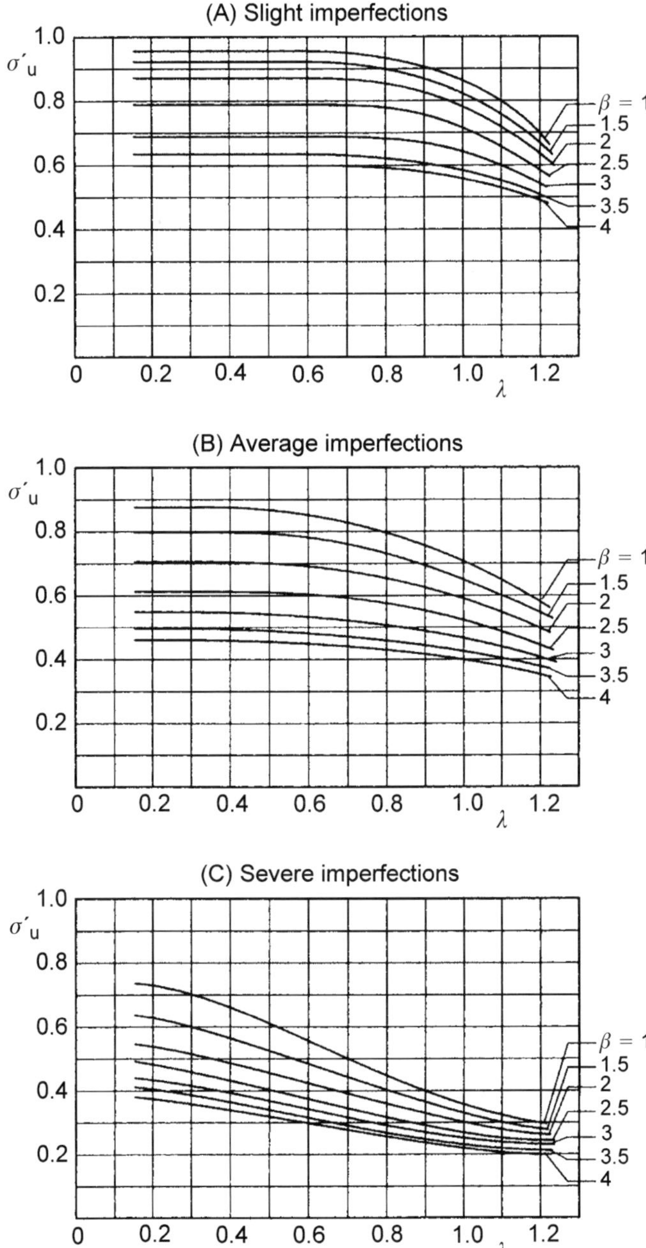

FIGURE 13.15 Compressive strength of panels. *Courtesy of RINA.*

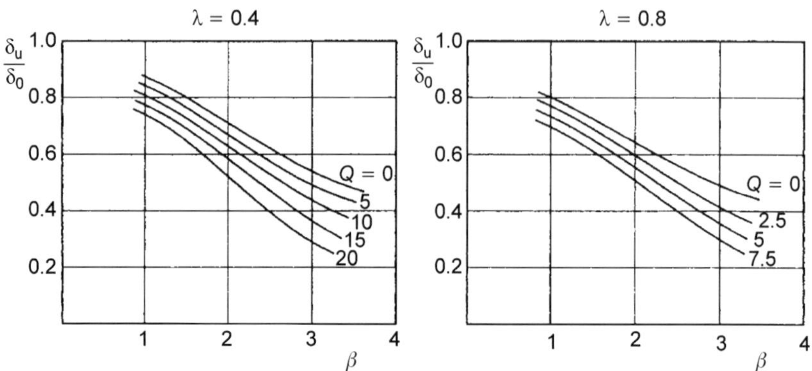

FIGURE 13.16 Influence of lateral pressure. *Courtesy of RINA.*

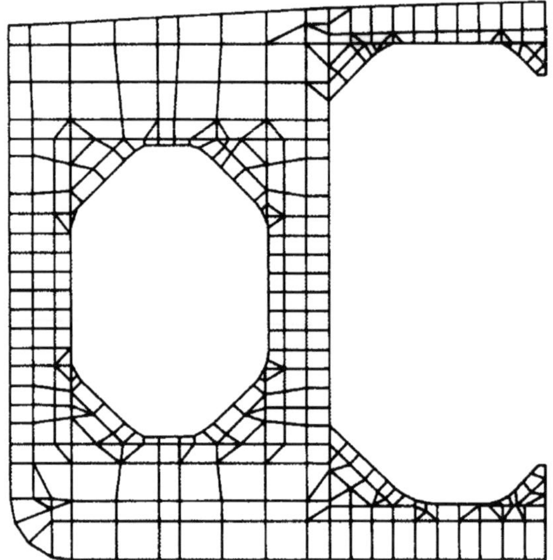

FIGURE 13.17 Transverse section elements.

mathematics of matrix algebra. It is only possible to give a simplified explanation of the principles involved. The structure is imagined to be split up into a series of elements, usually rectangular or triangular. The corners where the elements meet are called *nodes*. For each element an expression is derived for the displacement at its nodes. This gives strains and stresses. The displacements of adjoining elements are made compatible at each node and the forces related to the boundary forces. The applied loads and internal forces are arranged to be in equilibrium.

The finer the mesh the more accurately the stress pattern will be represented. In a structure such as that shown in Figure 13.17, elements of different shape and size can be used. Smaller elements would be used where it was suspected that the stresses would be highest and more variable.

The starting point in a comprehensive structural design approach would be an FEA of the complete hull using a relatively coarse mesh. The data from this global analysis would then be used to define the boundary conditions for more limited areas which would be studied using a finer mesh.

CORROSION

Corrosion Protection

The surface of all metalwork, inside and outside the ship, needs to be protected against the corrosive effects of the sea environment and of some of the cargoes carried. Most failures of marine structures are due to a combination of corrosion and fatigue. Both can be described as cumulative damage mechanisms. High-tensile steels are as liable to corrosion as mild steel. Hence when they are used to produce a lighter weight structure, corrosion can assume relatively greater significance.

Types of Corrosion

These can be classified as:

- *General corrosion*. This occurs relatively uniformly over the surface and takes place at a predictable rate.
- *Pitting*. Localised corrosion can occur under surface deposits and in crevices. Pits can act as stress raisers and initiate fatigue cracks, but the main concern with modern shipbuilding steels is penetration and subsequent pollution.
- *Differential aeration*. Debris and fouling on a surface can lead to different concentrations of oxygen which trigger local corrosion.
- *Galvanic action*. Sea water acts as an electrolyte so that electrochemical corrosion can occur. This may be between different steels or even between the same steel when subject to different amounts of working or when a partial oxide film is present. In the 'cell' that is created it is the anodic area that is eaten away. A few average values of electrical potential for different metals in sea water of 3.5% salinity and 25°C are listed in Table 13.5. If the difference exceeds about 0.25 V, significant corrosion of the metal with the higher potential can be expected.
- *Stress corrosion*. The combined action of corrosion and stress can cause accelerated deterioration of the steel and cracking. The cracks grow at a negligible rate below a certain stress intensity depending upon the metal composition and structure, the environment, temperature and strain rate. Above this threshold level the rate of crack propagation increases rapidly with stress intensity.

TABLE 13.5 Electro-Chemical Table

Material	Potential (volts)
Magnesium alloy sheet	− 1.58
Galvanised iron	− 1.06
Aluminium alloy (5% Mg)	− 0.82
Aluminium alloy extrusion	− 0.72
Mild steel	− 0.70
Brass	− 0.30
Austenitic stainless steel	− 0.25
Copper	− 0.25
Phosphor bronze	− 0.22

Protective Coatings

Painting can provide protection while the paint film is intact. If it fails in a local area serious pitting can occur. Careful preparation and immediate priming are needed. Classification societies specify a comprehensive range of protective coatings for a ship's structure depending upon the spaces concerned. Typical corrosion rates for different ship types against age of ship are presented in Figure 13.18.

Cathodic Protection

Two methods of protecting a ship's hull are commonly used under the term *cathodic protection*. The first, a passive system, uses a sacrificial anode placed near the area to be protected. Typically this might be a piece of zinc or magnesium. The corrosion is concentrated on the anode. A more effective system, an active one, is to impress a current upon the area concerned, depressing the potential to a value below any naturally anodic area. The potential is measured against a standard reference electrode in the water. Typical current densities required to be effective are 32 mA/m^2 for painted steel and 110 mA/m^2 for bare steel, but they vary with water salinity and temperature as well as the ship's speed and condition of the hull. The system can be used to protect the inner surfaces of large liquid cargo tanks.

OVERALL STRUCTURAL SAFETY

A designer must evaluate the probability of structural failure by each type and reduce it if possible. A suitable material must be chosen. In a steel ship this means a steel with adequate notch toughness in the temperatures and at the strain rates expected during service. Allowance must be made for

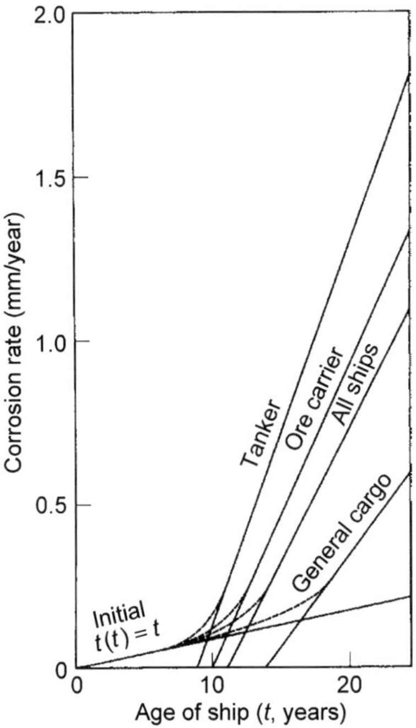

FIGURE 13.18 Corrosion rates. *Courtesy of RINA.*

residual stresses arising from the fabrication methods. Welding processes must be defined and controlled to give acceptable weld quality, to avoid undue plate distortion and defects in the weld. Openings must be arranged to reduce stress concentrations to a minimum. Allowance must be made for corrosion.

There will still be many reasons why actual stresses might differ from those calculated. There remain a number of simplifying assumptions regarding structural geometry made in the calculations. The plating will not be exactly the thickness assumed because of rolling tolerances. Material properties will not be exactly those specified. Fabrication will lead to small departures from the intended geometry. Intercostal structure will not be exactly in line either side of a bulkhead, say. Structure will become dented and damaged during service. All these introduce some uncertainty in the calculated stress values. Their effects can be assessed in statistical terms based on experience of the build and operation of ships. Rolling tolerances and metallurgical composition variation are set out in material specifications.

The loading experienced may differ from that assumed in the design. The ship may go into areas not originally planned. Weather conditions may not be as anticipated. Whilst many of these variations will average out over a ship's life it is always possible that a ship will experience some unusually severe combination of environmental conditions. It may even meet a freak wave.

Using the concept of load-shortening curves for the hull elements it is possible to determine a realistic value of the ultimate bending moment a hull can develop before it fails. The designer can combine information on the likelihood of meeting different weather conditions with its responses to those conditions, to find the loading that is likely to be exceeded only once in a ship's life. However, these values are subject to the uncertainties discussed above. Instead it is prudent to regard both loading and strength as probability distributions as in Figure 13.1 at the beginning of this chapter. In this figure load and strength must be expressed in the same way and this would usually be in terms of bending moment.

The area under the load curve to the right of point where it intersects the strength curve represents the probability that the applied load will exceed the strength available. The area under the strength curve to the left of that point represents the probability that the strength will be less than that required to withstand the load applied. The tails of the actual probability distributions of load and strength are difficult to define from recorded data unless assumptions are made as to their mathematical form. Many authorities assume that the distributions are Rayleigh or Gaussian so that the tails are defined by the mean and variance of the distributions. They can then express the safety in terms of a load factor based on the average load and strength. This may be modified by another factor representing a judgment of the consequences of failure.

Having ascertained that the structure is adequate in terms of ultimate strength, the designer must look at the fatigue strength. Again use is made of the stressing under the various weather conditions the ship is expected to meet. This will yield the number of occasions the stress can be expected to exceed certain values. Most fatigue data for steels relate to constant amplitude tests so the designer needs to be able to relate the varying loads to these standard data as was discussed earlier.

SUMMARY

The vertical bending moments and shearing forces a ship is likely to encounter can be assessed as can horizontal bending and torsion of the main hull. This loading is used, with estimates of the hull modulus, to deduce the stresses and deflections of the hull. The structure should be designed so that the maximum bending moment it can withstand is likely to be experienced only once in the life of the ship to minimise the chances that the hull will fail from direct overloading.

The main elements of the structure must contribute to overall strength and be able to withstand local forces. Stress concentrations, built-in stresses, fatigue and cracking must be allowed for. Superstructures can contribute to longitudinal strength if properly designed. Assessing the ability of grillages to carry load post buckling provides an ultimate load-carrying capability.

Associated with fatigue is the behaviour of steels in the presence of crack-like defects which act as stress concentrations and may cause brittle fracture below certain temperatures and at high strain rates, highlighting the need to use notch ductile (tough) steels. Corrosion can occur for a variety of reasons and must be controlled. The possible failure modes must be considered within the context of overall structural safety.

IACS have produced Common Structural Rules for a number of ship types, including bulk carriers and double hull oil tankers. These contain a wealth of data on structural matters and are available on the IACS website.

This page intentionally left blank

Chapter 14

Ship Design

INTRODUCTION

We now consider the synthesis of the design. A ship is large and complex; it is also expensive and must operate efficiently for a long time, often in excess of 25 years. There are no prototypes and even with a class of ships, the first is expected to be commercially viable from the date of acceptance. So the designer must 'get things right'. Particularly for warships, long life means that it is desirable that the design is adaptable to uncertain future roles as well as being suited to the initial roles identified. Design theorists call this the 'wicked problem' because establishing the true requirements can be more challenging than the design itself. Andrews (2011b) refers to a ship as being a physically large and complex (PLC) system.

Having no prototype, designers were forced for many years to use an evolutionary approach using a type ship. Weight and space breakdowns were a useful starting point depending upon whether a particular ship type was weight or space limited. One with a dense cargo needed a large hull to provide the buoyancy to support the weight leading to 'spare' space. A cruise ship requires more space for a relatively light 'cargo'. In warships the two extremes were represented by a battleship with its massive armour plating and a light aircraft carrier with large hangars. Whilst some novel aspects of a design, particularly hydrodynamic performance, could be checked by model tests, the absence of reliable analysis methods made a novel design a risky venture. Similar considerations meant that regulatory and classification bodies had to be generally reactive, rather than proactive, in thinking.

Not surprisingly the designer was often regarded as conservative. Today naval architects can be more adventurous in their approach because of the advances in understanding of a ship's behaviour – hydrodynamically and structurally – coupled with a tremendous increase in computer capacity. Novel designs can now be developed with much greater confidence.

This chapter gives an appreciation of what is involved in what is a very variable process. The approach to design is discussed in general terms, indicating what the naval architect must consider at each stage, and how relationships between different design features can be studied in a systematic way to produce a well-balanced design. The methodology can be applied in

Introduction to Naval Architecture. DOI: http://dx.doi.org/10.1016/B978-0-08-098237-3.00014-X

different ways but that application will usually be through the medium of a computer-aided design (CAD) system. The user must understand the basic approach and assumptions made in the chosen system, its strengths and weaknesses, arising perhaps from simplifying assumptions made in it.

Ship design is a complex undertaking requiring a methodical, disciplined approach, following a logical sequence. The detail steps will depend upon the type of ship and how novel it is. The owner must say what is needed so the starting point is a set of *requirements*.

THE REQUIREMENTS

A good set of requirements, the mission statement for the ship, will define the *functions* of the ship and the operational *capabilities* it should possess. Capabilities might be the ability to maintain a speed of 20 knots in average sea conditions, the ability to carry 500 standard containers or the ability to carry 1000 passengers. The statement of requirements should be couched in operational terms (it should be solution independent). It will be developed by informed dialogue between the owner and designer during the concept design phase.

The designer must work closely with the owner in developing the requirements as the owner may have a strong interest in some of the issues. In the case of the main propulsion plant, for instance, the new ship may be joining a fleet which to date has been exclusively diesel driven. If the new ship uses gas turbine drive, the owner will have to arrange for the retraining of engine room staff and will face additional logistics problems in providing spares. In addition to meeting an owner's specific requirements, the ship must meet the many international and national regulations discussed earlier.

In the case of warships, the government, as represented by the navy, is effectively the owner and the naval staff will specify what is needed to enable the navy to meet its commitments in support of the country's foreign policy.

THE DESIGN PHASES

General

The characteristics of the ship gradually evolve as decisions are made on how best to meet the requirements. Everything included in the ship must serve a useful purpose. For instance:

- The machinery must provide enough, but not too much, power for the desired speed.
- The hull, with its sub-division, must provide a safe vehicle for the intended service in the intact condition and be able to withstand a reasonable amount of damage.

- The various systems (e.g. electric, hydraulic and air conditioning) must be adequate to supply all needs, recognising that not all the connected equipment will be needed at the same time.

Margins, in weight, space, services and so on will allow for changes during the service life and provide the desired level of availability of any function.

The design process comprises three main stages/phases. Titles vary but those used here are considered descriptive of what is involved in each stage and they are commonly used (alternative titles in parentheses):

- Concept (Preliminary or Feasibility) design;
- Contract (Full) design;
- Detail (Build) design.

The purposes of these phases are quite distinct, and there may be breaks while approval is sought to proceed to the next phase. What is done in each phase can vary. Computers enable more comprehensive and detailed investigations to be carried out quickly at an earlier design stage helping the decision-making process by providing more extensive and reliable information. What is investigated, and how deeply, will be determined by what is possible and necessary at each stage.

Each design phase is variable in terms of what is studied and the amount of work associated with each design feature. It depends on the type and size of ship and the degree of novelty. Studying the movements of people within a ship involves more work in a cruise liner than in a crude oil carrier; vulnerability studies for a large aircraft carrier will be more complex than those for a frigate and so on. An innovative design requires more work than one closely resembling a previous ship.

The complex nature of a ship demands a systematic approach to ensure all aspects are considered at the appropriate time. The quality of the solutions that emerge will depend upon the ability and experience of the designer. Naval architecture is a blend of science and art, the former providing facts based on analysis and the latter leading to the right design balance between often conflicting demands.

Concept Design

It is generally agreed that this is the most important design phase as it establishes the true requirement the designer is to meet. It requires an experienced naval architect. It is innovative as all the different ways in which an owner's requirements can be met must be considered. Should the ship be a conventional displacement hull, a catamaran or a ground effect vessel? Interactions with other parts of the overall transport system must be considered – port facilities, linking road and/or rail systems. The introduction of container ships illustrates the possible widespread implications of some of the decisions made at this stage.

The naval architect will:

- determine from the owner the routes to be used. These define the environments in which the ship is to operate;
- establish the overall size, type of hull, capacity, speed, machinery and so on;
- create the basic design and analyse aspects of it in enough depth, to ensure that it meets the objectives of the owner, and
- make changes to meet the requirements more closely, effectively and economically.

The computer has probably had the greatest impact on the concept design phase. It enables more design variants to be considered quickly, each variant being studied in more depth with fewer approximations so providing more reliable estimates of cost and technical performance. As Andrews (2011a) argues, the decisions required can be effectively reached only if well-developed ship studies have been carried out to give solid, reliable data. He shows how this can be achieved using the Design Building Block (DBB) approach. As with any design method, this requires a wide ranging database of previous designs and equipment as well as an experienced designer.

To indicate the progressive nature of the concept design phase, Andrews and Pawling (2008) (and developed further in Andrews (2010)) split it into:

- Concept exploration – a wide ranging study of options. This stage may lead to a small number of different options needing more study to home in on a preferred solution.
- Concept studies to develop the preferred concept arising from concept exploration.
- Concept design which leads to decisions to commit the more substantial effort needed for the later design stages.

In concept exploration the designer starts with a very broad look at a wide range of options, studies the more promising and ends up with a single preferred option. The process is one of iteration. Intelligent figures, often based on a type ship but perhaps based on earlier more extensive studies, are needed in the early stages to ensure that the first solutions are reasonable. A type ship is an existing one similar in size and type to the new ship.

The iterative process has been likened to a spiral as each ship feature is considered more than once in progressively greater detail. Each cycle should approach the final design more closely. However, a spiral implies a steady progression and ignores the step functions that can occur. More power, than originally allowed, may be required to drive the ship at the desired speed and the next size of power plant available may be significantly larger and heavier.

A better way is to look at the design process as a sequence of selection, analysis and decision with feedback from later to earlier stages as illustrated in Figure 14.1 from Andrews (2011a).

The concept stage is also important as it is then that a large percentage of the total ship cost is committed, although the actual spend is small. Time spent at this stage can avoid expensive changes later. By the end of the concept design, it will have been established whether, or not, it is feasible to produce a ship to meet an owner's requirements within cost. The overall characteristics will have been agreed and the designer should be able to present the owner with a good estimate of the cost of acquisition and of the through-life cost of the ship.

Contract Design

The concept design must now be developed in enough detail to allow a contract to be negotiated for building the vessel. Some of the ideas will have been partially developed in the concept stage and again the amount of work involved depends upon the novelty of the design. A lot of calculations will be carried out, many using CFD and FEA methods, to establish the ship's main performance characteristics. These will be confirmed by model tests as the final form emerges.

The makeup of the ship includes the equipment, sub-systems and systems which contribute to its ability to perform its intended functions. Thus the design can be considered as a number of DBBs some of which will be distributed quite widely throughout the ship. The configuration (layout) of the ship will emerge as the building blocks are assembled to create the crew accommodation, main and auxiliary machinery spaces with their control areas, cargo spaces and cargo-handling equipment, storerooms and so on. The layouts of the various spaces will be produced in enough detail to confirm that adequate space has been allocated. The electrical, chilled water, air conditioning, hydraulic and compressed air system capacities will be defined and the associated cables and pipes sized and positioned. Duplicate supplies will be arranged for vital equipment such as the steering gear and cross-connections made to provide alternative supplies to various areas of the ship. The ship will be divided into zones which can be self-sustaining following damage. This outlines what the naval architect must do in collaboration with other engineers.

Not everything envisaged in the concept design will work out as planned, and discussions between the designer and owner will be an ongoing process. A record must be kept of decisions made, by whom and when, with safeguards to ensure that changes are made only by those with the necessary authority. The contract design may be carried out by the same group as produced the concept design or it may be contracted out, perhaps to a potential

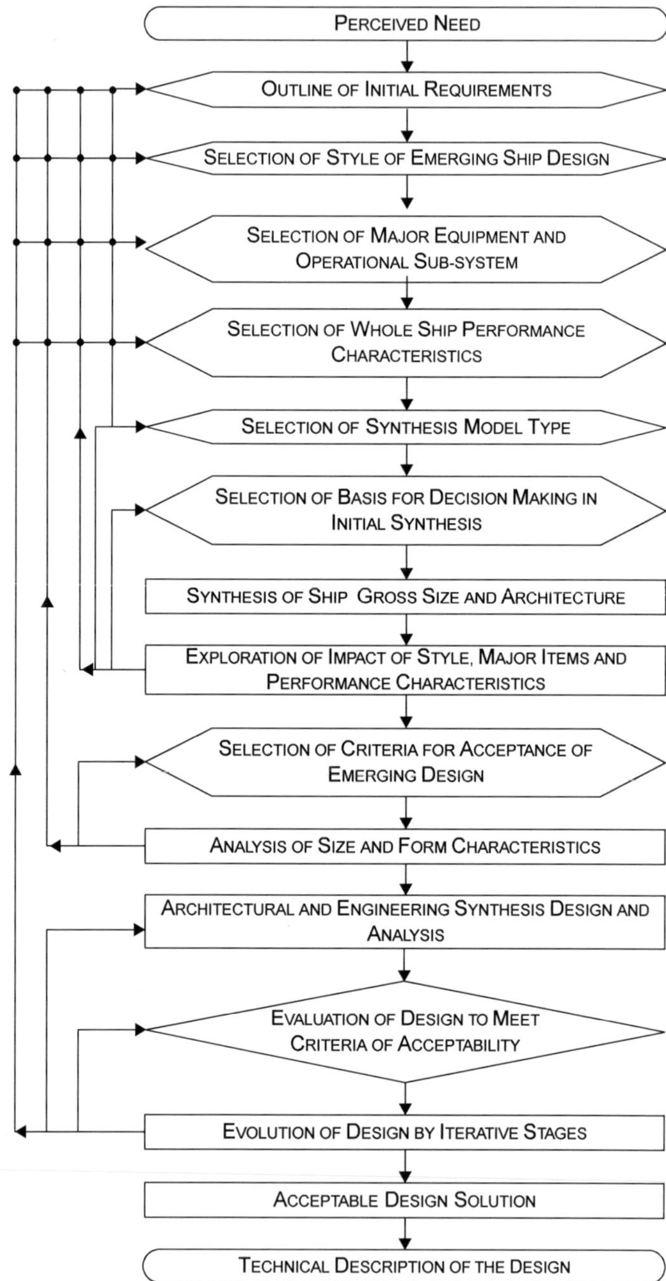

FIGURE 14.1 The design process. *Courtesy of RINA.*

builder or to some third party. Ideally the same CAD system will be used and this must include the safeguards controlling design changes.

The basic ship design definition will be supported by a mass of specifications, many international, which will control the development of the detail design.

Detail Design

Detail design is usually carried out by the builder's own design staff, working closely with the production department, but may be contracted out to another company as part of a consortium. The input is the output from the contract design and, again, ideally the same CAD system will be used to develop the detailed design. The output is the information needed by the production department to build the ship and order equipment and material, as and when needed to suit an agreed build programme. Included will be the specification of tests to be carried out as fabrication proceeds. Thus the testing of structure to ensure watertightness and structural integrity will be defined. Tests of pipe systems will lay down the test fluid and the pressures to be used, the time they are to be held and any permissible leakage. Tests will include an inclining experiment to check stability and sea trials to show that the ship meets the conditions of contract and the owner's requirements. For warships, contractor's sea trials are carried out to establish that the builder has fulfilled the contract. Then, after acceptance, the Ministry of Defence carries out further trials on weapon and ship performance in typical seagoing conditions.

There is a mass of documentation produced to define the ship for the users and maintainers – lists of spares and many handbooks. Much of these data are carried on the ship electronically or in microform.

A METHODICAL DESIGN APPROACH

Because it is so complex, with many features interacting with each other, the design must be approached in a systematic way. There are many ways to approach this problem and much will depend upon the designer's background experience. It is not possible to set out a detailed step-by-step procedure to be followed but the following sections discuss possible ways of achieving a well-balanced design.

Using a Type Ship

This is the traditional, evolutionary, process. Having selected a suitable successful ship of the right type, and with roughly the required performance abilities, the designer will note the following:

- Any experience with the type ship indicating how the design might be improved.

- Specific differences between it and the new ship – perhaps more shaft power, fewer crew, more cargo, better life-saving equipment and so on.
- Changes in regulations to be met in the new ship.

These changes will be incorporated into the old design to create a new design. The differences in size, weight and weight/space distributions may lead to an expanded hull form, with the original form being amended proportionately in all three directions. From hydrostatic data for the new form, the new draughts, trim and stability can be calculated. This will lead to further form changes. This revised form can be assessed for resistance. First the curve of areas will be looked at to ensure it is fair. Then form changes can be made in line with known effects on resistance.

The various hydrodynamic characteristics can be assessed using methodical series data, calculations and/or model tests. Layouts can be developed based on the type ship. And so the design develops in detail.

With No Suitable Type Ship

When there is no suitable type ship the designer must:

- Obtain a first estimate of overall size from the amount of cargo to be carried. Approximate formulae for the type of vessel can help with the first estimates of displacement, principal dimensions and form parameters.
- Obtain a first estimate of power needed to drive the ship at the required speed.
- Assess the size of crew, and hence the accommodation, required.
- Produce an upper deck arrangement to see if this dictates different dimensions to those initially estimated.
- Develop the general ship configuration reflecting how the ship is to operate and so on.

The designer must establish which ship features are most influential in determining the design characteristics – the design 'drivers'. The iterative nature of the process will be clear. It can be assisted by using DBBs as proposed by Andrews in several of his papers. Thus the machinery and accommodation blocks needed can be set within the ship layout.

Functions, Capabilities and Attributes

It is a good discipline to regard the ship as:

- being able to carry out a number of basic *functions*;
- possessing a number of *capabilities* enabling it to achieve the functions and
- possessing a number of *attributes* which underpin the capabilities.

Functions

These can be taken as the need to:

- **Float**. A ship must float at reasonable draughts with adequate freeboard, be stable in the intact condition and be able to accept a degree of damage. It must be seaworthy and strong enough to withstand the loads imposed on it in service.
- **Move**. It must be able to move in a controlled fashion at the intended speed and manoeuvre accurately relative to the ground in the wind and tide conditions likely to be met.
- **Trade**. It must carry the correct amount of cargo in specified conditions and be able to load and unload that cargo.

For the navy, it is said that a warship should float, move and fight.

Andrews and Pawling (2008) used DBBs for the top level functions of Float, Move and Fight (they were dealing with warship designs) and introduced infrastructure blocks to cover things such as accommodation.

Capabilities

A ship must possess a number of capabilities to be able to carry out the top level functions. The move function will require the capabilities of achieving the desired speed, following a straight course, manoeuvring in confined waters and coming to a stop in a reasonable distance. A capability supporting a tanker's ability to trade might be the ability to carry crude oil from A to B, loading and discharging it. Usually the capabilities will be multiple. For the fight function, a frigate may have to be able to detect and engage submarines, surface ships and aircraft, besides being able to assist in humanitarian missions. Some ships will need the capability of negotiating the Suez or Panama Canal.

It is the capabilities required that generally dictate the equipment and systems to be provided in the ship.

Attributes

To provide each capability, the designer must build in certain inherent attributes that enable it to carry out its functions economically and efficiently. Thus the ship must possess attributes such as stability, strength, manoeuvrability, seaworthiness, a good internal environment – air quality, temperature, ambient noise and vibration; adequate cargo-handling systems for loading, unloading and internal movement; adequate navigational systems – sensors and actuators; systems for anchoring and mooring and easy access around the ship.

A ship's attributes can involve a wide range of calculations to check that the required level of capability has been achieved. These were dealt with in earlier chapters. Some will be kept under constant review as the design develops. Stability and strength, in particular, are likely to be affected by progressive changes during design and must be monitored.

It is the attributes that give a ship the capability to carry out the required functions for which it was designed.

Inter-Dependencies

Having established the attributes of the ship, it is necessary to consider how individual systems and equipments contribute to them, and hence to the capabilities and functions. Everything in the ship must serve a purpose, possibly supporting several. The designer can produce diagrams showing how the various elements of a design combine and interact to give it a specific capability. These are known as *dependency diagrams* or *trees*. In CAD systems, the ship three-dimensional (3D) model can depict these dependencies in spatial terms as well, which otherwise would be shown in 2D in general arrangement (GA) drawings. Thus to meet the speed element of the mobility capability the ship may need, inter alia:

- A set of main machinery, say diesel engines.
- Uptakes and downtakes.
- Fuel, cooling water, lubricating oil and so on.
- Gear box(es).
- Shaft(s) with shaft bearings, stern tube(s) and shaft bracket(s) and propeller(s).

These major elements will entail supporting equipments/systems such as tanks with pumps and piping for fuel, structural supports and electrical supplies.

The dependency diagram will show how all these elements are linked and how failure of any one element would affect the overall speed capability. Thus in a single-screw ship, the loss of the shaft will remove the mobility capability completely. In a multi-shaft ship, the loss of one shaft only degrades the capability, and the degree of degradation can be assessed. The probability of loss, or degradation, of a capability can be calculated from the probabilities of failure of the individual components and how they interact with each other.

For the 'to float' function, the ship must have the capabilities to sit in the water at reasonable draughts, possess adequate freeboard and be stable. The external hull and internal watertight structure will contribute to these capabilities.

The dependency diagram can be a powerful design tool. Apart from showing inter-dependencies they:

- show how the design is configured to meet the requirements;
- can be used in availability assessments;

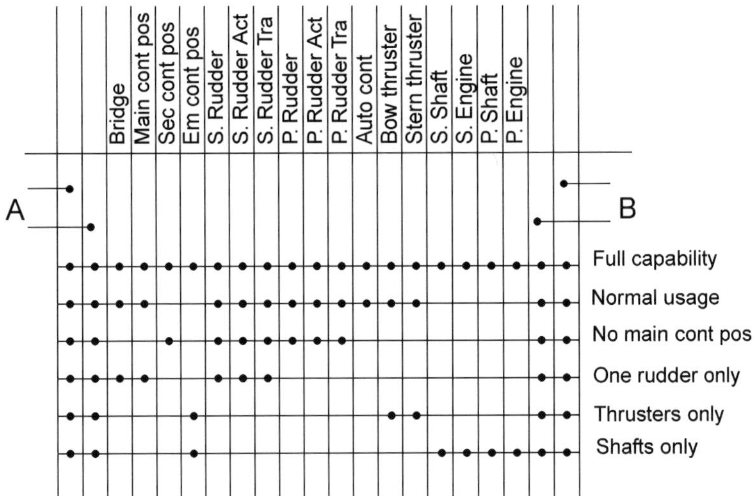

FIGURE 14.2 Simplified dependency diagram.

- can be used in vulnerability studies;
- provide one way of breaking a design down into its constituent parts – systems, sub-systems and equipment in terms of weight, cost, service demands and so on and
- enable costs to be allocated to capabilities so that the owner knows what each costs.

In using the diagrams in these ways, it is important that the interfaces are clearly defined to ensure that nothing is omitted or duplicated. Rules are needed on how those elements supporting more than one capability are to be dealt with. Going on one step, they provide a vehicle for defining packages of responsibility that can be delegated to individuals in the design and construction teams. Thus they provide a useful management tool.

To illustrate the concept, a simplified dependency diagram for the manoeuvring characteristic is shown in Figure 14.2. The ship may not carry all the items and there are more possible combinations of equipment for partial capabilities. A and B in the diagram represent the need for supporting services such as electricity, fuel, support, structure etc. These will be inputs from other sheets. 'Main Cont pos' and 'Sec Cont Pos' refer to the main and secondary control positions.

Availability

An owner wants a vessel to be available for use when needed. This is not necessarily all the time. Many ships have a quiet season when time can be found for refitting without affecting the planned schedules. Ferries are often refitted in winter months for that reason. *Availability* is a function of *reliability and maintainability*.

Reliability can be defined as the probability of an artifact performing adequately for the time intended under the operating conditions encountered. This implies that components must have a certain *mean time between failure* (MTBF). If the MTBF is too low for a given component, then that component will need to be duplicated so that its failure does not jeopardise the overall operation.

Maintenance is preferably planned with items refurbished or replaced before they fail. By carrying out *planned maintenance* in quiet periods, the availability of the ship is unaffected. The MTBF data can be used to decide when action is likely to be needed. To plan the maintenance requires knowledge of the *mean time to repair* (MTR) of components. Both MTBF and MTR data are assessed from experience with the components, or similar, in service. The other type of maintenance is *corrective* or *breakdown maintenance* which is needed when an item fails in service. Unless the item is duplicated, the system of which it is a part is out of action, or impaired, until repair is carried out. Some argue that a policy of planned maintenance is unnecessarily costly as items replaced still have some useful life left. It is a balance of risks – that of the extra cost against the possibility that a ship system will fail at some critical time.

The time taken to maintain can be reduced by adopting a policy of *refit or repair by replacement* (RBR) whereby complete units or sub-units are replaced rather than being repaired in situ. Frigates with gas turbine propulsion are designed so that the gas turbines can be replaced as units, withdrawal being usually through the uptakes or downtakes. The used or defective item can then be repaired as convenient without affecting the ship's availability, and the repairs can be carried out under better conditions, often at the manufacturer's plant. The disadvantage is that stocks of components and units must be readily available at short notice. To carry such stocks can be costly. But then an idle ship is a costly item. Again, it is a matter of striking the right balance between conflicting factors. To help in making these decisions, the technique of *availability modelling* can be used.

The dependency diagrams can be used in availability modelling of the various ship systems and capabilities. Some components of the diagram will be in series and others in parallel. Take the ability to move. The main elements were outlined above together with the supporting functions needed. Large items such as the main machinery can be broken down into their constituent components. For each item, the MTBF can be assessed together with the probability of a failure in a given time span. These individual figures can be combined to give the overall reliability of a system using an approach similar to the way the total resistance of an electric circuit is calculated from the individual resistances of items in series or parallel. High reliability of components is needed when many are used in a system. Ten components, each with a reliability of 99%, when placed in series lead to an overall reliability of $(0.99)^{10} = 0.905$. Ten units in parallel would have a reliability of $(1 - (0.1)^{10})$, effectively 100%.

Such analyses can highlight weaknesses which the designer can remedy by fitting more reliable components or by duplicating the unit. They also provide guidance on which spares should be stocked and in what quantities, that is the *range and scale* of spares.

DESIGN CONFIGURATION

Although a ship comprises many systems and equipments, design is not simply a matter of packing them all into the hull envelope. Due regard must be taken of the interactions between the various items and the efficient functioning of the end result. Indeed one can say that the layout, or configuration, is the real driving force behind the design and must be central to the design process. That is why in the older design process, the General Arrangement (GA) drawing was so important and useful. In CAD systems, it is the ship 3D model that provides the overall view. It helps to clarify the designer's thinking and aids communication of design intent to others. Figure 14.3 shows, for a frigate, some important layout considerations.

One feature not shown in Figure 14.3 (but which illustrates the importance of layout) is the provision for replenishment at sea which is important for warships which often need to operate for long periods away from base. There must be superstructure to take the high point connections, room on deck for landing stores and easy routes to the final storerooms, magazines and so on.

A ship is so large and complex that optimisation of the whole is not practical. Various optimisation methods have been developed but these generally relate to a sub-set of the totality of the variables – perhaps cost or powering. Therefore the designer needs to establish those main relationships which will

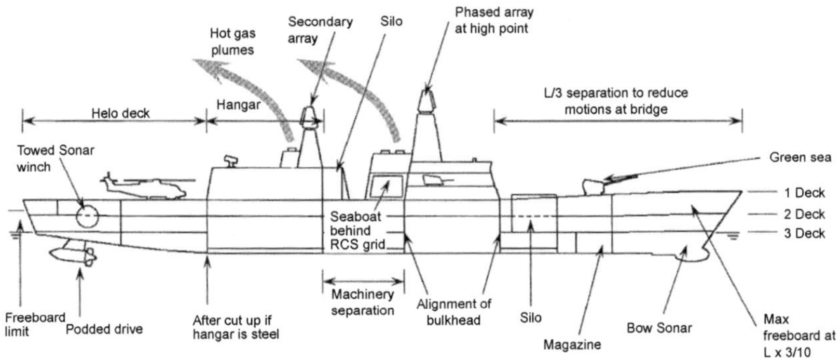

FIGURE 14.3 Frigate layout considerations. *Courtesy of RINA.*

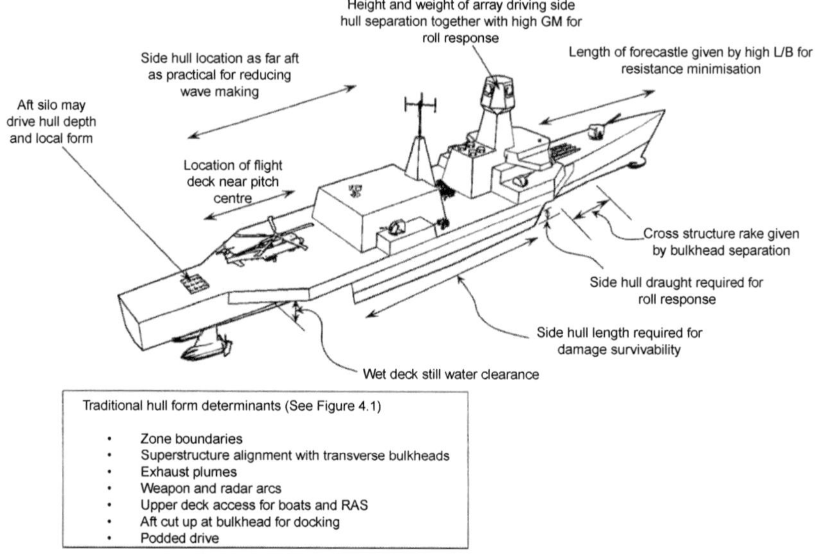

FIGURE 14.4 Trimaran configuration drivers. *Courtesy of RINA.*

determine the final design – what have become known as 'design drivers'. The design drivers for a trimaran frigate design are illustrated in Figure 14.4.

Andrews (2004), from which Figures 14.3–14.5 are taken, explained the importance of ship configuration and the DBB approach. The DBBs can be used in concept design to investigate significantly different ship configurations to meet a given set of requirements. The blocks represent various capabilities – accomodation blocks, weapon systems, machinery units and so on. Figure 14.5 shows how the blocks used in concept design can be used, with more detailed analyses within SURFCON, a program developed in the UK MOD. Such approaches became possible with the growing graphics capabilities of computers. The 2D and 3D ship representations enable the ship configuration to be developed and controlled. The designer uses the computer rather like a sketch pad. The use of these concepts within the PARAMARINE CAD system is discussed below.

CAD SYSTEMS

Computers were first used in naval architecture for individual, simple calculations and for producing lists of equipment and parts. Inputs to each calculation program and list were manual with the attendant risk of error. The next step was to arrange one program (e.g. defining the hull form) to produce an output suitable for feeding as input into other programs (e.g. a stability calculation). This led to integrated suites of programs with data

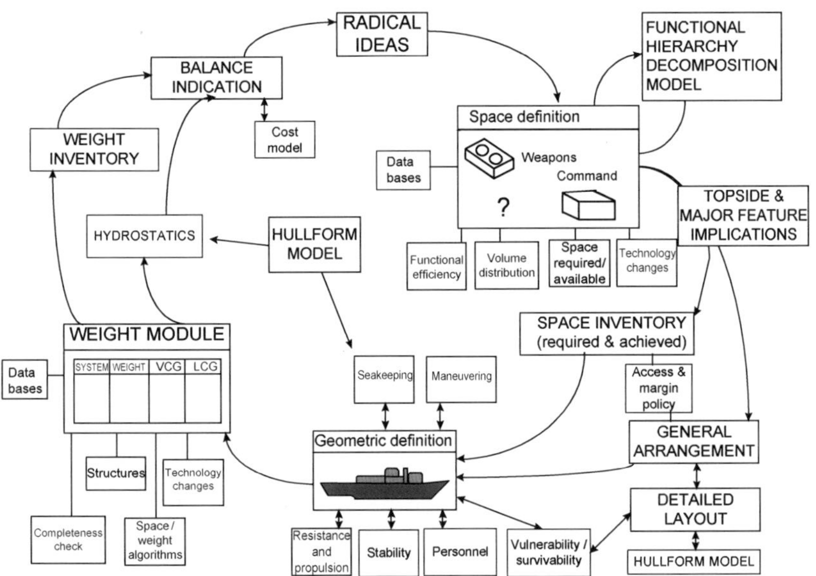

FIGURE 14.5 The SURFCON logic. *Courtesy of RINA.*

from later programs automatically updating the results of earlier programs. With increasing computer power, and the introduction of advanced graphics, designers can use 3D presentations in addition to conventional 2D. It is easy to position a 3D object in the ship either by typing in its coordinates relative to agreed reference axes or, less accurately, by pointing to a spot on the screen. Drawing upon a database of information, the computer will draw in the correct shape and show any space needed for installation, operation or repair. As a bonus the 3D pictures can be rotated for viewing from any aspect and can provide the means of 'moving' through a 'virtual' ship. So the modern CAD system was developed, bringing tremendous advantages to the designer.

Such systems eliminate errors in transcribing data from one program to another. They also provide the chance, if the software is well written, to 'check' the designer's work by, say, checking the consistency of input data. For instance, the geometry of a structure defined in an FEA program can be checked automatically for apparent inconsistencies relative to hull fairness or structural continuity. They can draw attention to potential interferences such as between structural elements and between structure and piping. Results of calculations can be automatically checked against required standards and attention drawn to cases where stress levels exceed acceptable limits. By holding information in a central database, the developing design definition can be worked on by staff at diverse locations, but limiting ability to change existing data to nominated key staff.

Understanding the CAD System

Some useful parallels can be drawn between CAD systems and word processing on the computer. Initially word processors were simply typewriters where the output was in a digital form that could be printed out. Many fonts were available instead of the one provided on the typewriter. Typists did not have to use paper and carbon copies; all copies (possibly hundreds) printed off were of the same quality. Apart from these advantages, the material could be amended in detail without having to retype the whole document. Special symbols were made available so that equations could be easily typed with super- or sub-scripts as necessary.

The printed words might be spelt wrongly but in due course the computer carried out automatic spelling and grammar checks drawing the writer's attention to suspected errors. Later the computer automatically corrected what it regarded as errors. Single key strokes could be used to represent a word, a sentence or a block of text. Pro formas could be created and called up when needed. A standard letter could be typed and then, with a suitable address list, sent out as a 'personal' letter to hundreds of people.

The aim was to increase efficiency with a user-friendly system. The person typing had no need to understand how the computer achieved what it did – merely what could be done. Whilst the aim of user friendliness has largely been achieved, the operator must be careful in its use. The following two examples can be considered:

- The appropriate language must be present within the software – e.g. Greek or English. Perhaps more subtly the correct English version must be used – United Kingdom or United States – otherwise the computer will produce 'wrong' spelling for many words.
- If an equation starts on a new line (or after a full stop) the letter p (for pressure, say) will be changed to P (which may represent power). The computer can be overridden but it must be watched or errors creep in.

In the same way, CAD systems have become ever more versatile over the years and the aim has been to make them user-friendly. But they do things based on specific assumptions and often introduce approximations to deal with complicated problems. Too often such features are not highlighted and sometimes, with the passage of time, even those developing the systems are unaware of the assumptions built in early on. Again let us take two examples:

- Use may be made of resistance data from methodical model tests. Are the characteristics of that particular series appropriate for a given new design?
- The wetted surface area is needed as part of the calculation. Is it assessed in a way consistent with that adopted in the methodical series used?

The lesson here is to be aware of possible problems in what is a very powerful and useful (indeed essential) tool.

Another parallel can be drawn with the humble slide rule. It provided a rapid means of carrying out calculations with reasonable accuracy for most engineering purposes. When greater accuracy was required, it could be provided by the cylinder slide rule. Then specialist rules were developed for specific tasks. The Froude slide rule was developed for resistance calculations; another rule was developed for calculations related to heating and insulation. It was important to use the correct 'tool' for the job. It is the same for CAD systems which have, in many cases, been developed with certain types of craft in mind or are related to a particular design phase.

Examples of CAD Systems

There are many computer aided design/computer-aided manufacture (CAD/CAM) systems on the market and it is not possible to say that one is the 'best'. This will depend upon the types of ship being designed and the design stage of most interest. The two systems discussed below are introduced to show the general features of such systems. Some systems are stronger on preliminary design and others on the detailed design leading into manufacture. The main systems are expanding to increase their coverage so that the same system can be used throughout the design, into build and then on into ship operations, maintenance and survey. A company's other systems (e.g. finance, procurement and management) must be able to link in smoothly with the chosen CAD/CAM system. Would-be users should carry out a study of all systems to find out which suits their needs best. Obviously one would not use a system developed specifically for small craft to design a supertanker but the differences between some programs are more subtle.

Two systems are outlined here briefly as illustrations. Both are well-developed systems widely used in industry. They provide the opportunity for complete ship design definition of hull form, general naval architectural calculations and outfitting. The two systems are:

- *PARAMARINE* which was developed by the Graphics Research Corporation (GRC) and is marketed by QinetiQ GRC (www.grc.qinetiq.com). Version 7 was issued early in 2011 and enables many designers to work concurrently on the same vessel design. It has links with FORAN which is the other CAD/CAM system considered here.
- *FORAN* developed by SENER Marine (www.sener.es). The latest version V70 was released in 2010 with an improved GA module working in both the 2D and 3D environments.

Over the years, both systems have undergone progressive improvement in terms of their capabilities and the user friendliness of their interfaces.

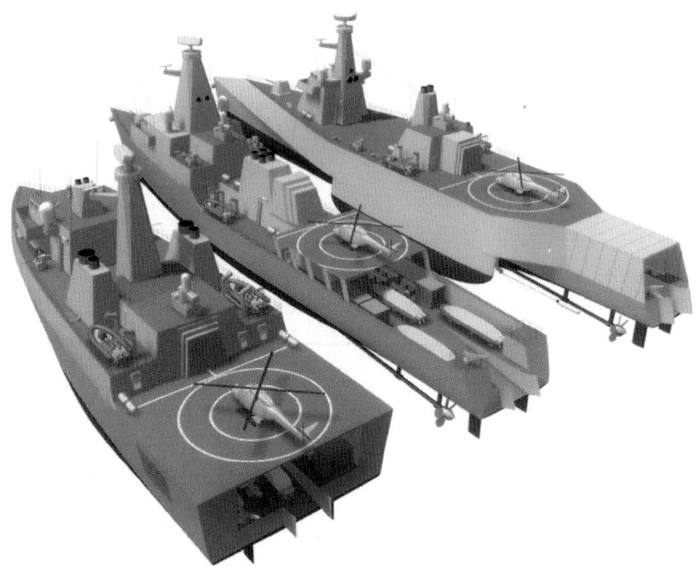

FIGURE 14.6 OPV configurations. *Courtesy of RINA.*

PARAMARINE, initially developed to replace a system used in warship design, was at first stronger on early design and FORAN was stronger in the detail design and interaction with CAM systems. Both now have more across the board capabilities. PARAMARINE is essentially a design and certification tool for ships and submarines that ventures into the production space through 'design for production' studies. It has been used by BAE in the design of the Type 45 destroyers and the Astute class of nuclear submarines. FORAN is being widely used by Babcock Marine in the build of the United Kingdom's two new aircraft carriers.

PARAMARINE was used by Andrews et al. (2005) in their 'DBBs' and Equipment Blocks approach to describe how a new ship design can be synthesised. (The DBBs should not be confused with 'Construction Building Blocks' used for ship assembly.) They used PARAMARINE for hull form hydrostatics, stability and so on. The baseline model for a Landing Platform Dock (LPD) used 99 DBBs and 118 equipment items (e.g. radars and engines) taken from the library of equipment. As an example of a presentation possible from the DBB approach, Figure 14.6, from Andrews (2011a), shows a visual comparison of three ship configurations of an Offshore Patrol Vessel (OPV) modular payload arrangement.

To enable blocks to be manipulated, each DBB carries information such as the weight and space needed for each component. Some of these items will be fixed; others will vary depending upon size and shape. Scaling algorithms are allocated to variable items. At any stage, the system checks that the volume

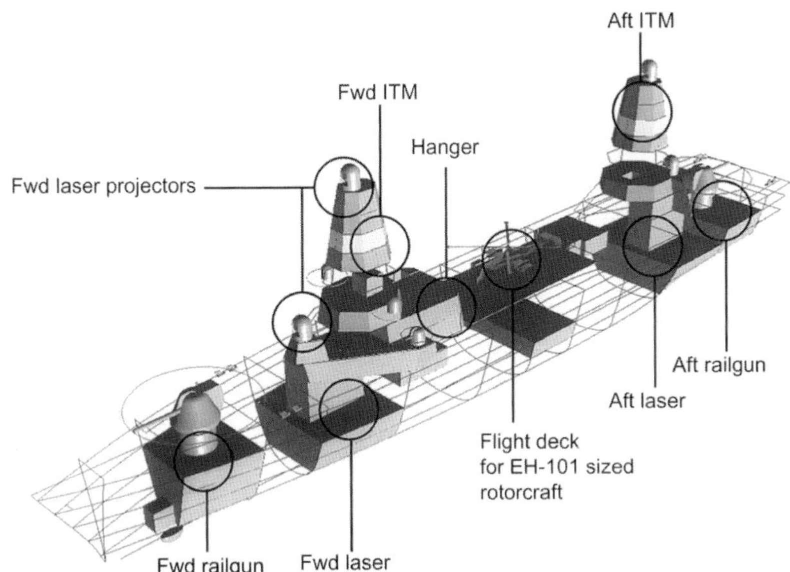

FIGURE 14.7 The Fight functional group. *Courtesy of RINA.*

within a block is adequate to accommodate all the items in it. Blocks are allocated to, and colour coded to represent, the basic functions. Once the model is set up, changes in layout, some possibly quite radical, can be studied. In this particular case, the aim of the studies was to 'design for production'.

Andrews and Pawling (2008) chose a trimaran option of the US Navy Littoral Combatant Ship they had produced as the result of an investigation for the US Navy's Office of Naval Research into the capability of the University College, London, DBB approach to Preliminary Ship Design and specifically its realisation in the early Stage Design module of the GRC's PARAMARINE ship design tool. The paper describes in depth the intermediate steps in the evolution of the detailed concept study as well as presenting the major design issues as they arose during the concept process.

The DBB approach is used in Andrews (2011a) to look at a wide ranging set of concept studies. The layout of the 'fight' function of a futuristic electric weapons destroyer is shown in Figure 14.7.

Foran

FORAN is a widely used and highly respected system which has been developed over many years. It is used for merchant ships, super yachts and warships. It uses a range of modules and is described briefly here to illustrate how the various modules interact to create a comprehensive product model (PM).

Common features relating to all the modules include the following:

- They draw upon extensive databases for equipment (size, shape, weight, cg position and service needs) and standard parts (e.g. plating elements and system fittings).
- User-friendly interfaces, developed over the years with extensive user experience feedback.
- They feed into and draw upon the PM to ensure consistency of design and configuration layout.
- Interaction between modules, as necessary, to avoid interface clashes and inconsistencies.
- The designer works in a 3D environment with outputs in 2D and 3D in the form of drawings, diagrams and parts lists.
- There are many automatic aids built in to ensure that items do not clash, that standard rules (e.g. for acceptable bend radii of piping) are applied and so on.
- They are updated regularly to take account of changing technology and to ensure that they reflect the latest international requirements.
- They are backed up by suites of calculation programs. These include programs for calculating the naval architectural characteristics of the design and others for calculating system requirements in terms of pipe/ducting sizing. These are also regularly updated – the stability programs, for instance, allow for the latest International Maritime Organisation (IMO) probabilistic methods of damage stability assessment.

Referring to Figure 14.8, the contents/capabilities of the various modules include:

- The PM contains the complete description of the design as it develops. Unconventional and asymmetric hulls can be accommodated.
- FBUILD allows a FORAN project to be organised taking account of the fabrication and assembly processes at the shipyard.
- FNORM contains the data on norms, standards and configuration parameters for workshop drawings.
- FDESIGN covers all the design disciplines, allowing the generation of drawings of selected elements of the 3D model. It facilitates large assembly management with interference analysis and the detection of conflicts between items.
- FHULL is the 3D structural model (e.g. flat and curved plates and stiffeners) updated during the whole design process. It includes a 'walk through' ability, parts lists and weight and cg data.
- FSURF has advanced tools to define the ship surface. Appendages can be included or treated separately. Transformation functions provide for the generation of new hull forms from a parent hull.

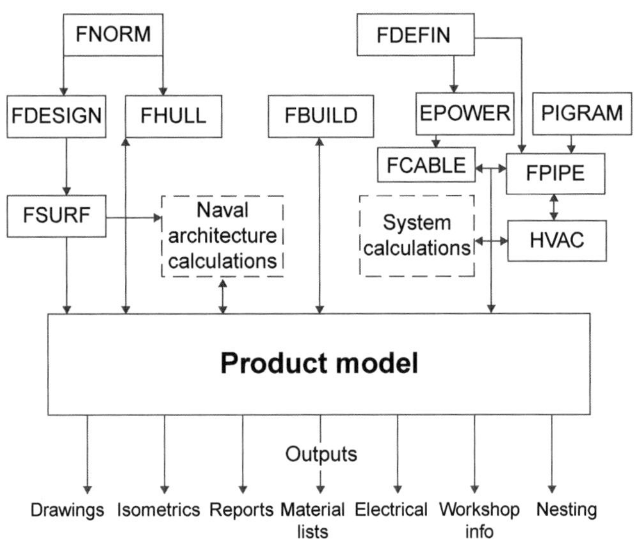

FIGURE 14.8 Schematic of FORAN.

- FDEFIN contains all outfitting standards including HVAC components, support and so on.
- EPOWER allows siting of electrical elements with calculation of cable sizes and runs. FCABLE defines 3D cable paths allowing for cable segregation and routeing rules.
- FPIPE allows the routeing of pipes, ventilation ducts and cable trays. PIGRAM allows the insertion of fittings and automatic control of the logic of connections.
- HVAC allows definition of trunk sections. Pressure losses, fan sizing and duct dimensions are calculated.
- Naval architecture calculations are in accord with the latest regulations. Sub-modules include:
  - Hydrostatics for hydrostatic curves, Bonjean curves, deadweight scale, cross curves of stability, freeboard and floodable lengths.
  - Draughts in still water in various load conditions. Static and dynamical stability for heeling angles can be found.
  - Damage stability. FLOOD deals with the deterministic methods of determining damage stability by the added weight or lost buoyancy methods. FSUBD uses the Safety of Life at Sea (SOLAS) probabilistic method, including the study of intermediate flooding, to calculate the A and R values for the ship.
  - Power prediction uses a number of modern powering prediction methods. It covers design of propeller, sternframe and rudder together with estimates of manoeuvrability.

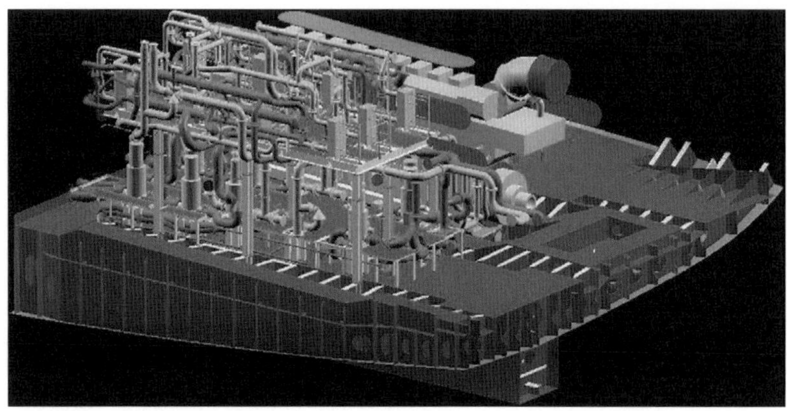

FIGURE 14.9 Screenshot, FORAN. Outfitting. *Courtesy of SENER.*

- General arrangement (GA). A module, FGA, supports the design and management of all ship spaces, setting up preliminary layouts of equipment and accommodation spaces. Capacities and tonnages can be calculated.
- Launching provides graphic representation with static and dynamical stability calculations for launch or floating from a dry dock.

- Systems calculations cover the range of calculations needed for designing the multitude of systems running through the ship.

In FORAN, a model (using both 2D and 3D representations) is built up and detailed as the design progresses. A typical screen presentation from FORAN, related to outfitting, is given in Figure 14.9. The system outputs can assist not only the designer but also the builder, operators and surveyors. If a computer model has been produced previously for a ship coming in for modification, it can be used to plan the desired changes. If no model exists, laser scanning of the actual ship can be used to create a 3D model.

It will be noted that the computer models within CAD systems contain the data that would be needed for the dependency diagrams and so, if desired, availability and vulnerability assessments could be carried out as the data defining the ship increases. Besides equipment inter-dependencies the computer-generated model shows spatial relationships. The model for an existing ship can be used as the starting point for the next similar design, or a completely new design can be synthesised from the appropriate components provided an adequate database is available.

COSTS

A ship must be able to carry out its intended functions economically, so costs are an important driver for the design. Unless the costs of acquiring and running a merchant ship are less than the revenue it can earn, the ship will be a liability. A prospective merchant ship owner/operator will have carried out market research to establish:

- the volume of trade (the goods to be transported annually);
- pick-up and delivery ports, the desired transit times;
- build and maintenance costs;
- loan repayments and interest, corporation and other taxes and
- freight rates, running costs for fuel, crew, harbour dues and so on.

Such data are used to model the economics of the investment and then, in collaboration with the designer, the owner will look at various operational scenarios in computer simulations to find the best size of ship(s) and the desirable speed.

In common with other data held in companies, cost data are often spread among many people and several departments. For ease of access, it is important that they are stored in a logical, codified way. This is often achieved by allocating costs to components of the ship which can be aggregated in various ways. Fischer and Holbach (2011) suggest using NATO's Expanded Ship Work Breakdown System which is a numeric coding system. They also discuss the allocation of direct and indirect costs, how to establish cost drivers and the use of sensitivity analysis to show how the value of a project is influenced by variations in assumed input values. All data are subject to uncertainty but particularly those related to what might happen during the life of the ship. Costs can be grouped to show the cost of providing a ship with a given attribute to support decision making on whether that attribute should be provided.

A difficulty in using data, to compare the profitability of different trading scenarios, is that various items of expenditure and income occur at different times spread over many years. Build costs arise early on, and then operating costs, including costs of crew, bunkering, port charges, refitting and repair, will be spread over the life cycle. To compensate for this, the owner works in terms of *net present value* (NPV). The concept is based on the fact that P invested in something that yields a compound interest rate of r will, after y years, be worth P_1, where:

$$P_1 = P(1+r)^y$$

Or put the other way:

$$P = P_1(1+r)^{-y}$$

Over a period of n years, if A_y is the annual cash flow in year y, the overall cash flow has a NPV of:

$$\text{NPV} = \sum A_y(1+r)^{-y}$$

where the summation is over n years.

In comparing different options, the owner will be looking for the highest NPV. Clearly the results are only as good as the assumptions made as to the values of the variables. Interest rates can be volatile and freight rates are highly so. The process can be inverted to provide a freight rate needed to give a NPV of zero. The sensitivity of the NPV to changes in the variables can be studied. Although not a factor in the cash flow calculations, the owner will keep in mind the depreciation in value of the ship.

Whilst a commercial ship owner can establish the cost effectiveness of an investment, it is more difficult for a government to decide the value of its warships. Warships do not 'earn' in the commercial sense, and it is necessary to compare the through-life costs of different ways of meeting the same operational need, taking into account its likely availability when required. The naval staff has then to decide whether this amount of money can be allocated from the defence budget against the competing bids of other requirements. If not, then the requirement must be reduced till an acceptable balance is achieved between ambitions and affordability.

One problem faced by designers is that whilst they can reasonably argue that a particular new layout or structure will be cheaper to build and fit out, the builder, not having experience of that way of working, may not be able to quantify the effect on cost. It is generally accepted that providing more space to facilitate access does reduce costs of build.

The NPV approach can be used to determine whether it is wise to build features into a ship to reduce operating costs, by showing whether the extra build cost is likely to be recouped during operations. Thus it might be better to use more automation to reduce crew size if the cost of automation is less than the associated crew costs over the life of the ship. These are not easy balances to assess. Besides being paid the crew must be trained, they need space on board and so on. Automation will bring with it initial and maintenance costs, with the need for maintainers offsetting in part other crew reductions.

SOME GENERAL DESIGN PROBLEMS

It has been seen that a ship will need to possess certain capabilities and attributes to meet an owner's requirements. Now consider some general attributes of design which apply to all, or most, ship types. Different ship types are discussed later.

Hull Form and Speed

The hull form is largely determined by the required speed – a finer hull for higher Froude numbers. The initial form will be chosen from historical data – a methodical model series, computer-generated forms or a type ship. It will be adjusted to enable it to carry its cargo more efficiently and to obtain the desired hydrodynamic characteristics. The influence of form changes on those characteristics has been discussed in the relevant chapters.

Some regard high speed as a status symbol, but it is expensive of power and fuel and if pitched too high can lead to an uneconomic ship. Faster ships can make more journeys in a given time period. Passengers like short passage times and are often prepared to pay a premium to get them as in the case of high-speed catamaran ferries. Some goods require to be moved relatively quickly. They may be perishable and a balance must be struck between refrigeration and a fast transit. For other products speed may be of little consequence. For instance, as long as enough oil is arriving in port each day, it does not matter to the customer how long it has been on passage. It is important to the ship owner who needs to balance speed, size, number of ships and capital locked up in ships and goods in transit to achieve the desired flow rate economically.

Capacity and Size

Usually there will be a certain volume of goods the ships of a fleet need to carry. The ‘goods’ may be cargo, people or weaponry. How many ships are needed and the amount to be carried in each individual ship will depend upon the rate at which goods become available. This will depend in turn, upon the supporting transport systems on land. Taking ferries as an example, one super ferry sailing each day from Dover to Calais, capable of carrying 1 day’s load of lorries, cars and passengers, would not be popular. Passage for most would be delayed, large holding areas would be needed at the ports and the ship would be idle for much of the time. Whilst such an extreme case is clearly undesirable, it is not easy to establish an optimum balance between size of ship and frequency of service. Computer modelling, allowing for the variability of the data, is used to compare different options and establish parameters such as the expected average waiting time, percentage of ship capacity used and so on.

Cargo Handling

In deciding what cargo-handling equipment to fit, a balance is needed between giving a ship the ability to load and discharge its own cargo and reliance upon the terminal port facilities. If the ship is to operate between well-defined ports, the balance may be clear. If the ship is to operate more flexibly, it cannot rely on specialist unloading equipment and must carry more of its own.

The development of the container ship was closely linked to the development of special container ports and the supporting road and rail networks for moving the containers inland. Similarly large crude oil carriers can expect good facilities at the loading port and the refinery terminal.

Influence of Nature of Goods Carried

Particularly for those goods where large volumes are to be shipped the nature of the cargo has come to dictate the main features of the ship. The wool clippers on the Australian run were an early example. More recently tankers have come to the fore, and with the growing demand for oil and its by-products, the size of tankers has grown rapidly. The major influences are the possible storage methods and the means of loading and discharging. Oil can be carried in large tanks and can be pumped out. Some particulate cargoes can be handled similarly or by conveyor belts and huge grabs. This has led to bulk carriers for grain, iron ore and coal.

Mixed cargoes are often placed in containers of a range of standard sizes. This improves the security in transit and reduces time in port. In other cases, the cargo is brought to the ship in the land transport system units. First came the train ferries and then the roll on/roll off ships. Cars can be driven on and off for the delivery of new cars around the world or for people taking their cars on holiday.

Transiting the World's Major Waterways

There may be limits imposed on the size of a ship by external factors such as the geographical features, and facilities, of the ports and waterways to be used. Many ships need to use rivers but the three waterways that are of particular interest are:

- *The Suez Canal* (The Suez Canal Authority) was built to reduce the passage time between Europe and the East by joining the Mediterranean to the Red Sea. It is 192 km long and, whilst it imposes no length limit, it limits draughts to 19 m. Speed limitations vary but average transit times are about 14 h. The ships must carry a canal pilot and are routed in convoys. There are no locks. There are dredged channels which mean that greater draughts are permitted at certain beams and less draught for wider ships. There are plans to deepen the central channel.
- *The Panama Canal* (The Panama Canal Commission) which links the Atlantic and Pacific oceans. It is 80 km long and uses three flights of dual parallel locks to raise ships from sea level to some 26 m above that level and then lower them again. The maximum allowable dimensions of ships are dictated by the size of locks and the current dimensions are as follows: length 294.1 m, beam 32.2 m and draught 12 m in tropical fresh water. The Panama Canal Authority is currently expanding the system. There are to be two new flights of locks, one at the Pacific and one at the Atlantic end of the canal, which will operate additionally to the current locks. Additionally parts of the waterway are being widened and deepened. When the expansion is completed in 2014/2015, the dimensions of

vessels that can transit the canal will be: length 366 m, beam 49 m and draught 15 m.

- *The Saint Lawrence Seaway*, providing access from the Great Lakes to the Atlantic. The maximum acceptable vessel dimensions are length 223 m, beam 23.2 m and draught 7.92 m. Access is limited by ice in winter. There is an air draught limit of 35.50 m.

The importance of these routes is shown by the number of ships using them each year – some 14,000 for the Panama Canal and 25,000 for the Suez Canal. The use of the waterways requires a ship to pay tolls and not to exceed certain critical dimensions. Both tolls and dimensions are subject to detailed conditions and special certificates are needed. A designer/operator should consult the relevant authority for those details but a lot of data can be found on associated websites. The limitations on dimensions have led to the terms *Suezmax* and *Panamax* being applied to bulk carriers just within the limits of dimension. Those not able to use the canals are referred to as *Capesize*.

Manning/Complement

A significant factor in determining the size of a ship is the complement needed for effective manning and the associated accommodation and catering requirements. The features dictating a ship's complement include the type of ship, machinery – main and auxiliary, policy on automation of systems, policy for maintenance of the ship and its equipment, and average passage time.

A passenger ship requires more crew to serve the passengers. Cargo ships require less people than a warship because of the nature and demands of the payload. Leaving aside passenger ships, merchant ships might typically have a crew of 20–30, a frigate will have about five times that number. The more automation fitted, the smaller the crew for operation but the number of maintainers may increase. Long passages demand more crew to provide domestic services. Over the years technological improvements have led to significant reductions in manning levels for both warships and merchant ships.

The numbers needed to man the ship must be considered for different operating conditions; in the case of a warship, the needs for cruising are quite different to those for action. Most ships work on a three watch system, so some positions will demand three people. A first approximation to the complement of a new ship can be made by scaling from an existing ship. A better approach, if the data are available, is to build up a complement from knowledge of the way a ship is to be operated and of the equipment and systems fitted, taking each department in turn.

In assessing the accommodation required, margins must be applied to manning estimates. These are to cover possible errors in initial estimation, training billets, advancement of crew and changes that may be made to the ship during the ship's life.

Merchant ship conditions of service and living arrangements are subject to various IMO and International Labour Organisation (ILO) regulations. As an instance an ILO Guide to International Labour Standards requires ships carrying 100 or more seafarers and ordinarily engaged on international voyages of more than 3 days' duration to have a medical doctor as a member of the crew.

Aesthetics

The benefits of good aesthetics are not quantifiable but it is important that ships should 'look good'. This applies not just to the external appearance but also to the interior design. Passengers and crew feel better in a pleasant environment and the former are more likely to form an attachment to an attractive ship. The appearance, however, must not detract from the ship's main functions. Thus a drill ship cannot avoid the drilling rig and so cannot be made 'sleek'. A number of things can be done to improve the aesthetics of a ship (some are matters of personal preference) including:

- raking the bow, masts, funnels and the fore end of the superstructure;
- streamlining the superstructures. This should also reduce air resistance;
- employing specialist interior designers for important passenger spaces and
- use of lighting to suit the work done or influence the 'mood' in various spaces.

Other actions include removing clutter, screening runs of cabling, ducting and pipes (bearing in mind the possible need to access them for maintenance or in an emergency) and keeping paintwork in good condition.

Looks are important even in warships which must provide a 'presence' in any part of the world where trouble is brewing. The ship needs to look as though it can take any action needed to deal with the situation. It needs to look powerful, and naval personnel will take more pride in a good-looking ship.

SAFETY

People are increasingly aware of safety issues in their daily lives and they are unwilling to accept levels of risk that might have been acceptable 50 or 100 years ago. The *Titanic* disaster brought home the fact that no ship is unsinkable, no matter how big. The loss of life in the *Herald of Free Enterprise* and the *Estonia* highlighted the potential dangers of designs with large open deck spaces. This showed the danger that changes in design, as technology develops, can get ahead of the regulations intended to promote safety.

Efforts by the international community are improving the situation but the naval architect must not become complacent. Too often the general public is unaware of the unexplained ship losses involving large and relatively new ships. The *MV Derbyshire* was such a case until the relatives and unions lobbied the government. Subsequent investigations led, amongst other things, to the finding of the hull on the sea bed, the tightening up of regulations concerning hatch covers and the acceptance that freak waves are not as rare as previously thought. In bulk carriers, water ingress alarms are being fitted and so are double hulls. Some ships are provided with hull stress monitoring systems involving strain gauges, pressure transducers and motion sensors. These help the master avoid undue straining of the hull at sea and during loading and unloading.

In recent years, bulk carriers, tankers and Ro-Ro ships have received quite a lot of attention from the maritime community. Whilst still more needs to be done in those areas, they are not the only causes for concern. Spouge (2003) pointed out that of losses of ships over 100 gross tonnage in the period 1995–2000, 42% were general cargo ships and 25% fishing vessels. The high rate of cargo ship loss is due in part to the greater number of these ships in service, but if the loss rate per 1000 ship years is taken, figures of about 5.4, 3.3 and 1.5 are obtained for cargo ships, dry bulk carriers and oil tankers respectively.

The lessons to be learnt are that:

- it is not that just high profile ships that need attention;
- it is essential to analyse the data available on losses to detect trends and potential reasons for the losses so that remedial action can be taken.

Many of the ships lost will have been built, maintained and manned in accord with the latest rules and regulations. Still some ships are not as safe as they could, and possibly should, be. This is partly due to the following reasons:

- Regulations are often a compromise between what is regarded by many as good practice and what others are prepared to accept for economic reasons.
- The time between failures being experienced and analysed, and the corrective action decided upon, agreed and implemented.
- Advancing technology and changing trade requirements can lead to ships with new features, and operating patterns, which have not been fully proven. Modern analysis methods, testing of hydrodynamic or structural models, and of materials in representative conditions, can all help but the final proof of the soundness of a design is its performance at sea.

Ships cannot be made completely safe against all eventualities. Some measures to improve safety might make the ship virtually unusable. Too great a level of internal watertight sub-division, carried up high in the ship, would make it very difficult to move around or to stow cargo effectively. Some are very costly, making the ship uneconomic to operate.

Survivability After Damage

A ship can be seriously damaged by or lost because of:

- water entering as a result of damage or human error in not having watertight boundaries sealed;
- fire or explosion;
- structural failure due to overloading, fatigue or fracture, possibly brittle in nature, or
- loss of propulsive power or steering, leading to collision or grounding.

Damage Scenarios

A ship's survivability after an accident depends upon many unpredictable factors and illustrates graphically the uncertainties that so often face the designer. For instance:

- What was the form of the accident? Was it a collision, grounding or fire?
  - If a collision, where, and how deeply, did the impacting ship strike? What were the speeds of the two vessels? What were their relative sizes? What was the state of watertight closures at the time of impact?
  - If a grounding, what was the nature of the seabed – soft or rocky? Where along the ship, and transversely, was the area of contact? Was the tide rising or falling?
  - If a fire, where did it originate? Was it able to feed on combustible material in the area? What was the state of fireproof boundaries at the time? Were there automatic sprinkler systems fitted in the area? Did they work straight away?
- Was the main hull structure weakened in any way? If so by how much?
- Where did the accident take place? Was it close to land? Are rescue services readily to hand in the area?
- What were the weather conditions at the time in terms of winds, waves and tide? Are these likely to deteriorate?
- What actions did the crew take? Were these the best ones and were they put in hand immediately?
- What was the state of loading at the time? Was the ship fully loaded or travelling in ballast? Had the ship been at sea some time using up fuel and stores?
- What services were available to help combat the accident – electrical supplies, pumps and so on? Can the ship move? Can it steer?
- Was any of the life-saving equipment damaged in the incident?

This is not intended as an exhaustive list but it illustrates the sort of factors to be considered in assessing a ship's ability to survive an accident. Some of these matters are discussed elsewhere.

The uncertainties, and the complexities of dealing with specific scenarios in a thorough way, have meant that over the years naval architects have had to make many simplifying assumptions in attempting to assess the survivability of a given design. The increasing power of computers, with more advanced programs, has meant that gradually more comprehensive treatments, and more of them in a given time, have become possible, such as those associated with damage stability.

Warships have to be designed to retain as much fighting capability as possible following attack. Knowing the destructive power of the torpedo, for instance, the designer can arrange the spacing of transverse bulkheads so that not more than two adjacent spaces will be directly open to the sea. Because a torpedo might strike on a bulkhead, there is always the prospect of two spaces being flooded at least. Thus the designer will:

- ensure that the ship will not sink wherever the torpedo strikes along its length. Flooding may be asymmetric leading to large heel angles and longitudinal bulkheads are avoided as far as possible.
- in a large ship with two or more separate main machinery units, arrange their spacing so that not all propulsive power will be lost due to one hit.
- separate essential services and provide cross-connections so that some electrical power and so on will remain after a single hit.

In the case of merchant ships, apart from those built with government subsidies on the understanding that in the event of war they will be used for military purposes, the designer does not need to consider deliberate enemy attack. However, there are the natural hazards to consider and the steps taken over the years to improve damage stability have already been discussed.

Vulnerability

A ship might be quite safe while it remains intact but suffers extensive damage, or loss, as a result of a relatively minor incident. For instance, a ship with no internal sub-division could operate safely until water entered by some means. It would then sink. Such a design would be unduly *vulnerable*.

An incident may involve another ship, in a collision say, or result from an equipment failure. Thus loss of the ability to steer the ship may result in grounding. It can arise from human error, the crew failing to close and secure watertight doors and hatches. Often it will be the result of several failures occurring at the same time.

For each way in which a ship may be damaged, the outcome of that damage on the ship and its systems can be assessed. The aim is to highlight any undue weaknesses in the design. Taking the steering system as an example, the various elements in the total system can be set out in a diagram showing their inter-relationships (see Figure 14.2). There will be the bridge console on which rudder angles or course changes are ordered, the system by which these orders are transmitted to the pumps/motors driving the rudder and the rudder itself. If two rudders are fitted, the two systems should be as independent as possible so that an incident causing one of the rudders to fail does not affect the other. If only one rudder is fitted, the system would be less vulnerable if duplicate motors/pumps were provided. Wiring or piping systems and electrical supplies can be duplicated. Each duplication costs money, space and weight and so it is important to assess the degree of risk and the consequences of failure. The consequences are likely to depend upon the particular situation in which the ship finds itself. Loss of steering is more serious close to a rocky coast than in the open ocean. It may be even more serious within the confines of a crowded harbour. Thus safety and vulnerability studies must be set within the context of likely operational scenarios.

It will be apparent that probabilities play a major role in these studies and the statistics of past accidents are invaluable. For instance, from the data on the damaged length in collisions and groundings, the probability of the ship being struck at a particular point along its length and of a certain fraction of the ship's length being damaged in this way, what is likely to happen in some future incident, can be assessed. This is the basis of the latest IMO approach to merchant ship vulnerability. The probability of two events occurring together is obtained from the product of their individual probabilities. Thus the designer can combine the probabilities of a collision occurring (it is more likely in the English Channel than in the South Pacific), that the ship will be in a particular loading condition at the time, that the impact will occur at a particular position along the length and that a given length will be damaged. The crew's speed and competence in dealing with an incident are other factors. IMO have proposed standard shapes for the probability density functions for the position of damage, length of damage, permeability at the time and for the occurrence of an accident. There is a steady move towards probabilistic methods of safety and vulnerability assessment, and passenger and cargo ships are now studied in this way.

It must be accepted, however, that no ship can be made absolutely safe under all possible conditions. Unusual combinations of circumstances can occur and freak conditions of wind and wave will arise from time to time. In 1973, the *Benchruachen*, with a gross tonnage of 12,000, suffered as a result of a freak wave. The whole bow section 120 ft forward of the break in forecastle was bent downwards at 7°. When an accident does occur, the question to be asked is whether the design was a reasonable one in the light of all the circumstances applying. No matter how tragic the incident, the design itself may have been sound. At the same time, the naval architect must be prepared to learn as a result of experience and take advantage of developing technology.

Accident Investigations

Even with well-designed, manned and operated ships, accidents can happen and their investigation often highlights useful points for designers to consider in the future.

In the United Kingdom, accident investigations are conducted by the *Marine Accident Investigation Branch (MAIB)* (www.maib.dft.gov.uk). The MAIB is a branch of the Department of Transport and examines and investigates all types of marine accidents to, or on board, UK vessels worldwide, and other vessels in UK territorial waters. It is independent of Maritime and Coastguard Agency and its head, the Chief Inspector of Marine Accidents, reports directly to the Secretary of State. The role of the MAIB is to determine the circumstances and causes of an accident with the aim of improving safety at sea and preventing future accidents. It is not the MAIB's purpose to apportion liability, nor, except when necessary to achieve the fundamental purpose, to apportion blame. They do not enforce laws or carry out prosecutions. Their powers are set out in the Merchant Shipping Act.

Another type of investigation is carried out by the *Salvage Association* (www.wreckage.org). The Salvage Association, founded in London in 1856, serves the insurance industry. When instructed, it carries out surveys of casualties to ascertain the circumstances, investigate the cause, the extent of damage and to assess the cost to rectify. On-site inspections are usually needed, although thought is being given to using remote video imaging fed back to experts at base. The Association aims to give a fast service, giving preliminary advice within 48 h.

Action by the Designer

Apart from meeting all the legal requirements, a designer should:

- Consider whether any novel features of a design require special consideration.
- Look for any potentially weak spots which can be improved. This will often be at little cost if addressed early enough in the design process.
- Use dependency diagrams, or the like, drawn up as part of the design process, to establish where duplication of critical equipment is desirable.
- Ensure that the builders and operators are aware of the reasons the design is configured in the way it is and ensure that this design intent is carried through into the ship's service life.
- Carry out *failure mode effect analyses* of critical equipment and systems. This calls for experience of failures and why they occur and requires a dialogue between the designer and users.
- Produce a *safety case*, identifying how a ship might suffer damage, the probability of occurrence and the potential consequences.

The Safety Case

The safety case concept consists of four main elements:

- The safety management system, including establishing, implementing and monitoring policies. It is these policies that set the safety standards to be achieved, i.e. the aims. It is the opposite of the prescriptive approach in which the system is made to adhere to a set of rules and regulations. The safety case is targeted at a particular ship, or installation, in a given environment with a specified function.
- Identification of all practical hazards.
- Evaluating the risk level of each hazard and reducing the level of hazards for which the risk is judged to be unacceptable. The risk of a hazard is the product of its probability of occurrence and the consequences if it does occur. The judgement of acceptability is a difficult one. It is usually based on what is known as the ALARP (As Low As Reasonably Possible) principle.
- Being prepared for emergencies that could occur.

Such studies can guide the designer as to the safety systems that should be fitted on board. Analysis might show a need for external support in some situations. For instance, escort tugs for tankers might be deemed desirable in confined waters or areas of special ecological significance. Many of the factors involved can be quantified, but not all, making good judgement an essential element in all such analyses. The important thing is that a process of logical thought is applied, exposed to debate and decisions monitored as the design develops. Some of the decisions will depend upon the master and crew taking certain actions and that information should be declared so that the design intent is understood.

Safety is no academic exercise and formal assessments are particularly important for novel designs or conventional designs pushed beyond the limits of existing experience. Thus following the rapid growth in size of bulk carriers, that class of ship suffered significant numbers of casualties. One was the *MV Derbyshire*, a British bulk carrier of 192,000 tonne displacement. From 1990 to 1997, 99 bulk carriers were lost with the death of 654 people. An IMO conference in 1997 adopted important new regulations which it was hoped would help prevent ship loss following an accident. These came into force in 1999.

The loss of a ship for some unknown reason is most worrying. To assist with these cases, and in accident investigations more generally, a new regulation was adopted by IMO in 2000 which requires many ships to be fitted with 'black boxes' similar to aircraft practice of many years standing. These *voyage data recorders* (VDRs), to give them their correct title, are to be fitted in all passenger ships, and in other ships of 3000 gross tonnage upwards. There is provision for retrospective fitting in some older ships. Recommendations on voyage recorders are contained in IACS Rec. 85 (2005).

The VDRs, whose use has previously been encouraged, but not mandatory, record pre-selected data relating to the functioning of the ship's equipment and to the command and control of the ship. They are put in a distinctive protected capsule with a location device to aid recovery after an accident.

Certain ships are also to be required to carry an *automatic identification system* capable of providing data about the ship such as identity, position, course and speed automatically to other ships and shore authorities.

SUMMARY

Design is a process of synthesis bringing together a wide range of disciplines and analysis methods. In the past, ship design was a process of evolution, the starting point for a new design being a type ship to which changes were made. With modern analysis tools and powerful computers, the naval architect can be more innovative. Ships, however, are complex and their design must be approached in a methodical manner. There are no prototypes so the designer must 'get it right' first time. There are three distinct phases – concept, contract and detail design. It is in the concept phase that the designer will establish the broad characteristics of the design in consultation with the owner. The actual design process within each phase varies with the type of ship and how novel it is. One approach is to regard the ship as possessing certain capabilities and attributes which confer upon it the ability to float, move and trade – the three key functions. Everything in the ship has a part to play and the design configuration is key to the development of a good design. Also an understanding of the relationships between equipment and systems, enabling them to contribute to the functions, is important. Embedded within the design process are assessments of the ship's characteristics such as stability, strength, powering, manoeuvrability and motions. The design must be cost-effective, require minimum manning, be available when needed and not be unduly vulnerable. The safety of the ship, the people on board and the environment in which it sails are all important.

CAD systems are widely used. Many have been developed each with its particular advantages.

This page intentionally left blank

Chapter 15

Ship Types

INTRODUCTION

Having described the design process and a number of general ship attributes, we now consider the characteristics of different ship types designed to meet the specific needs of owners. Even within a ship type, there will be considerable variation of size, propulsion plant and so on. Changes are always occurring due to developing technology and national and international regulations, particularly those associated with safety. The double hull tanker is an example. This chapter discusses broad aspects of the design of different types leaving the reader to go to the quoted sources for details of individual ships.

In the mid-twentieth century, it was generally true that if a developing technology made a thing possible it was advantageous to apply it. This is no longer true and much that could be achieved technically is not adopted for economic reasons. The phasing out of Concorde is an example from the world of air travel of economics overriding what technology has to offer.

MERCHANT SHIPS

The development of merchant ship types is dictated largely by the nature of the cargo and the trade routes. Broadly they can be classified according to whether the 'cargo' is human or material. Much information including typical structural details can be found in *IACS; REC 082 (2003)*.

Passenger Ships

General

Passenger ships include ferries, cruise ships and liners. Safety is a major concern and considerable thought has been given to achieving rapid and safe evacuation, the operation often being studied using computer simulation. For instance, quicker access is possible to lifeboats stowed lower in the ship's superstructure; chutes or slides can be used by passengers to enter lifeboats already in the water either by directly entering into the boat or by using a transfer platform. Such systems must be effective in adverse weather conditions and when the ship is heeled. Shipboard arrangements take account

Introduction to Naval Architecture. DOI: http://dx.doi.org/10.1016/B978-0-08-098237-3.00015-1

of the land-based rescue organisations covering the areas in which the ship is to operate. Allowance must also be made for the fact that passengers may panic and some may be unfit or partially disabled in some way. The Maritime and Coastguard Agency has issued guidance on the needs of disabled people and the Ferries Working Group of the Disabled Persons Transport Advisory Committee (DPTAC) has issued more detailed guidance.

Safe Areas and Safe Return to Port

In view of the increasing size of these ships, and the possible difficulties in evacuating large numbers of passengers, the 'safe area' concept was developed by the International Maritime Organisation (IMO) so that, in the event of a casualty, people can stay safely on board as the ship proceeds to a safe haven. A safe area in the context of a casualty, and from a perspective of habitability, is any area outside the main vertical zone(s) in which an incident has occurred, that can safely accommodate all persons on board (more than one zone may be needed) to protect them from hazards to life or health and provide them with basic services. Basic services needed within a safe area, include sanitation, water, food, space for medical care, shelter from the weather and so on. On-board safety centres are to be provided from which safety systems can be operated.

IMO's Safe Return to Port regulations apply to (post July 2010) new vessels of 120 m or more or having three or more main vertical zones. It recognises that a ship is often its own best lifeboat. The regulations require that, after a fire or flooding incident (levels of casualty threshold are defined), basic services are to be provided for all on board and that certain systems remain operational for a safe return to port. Thus essential propulsion, electrical, firemain, navigation and steering systems are to remain operable, and the ship must have internal and external communications. For fire casualties exceeding the casualty threshold but not exceeding one vertical zone, systems for orderly evacuation have to be available for 3 hours. Providing this return to port capability requires special design provisions. See the Germanischer Lloyd guidelines of 2009. Some aspects of safety, such as damage stability, have been considered elsewhere.

Ferries

The ferry provides a link in a transport system, and large ocean-going ferries are a combination of roll-on roll-off ships (Ro-Ro) and passenger vessel. Such vessels have three zones: the lower machinery space, the vehicle decks and the passenger accommodation. A large stern door and, often also, a bow door provide access for cars and lorries to the various decks which are connected by ramps. Great care is needed to ensure these doors are watertight and proof against severe weather. There is usually a secondary closure arrangement in case the main door should leak. The passenger accommodation varies with length of the journey. For short-haul or channel crossings, public rooms with aircraft-type seats are provided, and for long distance ferries, cabins and sleeping berths are provided. Stabilisers and bow thrusters are usually fitted to improve seakeeping and manoeuvring. Size varies according to route requirements and speeds are usually around 20–22 knots.

Vehicles usually enter at one end and leave at the other, speeding up loading and unloading. There has been considerable debate on the vulnerability of Ro-Ro ships, should water get on to their vehicle decks. Various means of improving stability in the event of collision and to cater for human error have been proposed since the loss of the *Herald of Free Enterprise*, and regulations have been tightened up progressively. The later loss of the *Estonia* gave an additional impetus to a programme of much needed improvements. In recent years many high-speed passenger ferries, based on catamaran designs, have appeared.

Cruise Ships

Cruise ships have been a growth area both in terms of number of passengers being carried and in numbers and size of ships. Vessels are now capable of carrying 3000 or 4000 passengers at 22 knots. However, the larger cruise ships cannot use some ports and harbours in the more attractive locations. They must anchor well out and ferry passengers ashore in smaller boats. This takes time, and small or medium-sized ships may be desirable to cater for passengers who want to visit smaller islands.

In a cruise ship passengers are provided with a high standard of accommodation and leisure facilities. This results in a large superstructure as a prominent feature of the vessel. The many tiers of decks are fitted with large open lounges, ballrooms, swimming pools and promenade areas. Stabilisers are fitted to reduce rolling and bow thrusters are used to improve manoeuvrability. The story of the evolution of the cruise ship will be found in Dawson (2000).

Liners

The advent of long distance air travel very much reduced the number of passengers taking scheduled sea voyages by ship, but in recent years there has been a renewed interest in liners, particularly for Atlantic crossings. The *Queen Mary 2* is a good example. The history of liner development will be found in Dawson (2005).

Cargo Ships

General

The industry distinguishes between break bulk cargo, which is packed, loaded and stowed separately, and bulk cargo, which is carried loose in bulk.

General Cargo Ships

The general cargo carrier shown in Figure 15.1 is a design having a length of 209 m, deadweight of 48,000 dwt and speed of 16 knots. Two 70 tonnes safe working load (SWL) gantry cranes provide self-unloading capability. The ship can carry a wide variety of cargoes. Usually general cargo shops have several large clear open cargo-carrying spaces or holds. One or more 'tween decks may be fitted within the holds to provide increased flexibility in loading and unloading and permitting cargo segregation. Access to the holds is by openings in the deck, fitted with hatches.

Hatches are made as large as strength considerations permit to reduce the amount of horizontal movement of cargo within the ship. Cranes and derricks are provided for cargo handling. Typically the hatch width is about a third of the ship's beam. Hatch covers are of various types. Pontoon hatches are common in ships of up to 10,000 dwt, for the upper deck and 'tween decks, each pontoon weighing up to 25 tonnes. They are opened and closed using a gantry or cranes. In large bulk carriers, side rolling hatch covers are often fitted opening and closing by movement in the transverse direction. Another type of cover is the folding design operated by hydraulics. The coamings of the upper or weather deck hatches are raised above the deck to reduce the risk of flooding in heavy seas. The coamings can provide some compensation for the loss of hull strength due to the deck opening. They are liable to distort a little due to movement of the structure during loading and unloading of the ship and this must be allowed for in the design of the securing arrangements.

A double bottom is fitted along the ship's length, divided into various tanks. These may be used for fuel, lubricating oils, fresh water or ballast water. Fore and aft peak tanks are fitted and may be used to carry ballast

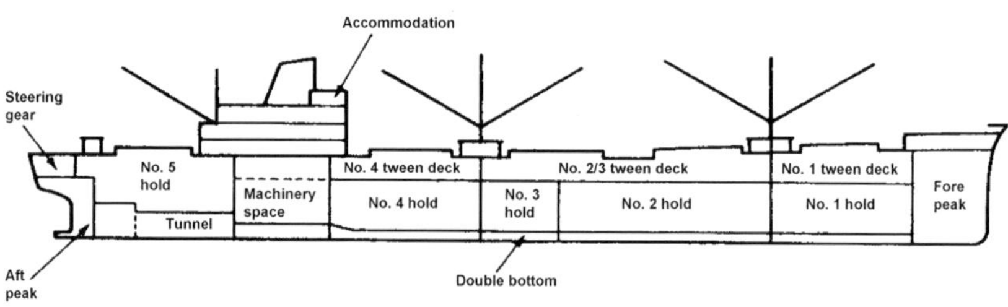

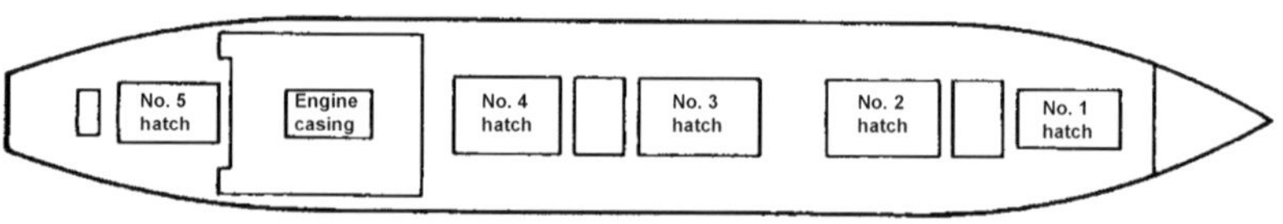

FIGURE 15.1 General cargo ship. *Courtesy of RINA.*

and trim the ship. Deep tanks are often fitted and used to carry liquid cargoes or water ballast. Water ballast tanks can be filled when the ship is only partially loaded in order to provide a sufficient draught for stability, better weight distribution for longitudinal strength and better propeller immersion.

The machinery spaces are usually well aft but with one hold aft of the accommodation and machinery spaces to improve the trim of the vessel when partially loaded. General cargo ships are generally smaller than the ships devoted to the carriage of bulk cargos. Typically their speeds range from 12 to 18 knots.

Refrigerated Cargo Ships (Reefers)

In these ships, a refrigeration system provides low temperature holds for carrying perishable cargoes in insulated holds. Cargo may be carried frozen or chilled at appropriate temperatures. The effect of low temperatures on surrounding structure must be considered. Refrigerated fruit is carried under modified atmosphere conditions, the cargo being maintained in a nitrogen-rich environment to slow the ripening process. The costs of keeping the cargo refrigerated, and the nature of the cargo, make a shorter journey time economic, and these vessels usually have speeds up to 22 knots. Up to 12 passengers are carried on some ships, this number being the maximum permitted without the need to meet full passenger ship regulations.

Container Ships

Container ships (Figure 15.2) are an example of an integrated approach to the problem of transporting goods. Goods placed in a container at a factory can be carried by road, rail or sea, being transferred from one to another. The container need not be opened until it reaches its destination, improving

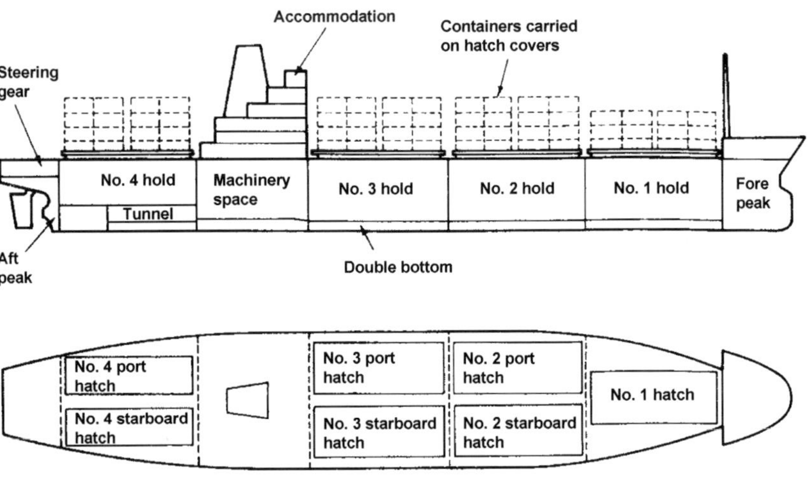

FIGURE 15.2 Container ship.

security. Any element of the overall system may impose restrictions on what can be done. The container height is likely to be dictated by the tunnels and bridges involved in land transport. Weight is likely to be dictated by the wheel loadings of lorries. The handling arrangements at the main terminals and ports are specially designed to handle the containers quickly and accurately. The larger container ships use dedicated container ports and tend not to have their own cargo-handling gantries.

Containers are reusable boxes made of steel, aluminium or FRP and come in a range of types and sizes, being of standard sizes (ISO specified) with a cross section 2.60 by 2.45 m and lengths of 6.10, 9.15 or 12.20 m. In Imperial units, containers are 8.5×8 ft with lengths of 20, 30 and 40 ft. Internal volumes and weight of goods that can be carried vary with the material. Details can be found on the websites of the operators.

The cargo-carrying section of the ship is divided into several holds with the containers racked in special frameworks and stacked one upon the other within the hold space. Containers may also be stacked on hatch covers and secured by special lashings. Some modern ships dispense with the hatch covers, and pumps deal with any water that enters the holds. Each container must be of known all up weight and stowage arrangements must ensure the ship's stability is adequate, as well as meeting the offloading schedule if more than one port is involved. The ship's deadweight will determine the total number of containers it can carry.

Cargo holds are separated by a deep web-framed structure to provide the ship with transverse strength. The structure outboard of the container holds is a box-like arrangement of wing tanks providing longitudinal and torsional strength. Wing tanks can be used for water ballast and used to counter the heeling of the ship when discharging containers. A double bottom is fitted which adds to the longitudinal strength and provides additional ballast space.

Accommodation and machinery spaces are usually located aft leaving the maximum length of full-bodied ship for container stowage. The overall capacity of a container ship is expressed in terms of the number of standard 20 ft units it can carry, i.e. the number of *twenty-foot equivalent units* (TEU). A 40-ft container is classed as 2 TEU. Container ships are of ever increasing size to take advantage of the economies of scale. They tend to be faster than most general cargo ships, with speeds up to 30 knots. The larger ships can use only the largest ports, but as these are fitted out to unload and load containers, the ship itself does not need handling gear. Smaller ships are used to distribute containers from large to smaller ports. Since the smaller ports may not have suitable handling gear, these ships can load and offload their own cargoes.

Some containers are refrigerated having their own independent cooling plant or being supplied with coolant from the ship's refrigeration system. Being insulated, refrigerated containers have less usable volume. Temperatures for a refrigerated unit would be maintained at about $-27°$C and for a freezer unit at about $-60°$C. They may be carried on general cargo ships or on dedicated refrigerated ships. A design without hatch covers enables the cell structure, in which the containers are stowed, to be continued above deck level giving greater security to the upper containers. An open hold allows easier dissipation of heat from the concentration of reefer boxes.

Barge carriers are a variant of the container ship. Standard barges are carried into which the cargo has been previously loaded. The barges, once unloaded, are towed away by tugs and return cargo barges are loaded. Minimal or even no port facilities are required and the system is well suited to countries with extensive inland waterways.

Roll-On Roll-Off Ships

These vessels (Figure 15.3) are designed for wheeled cargo, often trailers and are generally similar to Ro-Ro ferries.

The cargo may be driven aboard under its own power, towed on board or loaded by straddle carriers or fork lift trucks. One or more hatches may be provided for containers or general cargo served by deck cranes. Where cargo, with or without wheels, is loaded and discharged by cranes the term 'lift-on lift-off' (Lo-Lo) is used. In recent years, large numbers of completed cars have been moved from the country of manufacture to the importing country by specialised car carriers.

Bulk Carriers

The volume of cargoes transported by sea in bulk increased rapidly in the second half of the twentieth century, particularly oil-related products. These ships carry cargoes which do not need packaging and which can benefit from the economies of scale. Most bulk carriers are single-deck ships, longitudinally framed with a double bottom, with the cargo-carrying section of the ship divided into holds or tanks. The hold or tank arrangements vary according to the range of cargoes to be carried. Framing is contained within

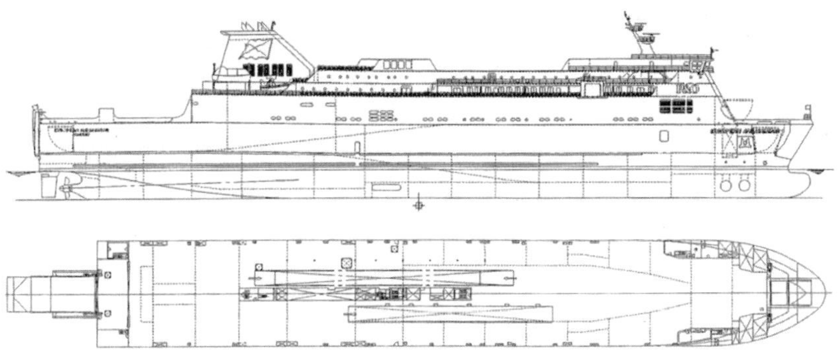

FIGURE 15.3 Ro-Ro ship. *Courtesy of RINA.*

the double bottom and wing tanks to leave the inner surfaces of the holds smooth. They may be designed to the maximum dimensions permitted in transiting the Suez or Panama canals.

Bulk carriers can be subdivided into tankers and dry bulk carriers. Requirements such as the permitted lengths of cargo holds vary with the size of ship, and the following comments are for general guidance only.

Tankers

Tankers are used for the transport of liquids. They include crude oil carriers, product tankers, gas tankers and chemical carriers.

Crude Oil Carriers

These carry the unrefined crude oil and they have significantly increased in size in order to obtain the economies of scale and to respond to the demands for more oil. Designations, such as Ultra Large Crude Carrier (ULCC) and Very Large Crude Carrier (VLCC), have been used for these huge vessels. The ULCC is a ship with a capacity of 300,000 dwt or more and the VLCC has a capacity of 200,000–300 000 dwt. Crude oil tankers with deadweight tonnages in excess of half a million have been built, although the current trend is for somewhat smaller (130,000–150,000 dwt) vessels.

The cargo-carrying section of the tanker is usually divided into tanks by longitudinal and transverse bulkheads. The size and location of these cargo tanks is dictated by the IMO Convention MARPOL 1973/1978. These regulations require the use of segregated ballast tanks and their location such that they provide a barrier against accidental oil spillage. The segregated ballast tanks must be such that the vessel can operate safely in ballast without using any cargo tank for water ballast.

Figure 15.4 shows a double-hulled VLCC of 335 m length, beam 58 m, draught 22.7 m and deadweight 310,000 dwt. It is now mandatory for tankers of 5000 tonnes deadweight and more to be fitted with double hulls or an

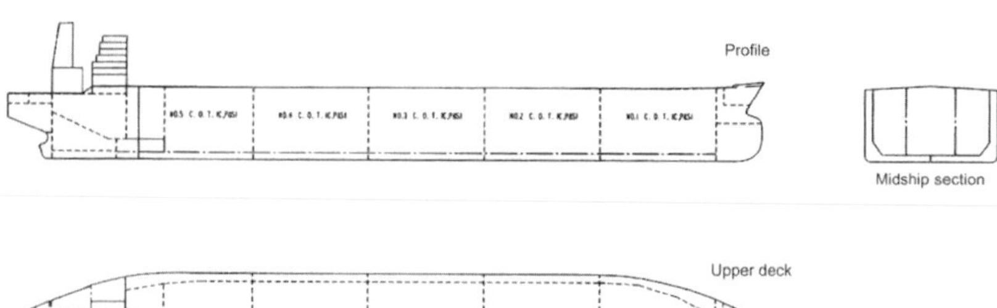

FIGURE 15.4 Layout of a double-hulled VLCC. *Courtesy of RINA.*

equivalent alternative design to comply with the MARPOL regulations. In the double-hull design, the cargo tanks are completely surrounded by wing and double-bottom tanks which can be used for ballast purposes. There has been debate on whether a double hull is the best way of reducing pollution following grounding or collision. IMO and classification societies are prepared to consider alternatives and one alternative is the mid-height depth deck design. In such ships, a deck is placed at about mid-depth which will be well below the loaded waterline, dividing the cargo tanks into upper and lower tanks. A trunk is taken from the lower tank through the upper tank and vented. The idea is that if the outer bottom is breached, the external water pressure will be greater than the pressure of oil in the lower tank and this will force oil up the vent trunk. Thus water enters the ship rather than oil escaping from it. Such tankers would still incorporate segregated ballast tanks outboard of the cargo tanks to safeguard against collision. For the detailed provisions, recourse should be had to the regulations of the authorities concerned.

Fire safety provisions are much more stringent for tankers than ordinary dry cargo ships, and other features include the following:

- Inert gas systems. 'Empty' tanks may still contain flammable gas which might explode. Tanks are normally filled with inert gas from the ship's boiler emissions which are cleaned and pumped into the empty tanks or into the spaces left above the oil in loaded tanks. An inert gas system is required on all new tankers and most existing tankers of more than 20,000 dwt.
- Equipment is duplicated to ensure that, in the event of mechanical failure, the ship can still be controlled. New tankers of 20,000 dwt and above must be fitted with an emergency towing arrangement fitted at either end of the ship.
- Ballast tanks, empty on the cargo-carrying leg of the voyage and loaded with water ballast for the return leg, are positioned so as to reduce the impact of a collision or grounding. Carrying oil in the forepeak tank, the ship's most vulnerable point in the event of a collision, is banned.
- The carriage of heavy grade oil (HGO) in single-hull tankers is now banned. HGO is:
  - crude oil having a density higher than 900 kg/m^3 at 15°C;
  - other oil having a density higher than 900 kg/m^3 at 15°C or a kinematic viscosity higher than 180 mm^2/s at 50°C and
  - bitumen, tar and their emulsions.
- Tankers and bulk carriers more than 5 years old are subject to a special inspection programme to detect any deficiencies resulting from age or neglect.

Cargo is discharged by pumps, each tank having its own suction arrangement, and a network of piping discharges the cargo to the deck for pumping ashore. Where piping serves several tanks, means are provided for isolating each tank. Accommodation and machinery spaces are aft and separated from the tank region by a cofferdam.

Tanks have a high-quality coating system, and a back-up anode system to give a coverage of 10 mA/m^2 should be included to control corrosion after coating breakdown. Once breakdown occurs, permanent ballast tanks can suffer severe corrosion and regular inspection is vital.

Some ships are fitted with equipment to measure actual strains during service, typically a number of strain gauges at key points in the structure together with an accelerometer and pressure transducer to monitor bottom impacts. Data are available on the bridge to assist the master in deciding whether a speed reduction or a change of course is needed. The information is stored and is valuable in determining service loadings and long-term fatigue data.

Product Carriers

After the crude oil is refined, the various products are transported in product carriers. The refined products carried include gas oil, aviation fuel and kerosene. Product carriers are smaller than crude oil carriers. The cargo tank arrangement is again dictated by MARPOL 73/78. Individual 'parcels' of various products may be carried at any one time which results in several separate loading and discharging pipe systems and greater subdivision of tanks. The tank surfaces are usually coated to enable a high standard of tank cleanliness to be achieved after discharge and prevent contamination. Ship sizes range from about 18,000 up to 75,000 dwt with speeds of about 14–16 knots.

Liquefied Gas Carriers

The most commonly carried liquefied gases are *liquefied natural gas* (LNG) and *liquefied petroleum gas* (LPG). They are kept in liquid form by a combination of pressure and low temperature. The combination varies to suit the gas being carried. Specialist ships are used to carry the different gases in a variety of tank systems, combined with arrangements for pressurising and refrigerating them. Natural gas is released as a result of oil-drilling operations and is a mixture of methane, ethane, propane, butane and pentane. The heavier gases, propane and butane, are termed 'petroleum gases'. The remainder, largely methane, is known as 'natural gas'. The properties, and behaviour, of these two basic groups vary considerably, requiring different means of containment and storage during transit.

Natural Gas Carriers

Natural gas is, by proportion, 75–95% methane and has a boiling point of −162°C at atmospheric pressure. Methane has a critical temperature of −82°C, which means that it cannot be liquefied by the application of pressure above this temperature. A pressure of 47 bar is necessary to liquefy methane at −82°C.

LNG carriers are designed to carry the gas in its liquid form at atmospheric pressure and at a temperature in the region of −164°C. The ship design must protect the steel structure from the low temperatures, reduce the loss of gas and avoid its leakage into the occupied regions of the vessel.

Tank designs are either self-supporting, membrane or semi-membrane. The self-supporting tank is independent of the hull. A membrane tank requires the insulation between the tank and the hull to be load bearing. Single or double metallic membranes may be used, with insulation separating the two membrane skins. The semi-membrane design has an almost rectangular cross section and the tank is unsupported at the corners which are usually rounded.

Figure 15.5 shows the cross section of a membrane tank of an LNG carrier of about 108,000 gross tonnage, with a liquid volume capacity of 158,000 m^3 (storage at −163°C). Service speed is 19.5 knots on just under 40,000 kW installed power. This particular design used electric propulsion with a dual fuel engine system. The engines can run on diesel oil only or burn mainly gas with diesel oil as a pilot fuel. The gas is from the boil-off of the liquid in the cargo tanks.

Generally the tank and insulation structure in LNG carriers is surrounded by a double hull. The double bottom and ship's side regions are used for oil or water ballast tanks whilst the ends provide cofferdams between the cargo tanks. The accommodation and machinery spaces are located aft and separated from the tank region by a cofferdam. LNG carriers have steadily increased in size and speeds range from 16 to about 20 knots.

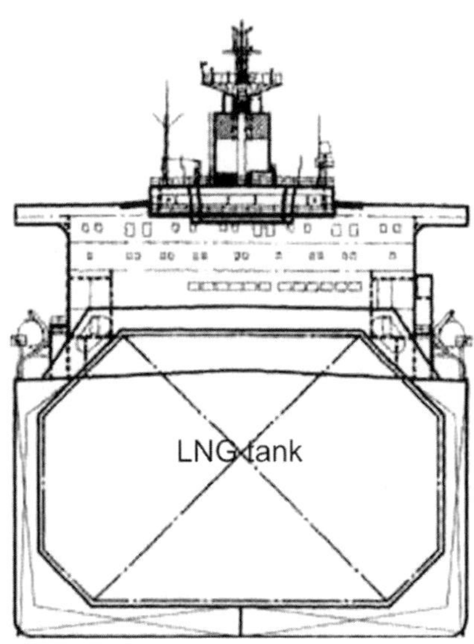

FIGURE 15.5 LNG carrier section. *Courtesy of RINA.*

Petroleum Gas Carriers

Petroleum gas may be propane, propylene, butane or a mixture. All three have critical temperatures above normal ambient temperatures and can be liquefied at low temperatures at atmospheric pressure, normal temperatures under considerable pressure, or some intermediate combination of pressure and temperature. The design must protect the steel hull where low temperatures are used, reduce the gas loss, avoid gas leakage and perhaps incorporate pressurised tanks. The fully pressurised tank operates at about 17 bar and is usually spherical or cylindrical in shape for structural efficiency.

Semi-pressurised tanks operate at a pressure of about 8 bar and at temperatures in the region of -7°C. Insulation is required and a reliquefaction plant is needed for the cargo boil-off. Cylindrical tanks are usual and may penetrate the deck. Fully refrigerated atmospheric pressure tank designs may be self-supporting, membrane or semi-membrane types as in LNG tankers. The fully refrigerated tank designs operate at temperatures of about -45°C. A double hull type of construction is used.

An LPG carrier of about 50,000 dwt is shown in Figure 15.6. It is a flushed deck vessel with four holds within which there are four independent, insulated, prismatic cargo tanks, supported by a load-bearing structure designed to take account of the interaction of movements and forces between the tanks and adjoining hull members. Topside wing, hopper side and double bottom tanks are mainly used for water ballast. Fuel is carried in a cross bunker forward of the engine room. Machinery and accommodation are right aft. It can carry various propane/butane ratios to provide flexibility of operation.

The double hull construction, cargo-pumping arrangements, accommodation and machinery location are similar to an LNG carrier. A reliquefaction plant is, however, carried and any cargo boil-off is returned to the tanks. LPG carriers exist in sizes up to about 95,000 m^3. Speeds range from 16 to 19 knots.

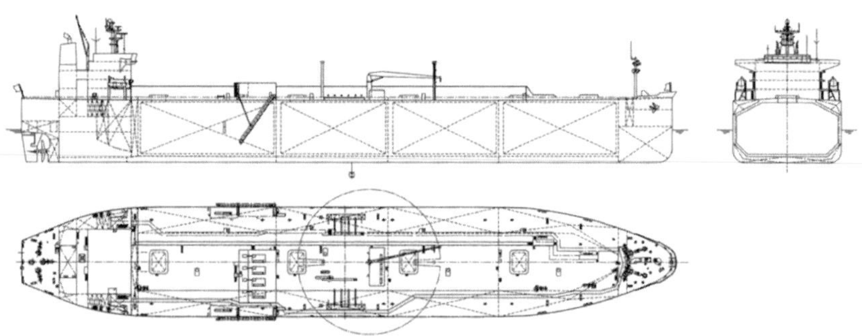

FIGURE 15.6 LPG carrier. *Courtesy of RINA.*

Chemical Carriers

A wide variety of chemicals is carried by sea. The cargoes is often toxic and flammable, so the ships are subject to stringent requirements to ensure safety of the ship and the environment. Different cargoes are segregated by cofferdams. Spaces are provided between the cargo tanks and the ship's hull, machinery spaces and the forepeak bulkhead. Great care is taken to prevent fumes spreading to manned spaces.

Dry Bulk Carriers

These ships carry bulk cargoes such as grain, coal, iron ore, bauxite, phosphate and nitrate. Towards the end of the twentieth century, more than 1000 million tonnes of these cargoes were being shipped annually, including 180 million tonnes of grain. Apart from saving the costs of packaging, loading and offloading times are reduced.

As the total volume of cargo increased so did the size of ship, taking advantage of improving technology. This growth in size has not been without its problems. In the 28 months from January 1990, there were 43 serious bulk carrier casualties of which half were total losses. Three ships, each of over 120,000 dwt, went missing. Nearly 300 lives were lost as a result of these casualties. To improve safety, IMO adopted a series of measures reflecting the lessons learned from the losses of these ships including the MV *Derbyshire* whose wreck was found and explored by remotely controlled vehicles. Among factors addressed were age, corrosion, fatigue, freeboard, bow height and strength of hatch covers. A formal safety assessment was carried out to guide future decisions on safety matters for bulk carriers.

In a general-purpose bulk carrier (Figure 15.7), only the central section of the hold is used for cargo. The partitioned tanks which surround the hold are used for ballast purposes. This hold shape also results in a self-trimming

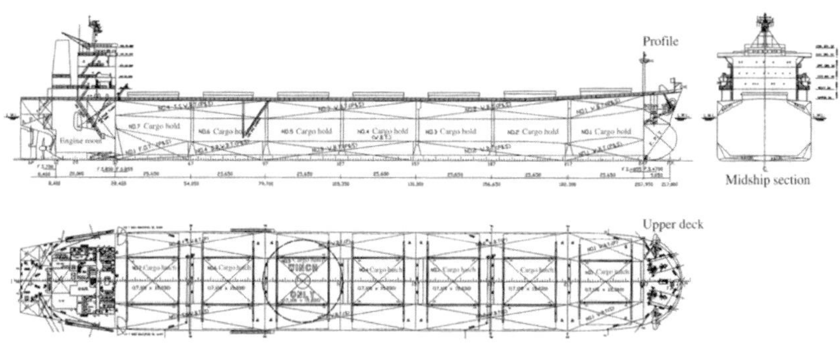

FIGURE 15.7 General-purpose bulk carrier. *Courtesy of RINA.*

cargo. During unloading the bulk cargo falls into the space below the hatchway facilitating the use of grabs or other mechanical unloaders. Hatchways are large, reducing cargo handling time during loading and unloading.

Combination carriers are bulk carriers which have been designed to carry any one of several bulk cargoes on a particular voyage, for instance ore, crude oil or dry bulk cargo.

Stability and loading manuals are provided to the master with information on operating the ship safely. Loading computer programs provide a full set of deadweight, trim, stability and longitudinal strength calculations. The very high loading rates, up to 16,000 tonnes/h, make the loading task one that needs careful attention.

An ore carrier usually has two longitudinal bulkheads which divide the cargo section into wing tanks and a centre hold which is used for ore. A deep double bottom is fitted. Ore, being a dense cargo, would have a very low centre of gravity if placed in the hold of a normal ship leading to an excess of stability in the fully loaded condition. The deep double bottom raises the centre of gravity and the behaviour of the vessel at sea is improved. The wing tanks and the double bottoms provide ballast capacity. The cross section would be similar to that for an oil/ore carrier shown in Figure 15.8.

An oil/ore carrier uses two longitudinal bulkheads to divide the cargo section into centre and wing tanks which are used for the carriage of oil. When ore is carried, only the centre tank section is used for cargo. A double bottom is fitted but used only for water ballast.

The ore/bulk/oil bulk (OBO) carrier is currently the most popular combination bulk carrier. It has a cargo-carrying cross section similar to the general bulk carrier but the structure is significantly stronger. Many bulk carriers do not carry cargo-handling equipment since they trade between specially equipped terminals. Combination carriers handling oil cargoes have their

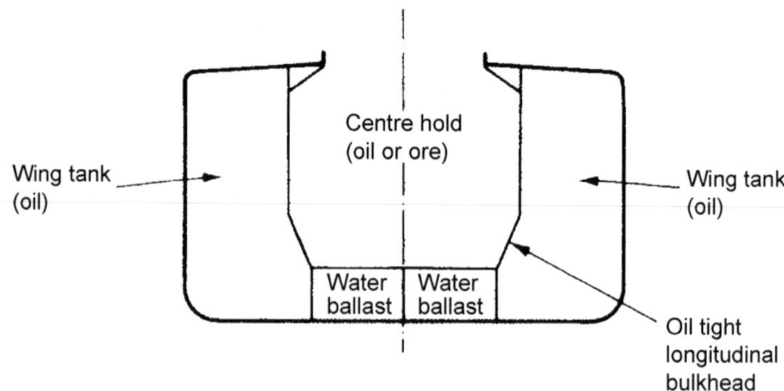

FIGURE 15.8 Section of oil/ore carrier.

own cargo pumps and piping systems for discharging oil. They are required to conform to the requirement of MARPOL. Deadweight capacities range from small to upwards of 200,000 tonnes. Taking a 150,000/160,000 tonne deadweight Capesize bulk carrier as typical, the ship is about 280 m in length, 45 m beam and 24 m in depth. Nine holds hold some 180,000 m^3 grain in total, with ballast tanks of 75,000 m^3 capacity. The speed is about 15.5 knots on 14 MW power. Accommodation is provided for about 30.

Preventing Shift of Bulk Cargoes

Regulations have existed for some time to minimise the movement of bulk cargoes and, in particular, grain. When a hold is filled with grain in bulk, it must be trimmed so as to fill all the spaces between beams and at the ends and sides of holds. Centreline bulkheads and shifting boards are fitted in the holds to restrict movement of the grain. Regulations require that the shifting boards or divisions extend downwards from the underside of deck or hatch covers to a depth determined by calculations related to an assumed heeling moment of a filled compartment.

The centreline bulkheads are fitted clear of the hatches and are usually of steel. Besides restricting cargo movement, they can support the beams if they extend from the tank top to the deck. Shifting boards can be removed when bulk cargoes are not carried.

Even with centreline bulkheads and shifting boards, spaces will appear at the top of the cargo as it settles down. To fill these spaces, feeders are fitted to provide a head of grain which will feed into the empty spaces. These feeders are usually trunks in part of the hatch in the 'tween decks above. Feeder capacity is 2% of the volume of the space it feeds. Such precautions permit grain cargoes to be carried with a high degree of safety.

SPECIALIST VESSELS

Tugs

Tugs perform a variety of tasks – moving dumb barges, helping large ships in confined waters, towing vessels on ocean voyages and engaging in salvage and firefighting operations. In most cases, they apply an external force to the ship they are assisting. This force may be applied in the direct or the indirect mode. In the former, the major component of the pull is provided by the tug's propulsion system. In the latter, most of the pull is provided by the lift generated by the flow of water around the tug's hull, the tug's own thrusters in this case maintaining its attitude in the water. To provide the direct force, tugs must develop a high thrust at low or zero speed. The pull it can exert at zero speed is known as the *bollard pull*. Tugs can be categorised broadly as inland, coastal or ocean going when they have a free-running speed of about 20 knots.

The tug shown in Figure 15.9 is a harbour tug of gross tonnage 203, length 25.4 m overall with a bollard pull of 45.45 tonnes. The general requirements for a tug are an efficient design for free running, a high thrust at zero speed (the *bollard pull*), an ability to get close alongside other vessels, good manoeuvrability and stability. Flat hull sections aft provide physical protection for the propellers as well as reducing the possibility of air being drawn into them.

Tugs can also be classified by the type and position of the propulsor units and are as follows:

- *Conventional tugs* have a normal hull and propulsion system using shafts and propellers. These last may be of fixed or controllable pitch, open or nozzled. Steerable nozzles or vertical axis propellers may be used. They usually tow from the stern and push with the bow.
- *Stern drive tugs* have the stern cut away to accommodate twin azimuthing propellers in nozzles. These propellers, of fixed or controllable pitch, can be turned independently through 360° to give good manoeuvrability. The main winch is usually forward and they tow over the bow or push with the bow.
- *Tractor tugs* are of unconventional hull form, with propulsors sited under the hull about one-third of the length from the bow, and with a stabilising skeg aft. Propulsion is by azimuthing units or vertical axis propellers. They usually tow over the stern or push with the stern.

In most tug-assisted operations, the ship moves at low speed. Concern for the environment, following the *Exxon Valdez* disaster, led to the US Oil Pollution Act of 1990. Escort tugs were proposed to help tankers, or any

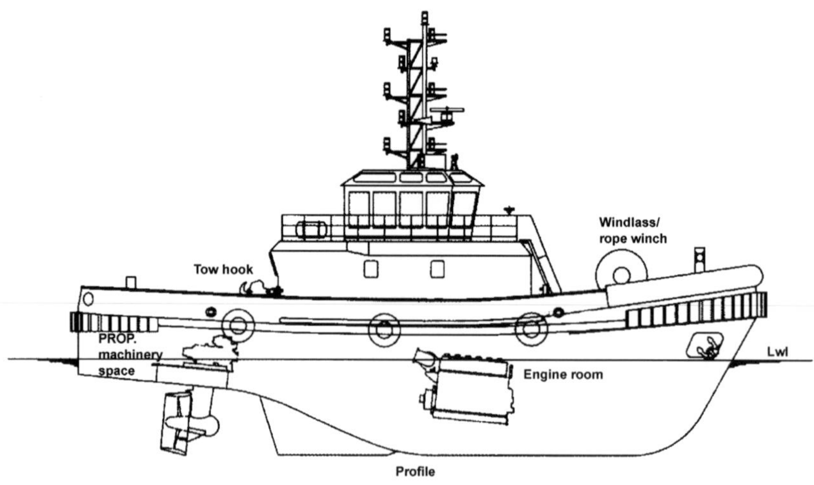

FIGURE 15.9 Tug. *Courtesy of RINA.*

ship carrying hazardous cargo, which found itself unable to steer. In this concept, the assisted ship may be moving at 10 knots or more. Some authorities favour a free-running tug so as not to endanger ship or tug in the majority (incident free) of operations. In this case, the tug normally runs ahead of the ship. It has the problem of connecting up to the ship in the event of an incident. For this reason, other authorities favour the tug being made fast to the ship when starting escort duties either on a slack or taut line.

Tugs can be used as part of an integrated tug/barge system. This gives good economy with one propelled unit being used with many dumb units.

The trends in tug design in the last decade of the twentieth century included the following:

- Tugs with azimuthing propulsion became most common for harbour work.
- Lengths range up to 45 m but tugs of 30–35 m dominate, with powers generally 2500–3000 kW with a few as high as 5000 kW.
- B/D and B/L ratios have increased to provide greater stability.
- Bollard pull is generally 60–80 tonnes at 5000 kW installed power.
- Free-running speeds ranging from 10 to 15 knots increasing linearly with the square root of the length.

Icebreakers and Ice Strengthened Ships

Ice as a hazard was discussed in the section on the ocean environment. The main function of an icebreaker is to clear a passage through ice so that other ships can use areas which would otherwise be denied to them. Icebreakers are vital to the economy of nations with ports that are ice bound for long periods of the year and those wishing to develop the natural resources within the Arctic. Icebreakers need to use special steels which remain tough at low temperature, extra structure in the bow and along the waterline, high propulsive power and manoeuvring devices which are not susceptible to ice damage.

The hull form enables them to ride up over the ice, this being one way of forcing a way through ice. The ship uses its weight to break the ice. The ship may be ‘rocked’ by transferring ballast water longitudinally. The hull is well rounded and may roll heavily as protruding stabilisers are unacceptable. Good hull subdivision and special hull paints are used. They are expensive to acquire and operate.

Other ships which need to operate in the vicinity of ice are strengthened to a degree depending upon the perceived risk. Typically they can cope with continuous 1-year-old ice of 50–100 cm thickness, have a double hull with thicker plating forward and in the vicinity of the waterline, have extra framing and a rounded bow form. Rudders and propellers are protected from ice contact by the hull shape. Engine cooling water inlets must not be allowed to become blocked.

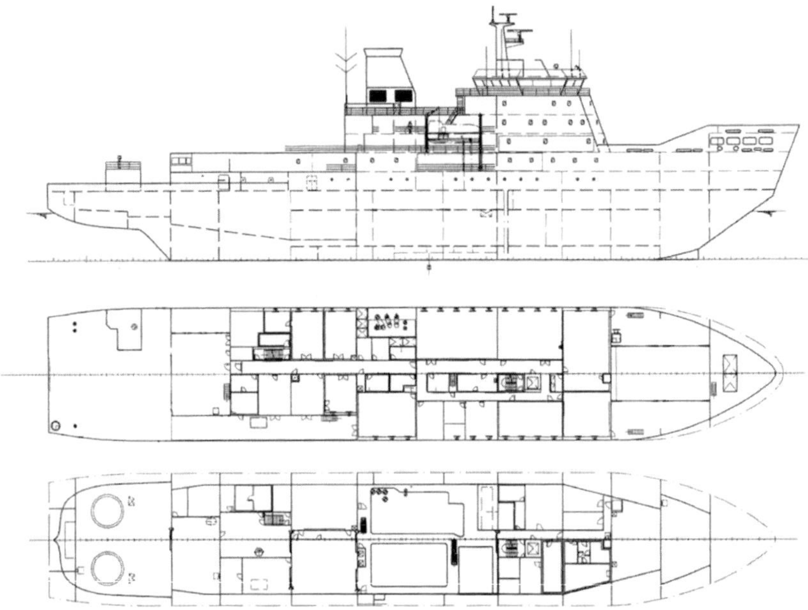

FIGURE 15.10 Ice-breaking research vessel. *Courtesy of RINA.*

Figure 15.10 shows an ice-breaking research ship of length 110 m and about 9000 tonnes displacement, with a block coefficient of 0.616. It has a cruising speed of 16 knots and endurance of 20,000 nm.

HIGH-SPEED CRAFT

A number of hull configurations and propulsion systems are available, each designed to overcome specific problems or to confer some desired advantage. Thus catamarans avoid the loss of stability at high speed suffered by round bilge monohulls. They also provide large deck areas for passenger use or deployment of research or defence equipment. Hydrofoil craft benefit from reduced resistance by lifting the main hull clear of the water. Air cushion vehicles give the possibility of amphibious operation. The effect of waves on performance is minimised in the Small Waterplane Area Twin Hull (SWATH) concept. Some craft are designed to reduce wash so that they can operate at higher speeds in restricted waters.

The choice of design depends upon the intended service. In some cases a hybrid is used. Although most applications of these concepts have been initially to small craft, some are now appearing in the medium size, especially for high-speed ferry services.

In commercial applications, one of the special characteristics, such as those mentioned above, may be the deciding factor in the adoption of a particular hull form. In other cases, particularly for ferries, it may be the extra speed possible. One way of assessing the relative merits of different forms is what is termed the *transport efficiency factor* which is the ratio of the product of payload and speed to the total installed power.

Monohulls

Monohulls can achieve high speed by suitable hull design. Round bilge forms at higher speeds can have stability problems and many small high-speed monohulls have hard chines. The chine is the intersection of the vessel's bottom and sides. In a hard chine form, the chine marks a distinct change in slope of the outer bottom. It is used for planing craft in which, at speed, a significant portion of the craft's weight is supported by hydrodynamic forces acting on the flatter section of the bottom. Hard chine forms with greater beam and reduced length give improved performance in calm water but experience high vertical accelerations in a seaway. Their ride can be improved by using higher deadrise angles leading to a 'deep vee' form. Current practice favours round bilge for its lower power demands and its sea kindliness at cruising speed, with the adoption of hard chines for Froude numbers above unity for better stability. The round bilge form can be fitted more readily with bilge keels to reduce rolling.

Surface Effect Ships

The earliest form of surface effect ship (SES) was the hovercraft in which the craft was lifted completely clear of the water on an air cushion, created by blowing air into a space under the craft and contained by a skirt. For these craft propulsion is by airscrew or jet engines. In some later craft, rigid sidewalls remain partially immersed when the craft is raised on its cushion and the skirt is only needed at the ends. The sidewalls mean that the craft is not amphibious and cannot negotiate very shallow water. They do, however, improve directional stability and handling characteristics in winds. They limit the leakage of air from the cushion reducing the lift power needed and they enable more efficient water propellers or waterjets to be used for propulsion. The sidewalls may provide sufficient buoyancy to keep the cross structure clear of the water when at rest with zero cushion pressure.

The effect of the air cushion is to reduce the resistance at high speeds. For ferries, which operate close to their maximum speed for the major part of their passage, it is desirable to operate at high Froude number to get beyond the wavemaking hump in the resistance curve. This, and the wish to reduce the cushion perimeter length for a given plan form area, means that most SESs have a low length to beam ratio. In other applications, the craft may be required to operate efficiently over a range of speeds. In this case, a somewhat higher length to beam ratio is used to give better fuel consumption rates at the lower cruising speeds.

SESs are employed as ferries on a number of short-haul routes. Passenger seating is located above the central plenum chamber with the control cabin one deck higher. Ducted air propellers and rudders are located aft to provide forward propulsion and lateral control. Centrifugal fans driven by diesel engines create the air cushion. Manoeuvrability is helped by air jet driven bow thrusters.

Early SESs were relatively high cost, noisy craft requiring a lot of maintenance, particularly of the skirts which quickly became worn. Later versions are considerably improved in all these respects. Naval applications include landing small numbers of covert forces and in mine hunting. In the former, an amphibious craft can cross an exposed beach quickly. The latter use arises from the relative immunity of SESs to underwater explosions.

Hydrofoil Craft

Hydrofoil craft make use of hydrodynamic lift generated by hydrofoils attached to the hull. As the craft accelerates, the resistance increases due to increasing wave resistance and then drops off as the main hull leaves the water. It is the forward movement through the water that causes a lift force on the foils counteracting the craft's weight. Once the hull is clear of the water, the resistance is reduced to that of the foils. With high lift to drag sections, high speeds are possible at relatively low powers. Once the hull is clear of the water, the lift required of the foils is effectively constant. As speed increases either the submerged area of foil will reduce or their angle of incidence must be reduced. This leads to the following two foil systems:

- Completely submerged, incidence controlled: these foils remain completely submerged, reducing the risk of cavitation, and lift is varied by controlling the angle of attack of the foils to the water. This is an 'active' system and can be used to control the way the craft responds to oncoming waves. It can be made to contour the waves.
- Fixed surface-piercing foils: these foils may be arranged as a ladder either side of the hull or as a large curved foil passing under the hull. As speed increases the craft rises thereby reducing the area of foil needed to create the lift. When meeting a wave, the forward foil is submerged more deeply, generates more lift and raises the bow. This is a 'passive' system.

Foils are provided forward and aft, the balance of area being such as to provide the desired ride characteristics. The net lift must be in line with the centre of gravity of the craft. Usually there is a large foil close to the LCG. This provides most of the lift and the other, smaller, foil maintains the trim angle of the craft. The surface-piercing foil will automatically provide a degree of transverse stability due to the greater foil area immersed on the lower side. With fully submerged foils ailerons either side of the foil can be moved to provide a moment opposing heel. Like the SES, the hydrofoil has been used for service on relatively short-haul journeys. Both types of craft have stability characteristics which are peculiarly their own.

Multi-Hulled Vessels

The twin hulls of the catamaran provide large upper deck areas for passenger facilities in ferries or for helicopter operations, say. Layout within the main box connecting the hulls can be very flexible. The greater wetted hull surface area leads to increased frictional resistance, but the relatively slender hulls can have reduced wave resistance at higher speeds, sometimes assisted by interference effects between the two hulls. A hull separation of about 1.25 times the beam of each hull is reasonable in a catamaran. Manoeuvrability is good.

High transverse stability and relatively short length mean that seakeeping is not always good. This is improved in the wave-piercing catamarans developed to reduce pitching, and in SWATH designs where the waterplane area is very much reduced and a large part of the displaced water volume is well below the waterline. See Figure 15.11. This confers much improved seakeeping performance but at the expense of increased draught and cost. Because

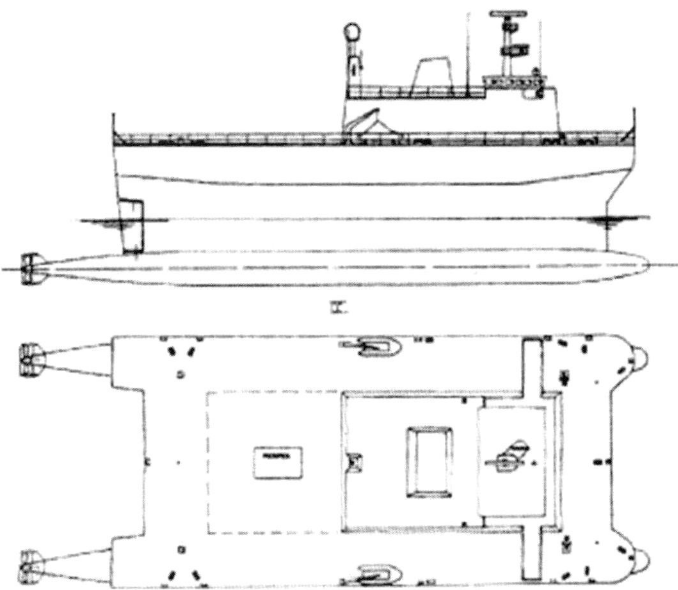

FIGURE 15.11 Concept study of a SWATH. *Courtesy of RINA.*

of their small waterplane areas, SWATH ships are very sensitive to changes in load and its distribution, so weight control is vital.

A development of twin hull vessels is the trimaran. Many design studies indicated many advantages with no significant disadvantages. To prove the concept, and particularly to prove the viability of the structure, a 98 m, 20 knot, demonstrator – *RV Triton* – was completed in 2000. Its structure was designed in accordance with the High Speed and Light Craft Rules of Det Norske Veritas (DNV). The main hull is of round bilge form. The side hulls are of multi-chine design on the outboard face with a plane inboard face. The main hull structure is conventional and integrated with a box girder like cross deck from which the side hulls extend. Propulsion is diesel electric with a single propeller, and rudder, behind the main hull with small side hull thrusters. The trials were extensive and in most cases successfully vindicated the theories. Studies have also been made of a pentamaran form with a slender main hull and two small hulls each side.

Comparisons of monohulls with multi-hull craft are difficult. Strictly designs of each type should be optimised to meet the stated requirements. Only then can their relative merits and demerits be established. For simpler presentations, it is important to establish the basis of comparison be it equal length, displacement, speed or carrying capacity.

Multi-hull designs have a relatively high structural weight and often use aluminium to preserve payload. Wave impact on the cross structure must be avoided or minimised; so high freeboard is needed together with careful shaping of the undersides of the cross structure.

Rigid Inflatable Boats

Inflatable boats have been in use as leisure craft for many years and, with a small payload, can achieve high speed. The first rigid inflatables came into being in the 1960s with an inflatable tube surrounding a wooden hull. Much research has gone into developing very strong and durable fabrics for the tubes to enable them to withstand the harsh treatment these craft get. Later craft have used reinforced plastic and aluminium hulls. Rigid inflatable boats (RIBs) come in a wide range of sizes and types. Some are open, some have enclosed wheelhouse structures, some have outboard motors and others have inboard engines coupled to propellers or waterjets. Lengths range from about 4–16 m and speeds can be as high as 80 knots.

Users include the military, coastguards, customs and excise, the Royal National Lifeboat Institution (RNLI), oil companies and emergency services. Taking the RNLI use as an example, the rigid lower hull is shaped to make the craft more sea kindly and the inflatable collar safeguards against sinking by swamping. Their 'D' Class is 5.12 m long, 1.9 m breadth and weighs 655 kg with a crew of three. It has a rigid deck, an inflated collar surrounding the deck and an inflatable keel.

WARSHIPS

Often merchant ship owners operate one type of vessel – cruise ships, tugs, ferries or fishing vessels. The military are responsible for a wide range of vessels including merchant ship types for replenishing fuel and stores. Warships should be considered in terms of the role they are to play as part of a task force as well as the individual roles they may have. The task force will have a mission that might be to land army units across a defended beach, as part of a multi-national (perhaps United Nations) force. To do this requires:

- producing local air superiority possibly outside the range of friendly shore based aircraft;
- defending the whole force against enemy aircraft, submarines, motor torpedo boats and mines;
- bombardment of the enemy ashore using guns and missiles and
- transporting the troops and their equipment to the area and then landing them on/across the beach which is likely to be well defended.

In discussing the features of individual ships, their role as part of a large task force must be borne in mind. All will need sensors to detect enemy ships and weapons to be able to defend themselves and attack others. They must be stealthy, i.e. difficult for an enemy to detect. Their ability to survive depends upon their *susceptibility* to being hit and *vulnerability* to the effects of a striking weapon.

Susceptibility

A ship's susceptibility depends upon its ability to avoid detection by an enemy and, failing that, to foil the enemy attack. Inevitably a warship creates a number of *signatures* which can betray its presence and provide signals on which weapons can home or which may trigger mines. All must be kept as low as possible to make detection more difficult. The signatures include:

- *Noise* from the propulsor, machinery or the flow of water past the ship. An attacking ship can detect noise by passive sonars without betraying its own presence. Noise levels can be reduced by special propulsor design, by mounting noisy machines and by applying special coatings to the hull. Creating a very smooth hull reduces the risk of turbulence in the water.
- *Radar cross section*. When a ship is picked up by enemy radar, the incident pulses are reflected. Although scattered to a degree, the strength of the returning pulse depends upon the ship's size and geometry. Arranging the structural shape so that the returning pulses are scattered over a wider arc can render the returning signal much weaker. Additionally, much of the incident signal can be absorbed by applying radar absorbent materials to the outer surfaces of the ship.

- *Sonar detection*. Whilst less can be done in the way of changing hull shape below water, sonar absorbing materials can be used to reduce the returning pulse when illuminated by active sonar. Hull coating can also reduce the ship's noise transmissions to make it more difficult to detect by passive sonars.
- *Infrared* emissions from areas of heat. The principle is the same as that in the instruments used by rescue services to detect the heat from human bodies buried in debris. The ship will be warmer than its surroundings but the main heat concentrations can be reduced. The funnel exhaust can be cooled. Heat transmissions into the air can be reduced by insulating hot spaces.
- *Magnetic*. Many mines are triggered by the changes in the local magnetic field caused by the passage of a ship. All steel ships have a degree of in-built magnetism which can be countered by creating opposing electromagnetic fields with special (degaussing) coils. In addition the ship distorts the earth's magnetic field. This effect can be reduced in the same way, but the ship needs to know the strength and direction of the earth's field in its current location in order to determine the corrections to apply.
- *Pressure*. The ship causes a change in the pressure field as it moves through the water and mines can respond to this. The effect can be reduced by the ship going slowly and this is the usual defensive measure adopted when, say, entering a port that is likely to be mined. By going astern the mine is likely to explode under the bow rather than under the stern with less significant damage to the ship.

It is impossible to remove the signatures completely. Indeed there is no need in some cases as the sea has a background noise level, and waves generate pressure variations, which help to 'hide' the ship. The aim of the designer is to reduce the signatures to levels where the enemy must develop more sophisticated sensors and weapons, usually at the expense of smaller warheads. Low signatures improve the chances of countermeasures being effective and of decoys seducing weapons aimed at the ship. As part of its countermeasures, a ship can jam an enemy's radars or tracking systems but acts such as this can themselves betray the presence of a ship. Passive protection methods are to be preferred.

Sensors

Sensors require a good field of view and their performance must not be degraded by the ship's own transmissions. Search radars require complete 360° coverage and are placed high in the ship. Hull-mounted sonars are usually fitted below the keel forward where they are remote from major noise sources and where the boundary layer is still relatively thin. Some ships carry sonars that can be towed astern to isolate them from ship noises and to enable them to operate at a depth from which they are more likely to detect a submarine.

Weapon control radars need to be able to match the arcs of fire of the weapons they are associated with. Increasingly this means 360° as many missiles are launched vertically and then turn in the direction of the enemy. Some weapon guidance systems depend upon the parent ship illuminating the enemy to enable weapons to locate and home in on the target. Others rely upon systems in the weapon itself – particularly for the final stages of engagement.

Own Ship Weapons

A ship's own weapons can present problems for it, apart from those of weight, space and supplies. They require good arcs of fire allowing for the shell/missile trajectory. Missiles create an efflux which can harm protective coatings on structure, exposed personnel or sensitive equipment. Weapons carry a lot of explosive material and precautions are needed to reduce the risk of premature detonation due, say, to the ship's electromagnetic transmissions. Magazines are protected as much as possible from penetration by enemy light weapons and special firefighting systems are fitted. Venting is provided to prevent high pressure build-up in the magazine in the event of a thruster motor igniting. Magazine safety is covered by special regulations and trials are carried out.

Enemy Weapons

Task forces adopt a policy of layered defence. The aim is to detect an enemy target or incoming weapon, at the greatest possible range and engage it with a long-range defence system. This may be a hard kill system to take out the enemy vehicle or weapon, or one which causes the incoming weapon to become confused and unable to press home its attack. If the weapon penetrates the first line of defence, a medium range system is used and then a short range one. It is in the later stages that decoys may be deployed. The incoming weapon's homing system locks on to the decoy and is diverted from the real target, although the resulting explosion may still be uncomfortably close. The shortest range systems are the *close-in weapon systems*. These are extremely rapid firing guns which put up a veritable curtain of steel in the path of the incoming weapon. At these very short ranges, even a damaged missile, may still hit the ship and cause considerable damage.

Vulnerability

Even good defence systems can be defeated, if only by becoming saturated. The ship, then, must be able to withstand a measure of damage before it is

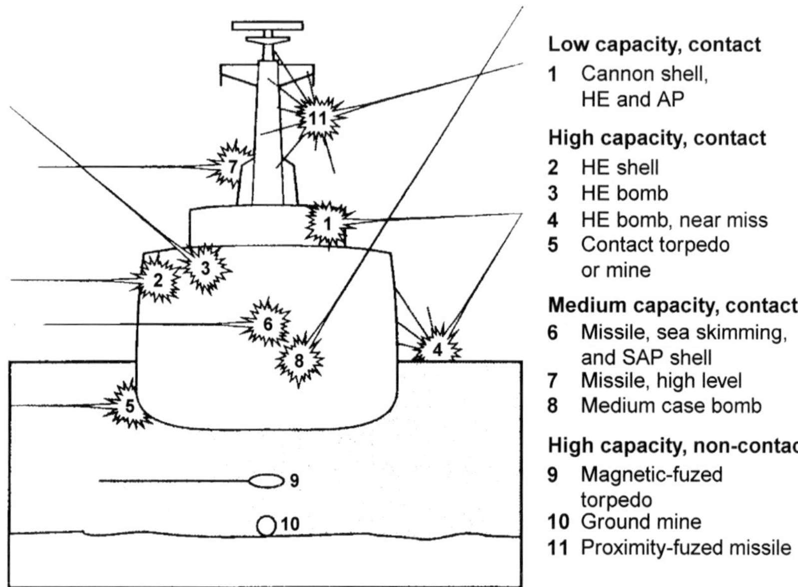

FIGURE 15.12 Conventional weapon attack. *Courtesy of RINA.*

put out of action completely and even more before it sinks. The variety of conventional attack to which a ship may be subject is shown in Figure 15.12.

The effects on the ship generally involve a combination of structural damage, fire, flooding, blast, shock and fragment damage. The ship must be designed to contain these effects within as small a space as possible by *zoning*, separating out vital functions so that not all of a given capability is lost as a result of one hit, providing extra equipment (redundancy) and protection of vital spaces. This latter may be by providing splinter proof plating or by siting well below the waterline. An underwater explosion is perhaps the most serious threat to a ship, particular if it occurs just below the keel producing whipping and shock.

Vulnerability Studies

General ship vulnerability was discussed earlier. Whilst important for all ships, it is especially significant for warships because they can expect to receive damage in action. A vulnerability assessment is carried out for each new design to highlight any weaknesses. The designer considers the probability of each of the various methods of attack an enemy might deploy, their chances of success and the likely effect upon the ship's capabilities. A fighting capability could be the ability to destroy an incoming enemy missile. The contribution of each element of the ship and its systems to each capability is noted. For instance, to destroy a missile would require detection and classification radar, a launcher and weapons, as well as electrics and chilled water services and a command system. Some elements will contribute to several capabilities. For each form of attack, the probability of the individual elements being rendered non-operative is assessed using a blend of historical data, calculation, model and full-scale tests. If one element is particularly susceptible to damage, or especially important, it can be given extra protection or it can be duplicated to reduce the overall vulnerability. This modelling is similar to that used for reliability assessments. The assessments for each form of attack can be combined, allowing for the probability of each, to give an overall vulnerability of the design. The computations can become quite lengthy and as some judgments are difficult to make, the results must be interpreted with care. For instance, reduced general services such as electricity may be adequate to support some but not all fighting capabilities. What then happens, in a particular engagement, will depend upon which capabilities the command needs to deploy at that moment. For this reason the vulnerability results are set in the context of various engagement scenarios. In many cases, the consequences of an attack will depend upon the actions taken by the crew in damage limitation. For instance, how effectively they deal with fire and how rapidly they close doors and valves to limit flooding. Recourse must be made to exercise data and statistical allowances made for human performance.

Whilst such analyses may be difficult, they can highlight design configuration weaknesses early in the design process when they can be corrected at little cost.

Types of Warship

Warships are categorised by their function – aircraft carrier, guided missile destroyer, frigate, mine countermeasure vessel, submarine support ship or amphibious operations ship. It is not possible to describe all these types but the following notes provide a feel for what is involved in their design. The task force will exercise defence in depth and if a carrier is present, its aircraft will provide a long range surveillance – airborne early warning (AEW) – and attack capability. Then there will be long- and medium-range missile systems on the cruisers and destroyers. All ships are likely to have decoy systems to seduce incoming weapons and a close-in weapon system. The range of sensors provided must match the requirements of the weapons fitted.

Aircraft Carriers

These are among the largest and most complex vessels in any navy. They carry aircraft which can act in support of amphibious operations, detect and attack enemy submarines using helicopters, provide early warning of enemy aerial attack, provide defence for the task force against aerial attack and use their own missile and gun systems to defend themselves against attack. The carrier design will be greatly affected by the type(s) of aircraft carried.

Carriers deploying fixed wing aircraft use catapults to assist take-off and arrester wires to assist recovery. The catapults are usually steam driven. They are at the bow and the fore end of the angled deck and must be long in order to keep accelerations to an acceptable level. For the same reason, the arresting wires have a long pull-out. These carriers are large and complex with, perhaps more than 2000 individual compartments and a large ship's complement augmented by the air crews and aircraft maintainers. Smaller carriers can operate vertical or short take-off and landing aircraft without these aids. Aircraft are stowed in large hangars just below the flight deck and/or in deck parks. The hangars must be at least two decks high as well as occupy most of the width of the ship. The bridge structure is on the starboard side of the flight deck which is angled to port. This means that an aircraft that 'overshoots' does not crash into aircraft parked forward on the deck. Lifts forward and aft move aircraft between the flight deck and the hangar. Aircraft usually have folding wings to facilitate stowage. In addition carriers have a deck landing mirror sight on the port side aft for guiding aircraft on to the deck, facilities for aircraft maintenance and large quantities of aviation fuel and weaponry.

Figure 15.13 shows a typical cross section of a carrier. This illustrates a number of problems faced by the carrier designer including running uptakes

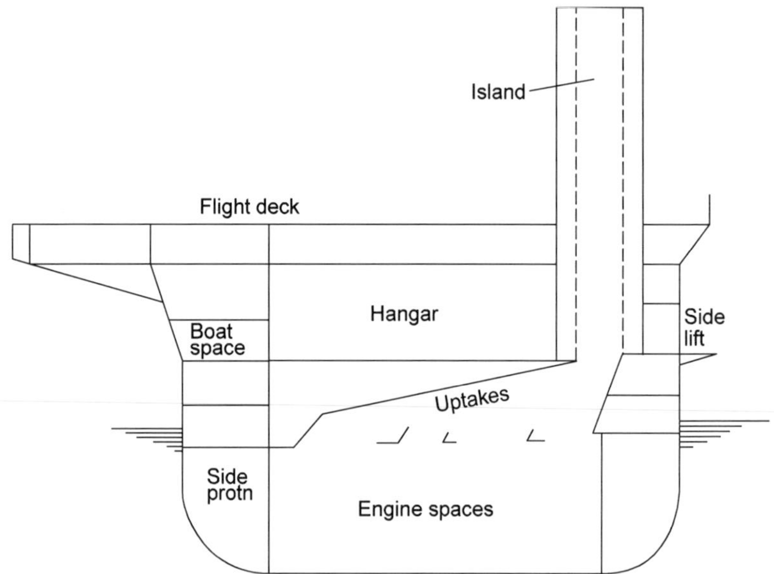

FIGURE 15.13 Aircraft carrier section.

to the funnel in the island, stowing boats and providing side protection against torpedo attack.

Amphibious Warfare Vessels

These fall into two general types:

- Largish vessels transporting troops with their vehicles, weapons and stores.
- Smaller craft to land the troops and their equipment. Helicopters would also be deployed for this part of the operation.

The larger vessels will include:

- Carrier type ships carrying troops and helicopters.
- Landing ships (dock) which are basically mobile floating docks in which smaller landing craft can be carried. They also have a helicopter operating deck over the dock and carry the troops, tanks and so on that are to go in the small craft and helicopters.
- Landing ships with bow doors and ramps which can disgorge troops and vehicles directly on the beach.
- Troopships for the main body of troops with smaller craft in davits for landing the troops on the beach.
- Ships carrying fuel, ammunition and stores to support the early stages of the landing.

The smaller craft will include:

- Craft for landing troops.
- Craft for landing tanks and other vehicles.
- Duplex drive vehicles which can be driven through the water and then across dry land.
- Craft with guns and rockets to give the force an increased shore bombardment capability.

Frigates and Destroyers

These cover a range of lengths and displacements, from 1500 to 6000 tonnes and from 80 to 150 m length. The UK Royal Navy uses the term frigate for smaller of the two, whereas in the US Navy, destroyers are the smaller. They have relatively slender hull forms because they operate at high speeds from 25 to 30 knots. They tend to be maids of all work but with a main function which may be anti-submarine or anti-aircraft. Their weapon and sensor fits and other characteristics reflect this. Most carry a gun for shore bombardment or to engage smaller surface vessels, but usually the main armament is some form of missile system designed to engage the enemy at some distance from the ship. The missile, having been fired from a silo or specialist launcher, may be guided all the way to the target by sensors in the ship or may be self-directing and homing. In the latter case, having been fired in the general direction of the target, the weapon's own sensors acquire the target and control the final stages of attack, leaving the ship free to engage other targets. Helicopters greatly extend the area of ocean over which the ship can exert an influence making them very useful in policing an area in anti-pirate or anti-smuggling operations. They are used for reconnaissance and for attacks on submarines or surface vessels. In acting as an escort to a task force, the ships must be sufficiently fast and manoeuvrable to hold and change station as the group changes course.

A key design driver for these vessels is the layout of the upper deck which typically must provide for:

- anchoring and mooring facilities forward;
- a gun forward plus a missile silo/launcher;
- a bridge/mast structure for command and for carrying sensors, with adequate arcs of view;
- funnels/uptakes;
- communication aerials. To give the required frequency range, some roof aerials must be long and this often leads to a second superstructure block and mast aft of the funnel(s). Other frequencies are provided by exciting masts and other structural elements;
- hangar and helicopter deck and
- a towed sonar system aft.

Many of these features impact upon the internal ship layout which limits the freedom the designer has in siting them. The funnel(s) in relation to the machinery spaces and the missile launchers relative to the magazines are two examples.

Mine Countermeasures Vessels

Mine countermeasure ships may be either sweepers or hunters of mines, or combine the two functions in one hull. Modern mines can lie on the bottom and only become active when they sense a target with quite specific signature characteristics. They may then explode under the target or release a homing weapon. They may only react after a selected number of ships have passed nearby or only at selected times. All these features make them difficult to render harmless.

Sweeping mines depend upon either cutting their securing wires or setting them off by simulated signatures to which they will react. The latest mines have been developed to the point where they are virtually unsweepable. They need to be hunted, detection being usually by a high-resolution sonar. They can then be destroyed by placing a small charge alongside the mine, usually laid by a remotely operated underwater vehicle, and setting it off. Because mine countermeasure vessels themselves are a target for the mines they are trying to destroy, the ship signatures must be extremely low and the hulls very robust. Nowadays hulls are often made from glass reinforced plastic and much of the equipment is specially made from materials with low magnetic properties.

Submarines

In spite of its limited range of military capabilities at that time, the submarine with its torpedoes proved a very potent weapon during the two world wars in the first half of the twentieth century. It could fire torpedoes, lay mines and land covert groups on an enemy coast. It could not remain submerged for very long and was more strictly a submersible. Since then its fighting capabilities have been extended greatly by fitting long-range missile systems which can be deployed against land, sea or air targets. The intercontinental ballistic missile, with a nuclear warhead, enabled the submarine to become the principal deterrent system of the major powers; cruise missiles can be launched against land targets well inland and without the need for the vessel to come to the surface. It is a difficult target for an enemy to locate and attack. The advent of nuclear propulsion enabled submarines to remain submerged for weeks on end, although this entails special air treatment systems. There are now a number of air independent propulsion (AIP) systems giving long underwater endurance. Closed-cycle diesel engines, fuel cells and Stirling engines are possibilities. The systems still require a source of oxygen such as high-test peroxide or liquid oxygen. Fuel sources for fuel cell application include sulphur-free diesel fuel, methanol and hydrogen. Thus today the submarine is a powerful, versatile, multi-role vessel. Clancy (1993) gives a comprehensive description of two nuclear submarines.

Submarines present a number of special challenges to the naval architect including:

- Although intended to operate submerged for most of the time, they must be safe and manoeuvrable on the surface.
- They are unstable in depth. Going deeper causes the hull to compress reducing the buoyancy force.
- Underwater, they manoeuvre in three dimensions, often at high speed. They must not betray their presence to an enemy by breaking the surface. Nor can they go too deep or they will implode. Thus manoeuvres are confined to a layer of water only a few ship lengths in depth.
- Machinery must be able to operate independently of the earth's atmosphere. The internal atmosphere must be kept fit for the crew to breathe and free of offensive odours.
- Escape and rescue arrangements must be provided to assist in saving the crew of a stricken submarine.

The layout of a typical conventional submarine is shown in Figure 15.14. Its main feature is a circular pressure hull designed to withstand high

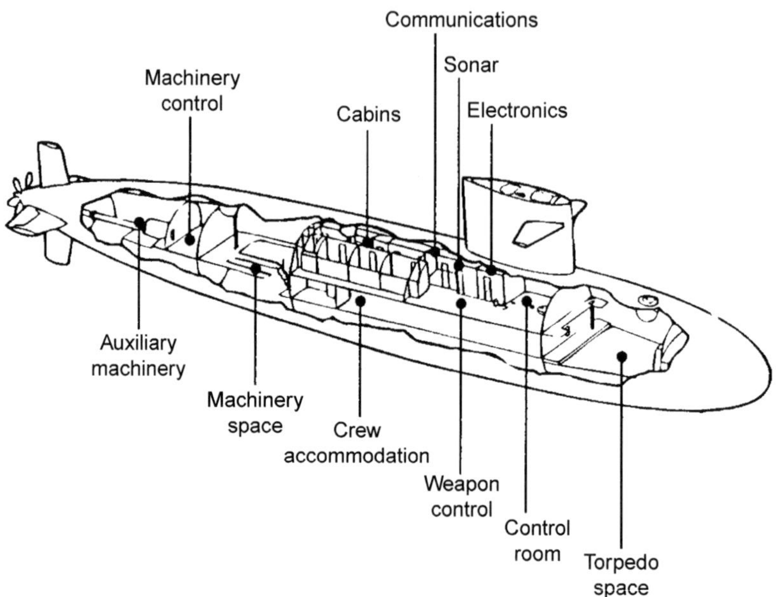

FIGURE 15.14 Submarine. *Courtesy of RINA*.

hydrostatic pressure. Since it operates in three dimensions, the vessel has hydroplanes for controlling depth as well as rudders for movement in the horizontal plane. Large tanks, mainly external to the pressure hull, are needed which can be flooded to cause the ship to submerge or blown, using compressed air, for surfacing. 'Conventional' submarines use diesels for surface operations and electric drive, powered by batteries, when submerged. An air intake pipe or 'snort' mast can be fitted to enable air to be drawn into the boat at periscope depth. Batteries are being constantly improved to provide greater endurance. Nuclear propulsion is expensive and brings with it problems of disposing of spent reactor fuel. For these reasons, increasing interest is being taken in AIP systems.

The hydroplanes, fitted for changing and maintaining depth, can only exert limited lift, particularly when the submarine is moving slowly, so the vessel must be close to neutral buoyancy when submerged, and the longitudinal centres of buoyancy and weight must be in line. The weight distribution before diving, and the admission of ballast water when diving, must be carefully controlled. The first task when submerged is to 'catch a trim', i.e. to adjust the weights by the small amounts needed to achieve equilibrium. Since there is no waterplane, when submerged the metacentre and centre of buoyancy will be coincident and BG will be the same for transverse and longitudinal stability. On the surface the usual stability principles apply, but the waterplane area is relatively small. The stability when in transition from the submerged to the surfaced state may be critical and needs to be studied in its own right. The usual principles apply to estimating the powering of submarines except that for deep operations there will be no wavemaking resistance. This is offset to a degree by the greater frictional resistance due to the greater wetted hull surface.

The pressure hull, with its transverse bulkheads, must be able to withstand the crushing pressures at deep diving depth. Design calculations usually assume axial symmetry of structure and loads. This idealisation enables approximate and analytical solutions to be applied with some accuracy. Subsequently detailed analyses can be made of non-axisymmetric features such as openings and internal structure. The dome ends at either end of the pressure hull are important features subject usually to finite element analysis and model testing. Buckling of the hull is possible but to be avoided. Assessments are made of *interframe collapse* (collapse of the short cylinder of plating between frames under radial compression), inter-bulkhead collapse (collapse of the pressure hull plating with the frames between bulkheads) and frame tripping.

The structure is developed so that any buckling is likely to be in the interframe mode and by keeping the risk of collapse at 1.5 times the maximum working pressure acceptably small. The effects of shape imperfections and residual stresses are allowed for empirically. Small departures from circularity can lead to a marked loss of strength and the pressure causing yield at 0.25% shape imperfection on radius can be as little as half that required for perfect circularity.

If a stricken submarine is lying on the seabed, the crew would await rescue if possible. For rescue at least one hatch is designed to enable a rescue submersible to mate with the submarine. The crew can then be transferred to the surface in small groups without getting wet or being subject to undue pressure. The first such rescue craft, apart from some early diving bells, were the two Deep Submergence Rescue Vessels (DSRV) of the US Navy. However, deteriorating conditions inside the damaged submarine may mean that the crew cannot await rescue in this way; the pressure may be rising due to water entry or the atmosphere may become polluted due, say, to water entering the batteries. In such cases, the crew can escape from the submarine in depths down to 180 m. One- and two-man escape towers are fitted to allow rapid compression thereby limiting the body's uptake of gas which would otherwise lead to the 'bends'. A survival suit is worn to protect against hypothermia and a hood holds a bubble of gas for breathing. In Russian submarines emergency escape capsules are provided.

Commercial applications of submarines have been generally limited to submersibles some of which have been very deep diving. Many are unmanned, remotely operated vehicles. Most of these applications have been associated with deep ocean research, the exploitation of the ocean's resources, rescuing the crews of stricken submarines or for investigations of shipwrecks.

A growing use is in the leisure industry for taking people down to view the colourful sub-surface world. In some cases, the submersible may be the only way of tackling a problem such as the servicing of an oil wellhead in situ which is too deep for divers. The search for, location and exploration of the wreck of MV *Derbyshire* used the capabilities of the Woods Hole Oceanographic Institution (WHOI) which operates several research ships together with a number of submersibles.

SUMMARY

The size, appearance and characteristics of ships vary with their intended function, leading to many diverse ship types – both commercial and military. For merchant ships the 'cargo' may be people or material. Passenger vessels include ferries, cruise ships and liners, some carrying several thousand passengers. Ferries may carry a mixture of passengers and vehicles – cars and lorries. Material cargoes may be carried in bulk, in containers or as discrete items. Each brings its own problems of loading, stowing and unloading. Specialist port facilities may be used or vessels may provide their own cargo-handling systems. Some vessels carry out specialist activities such as towing, fishing, lifesaving or offshore support. Again they require special features to enable them to carry out the intended function.

Warships include large aircraft carriers, cruisers, destroyers, frigates, amphibious assault ships and mine countermeasure vessels. Submarines must be able to operate safely for long periods submerged, manoeuvring in three dimensions. AIP systems may be provided to extend the underwater endurance and the atmosphere must be maintained at a high quality.

Only the broad characteristics of the various ship types are covered here. Reference can be made to other texts for details of individual types.

Appendix A

Presentation of Data

UNITS

The units used are those endorsed by the International Organisation for Standardisation, the *Système International d'Unites* (SI). See BS 5555; ISO 1000. The base units and some derived and supplementary units are given in Tables A.1 and A.2. Because some references the reader will need to use are in the older Imperial units, some useful equivalent units are given in Table A.3. In addition the mass unit of tonne (t) is recognised, with 1 tonne = 10^3 kg.

NOTATION

This book adopts the notation used by the international community, in particular by the International Towing Tank Conference and the International Ships Structure Congress. (Go to ITTC Symbols and Terminology on the ITTC website - http://ittc.sname.org.) It has been departed from in some simple equations where the full notation would be too cumbersome. Where there is more than one meaning of a symbol that applying should be clear from the context.

TABLE A.1 SI Base Units

Quantity	Unit Name	Unit Symbol
Length	metre	m
Mass	kilogram	kg
Time	second	s
Electric current	ampere	A
Thermodynamic temperature	kelvin	K
Amount of substance	mole	mol
Luminous intensity	candela	cd

TABLE A.2 Some Derived and Supplementary Units

Quantity	SI Unit	Unit Symbol
Plane angle	radian	rad
Force	newton	$N = kg\ m/s^2$
Work, energy	joule	$J = Nm$
Power	watt	$W = J/s$
Frequency	hertz	$Hz = s^{-1}$
Pressure, stress	pascal	$Pa = N/m^2$
Area	square metre	m^2
Volume	cubic metre	m^3
Density	kilogram per cubic metre	kg/m^3
Velocity	metre per second	m/s
Angular velocity	radian per second	rad/s
Acceleration	metre per second squared	m/s^2
Surface tension	newton per metre	N/m
Pressure, stress	newton per square metre	N/m^2
Dynamic viscosity	newton second per square metre	Ns/m^2
Kinematic viscosity	metre squared per second	m^2/s
Thermal conductivity	watt per metre kelvin	W/(m K)
Luminous flux	lumen (lm)	cd.sr
Illuminence	lux (lx)	lm/m^2

Where a letter is used to denote a 'quantity' such as length it is shown in *italics*. For a distance represented by the two letters at its extremities the letters are in italics. Where a letter represents a point in space it is shown without italics.

TABLE A.3 Some Equivalent Values

Quantity	UK Unit	Equivalent SI Unit
Length	foot	0.3048 m
	mile	1609.34 m
	nautical mile (UK)	1853.18 m
	nautical mile (International)	1852 m
Area	sq. ft	0.0929 m^2
Volume	cub. ft	0.0283 m^3
Velocity	ft/s	0.3048 m/s
	knot (UK)	0.51477 m/s
	knot (International)	0.51444 m/s
Standard acceleration, g	32.174 ft/s^2	9.80665 m^2/s
Mass	ton	1016.05 kg
Pressure	lbf/in^2	6894.76 N/m^2
Power	hp	745.7 W

Symbols

a resistance augment fraction
A area in general
B, b breadth in general
b span of hydrofoil or aerofoil
C coefficient in general, modulus of rigidity, wave velocity
D, d diameter in general, drag force, depth of ship
E modulus of elasticity, Young's modulus
f frequency
F force in general, freeboard
F_n Froude number
g acceleration due to gravity
h height in general
I moment of inertia in general
J advance coefficient of propeller, polar second moment
k radius of gyration
K_Q, K_T torque and thrust coefficients
K, M, N moment components on body
L length in general, lift force
m mass

M	bending moment in general
n	rate of revolution, frequency
p	pressure intensity
p, q, r	components of angular velocity
P	power in general, propeller pitch, direct load
Q	torque
R, r	radius in general, resistance
R_n	Reynolds' number
S	wetted surface
t	time in general, thickness, thrust deduction factor
T	draught, time for a complete cycle, thrust, period
u, v, w	velocity components in direction of x-, y-, z-axes
U, V	linear velocity
w	weight density, Taylor wake fraction
W	weight in general, external load
x, y, z	body axes and Cartesian coordinates
X, Y, Z	force components on body
α	angular acceleration, angle of attack
β	leeway or drift angle
δ	angle in general, deflection, permanent set
θ	angle of pitch, trim
μ	coefficient of dynamic viscosity
ν	Poisson's ratio, coefficient of kinematic viscosity
ρ	mass density
φ	angle of loll, heel, list
ω	angular velocity, circular frequency
V, ∇	volume
Δ	displacement force
η	efficiency in general
σ	cavitation number, direct stress
Λ	tuning factor

Subscripts

Much of the notation above is qualified by a subscript in particular applications. The subscripts used are:

B	block
D	developed (area), drag, delivered (power)
E	effective (power), encounter (waves)
F	frictional (resistance), Froude
H	hull
L	longitudinal, lift
M	midship section
O	open water (propeller)
OA	overall
P	longitudinal prismatic

PP	between perpendiculars
Q	torque
R	residuary (resistance), relative rotative (efficiency), rudder
S	shaft
T	transverse, total, thrust
VP	vertical prismatic
W	waterline, waterplane, wavemaking
WP	waterplane
y	yield (stress)
Θ	pitching, trimming
φ	rolling, heeling
ζ	wave elevation

PROBABILITY DISTRIBUTIONS

Many of the factors a naval architect deals with are variable and need to be dealt with in statistical terms. Wave heights in an irregular sea are an example. Any particular sequence of waves could be represented by measuring successive heights between crests and troughs and plotting the results as a series of rectangles the width of each being a height band (1.5–2.0 m, say) and the height representing the number of measurements in that band. This type of plot is known as a *histogram*. Histograms are not suitable for mathematical manipulation and for mathematical treatment they are replaced by a curve which follows the shape of the histogram.

The variables with which the naval architect deals (and many other random events) can be represented by two such curves, called *probability distributions*:

- The *normal* or *Gaussian* distribution in which the probability of occurrence of a value, x, of the variable under study is defined by the equation:

$$P(x) = [1/\sigma(2\pi)^{0.5}] \exp[-(x-\mu)^2/2\sigma^2]$$

- The *Rayleigh* distribution:

$$P(x) = [x/a] \exp[-x^2/2a]$$

when x is positive

In these equations μ = mean value of x, σ^2 = variance, σ = standard deviation, $2a$ = mean value of x^2 and $P(x)$ is the probability density.

The normal distribution is bell shaped and symmetrical about the mean, μ. The degree of spread of the curve is represented by the value of σ. Half the results will lie within a band between $\mu - 0.675\sigma$ and $\mu + 0.675\sigma$.

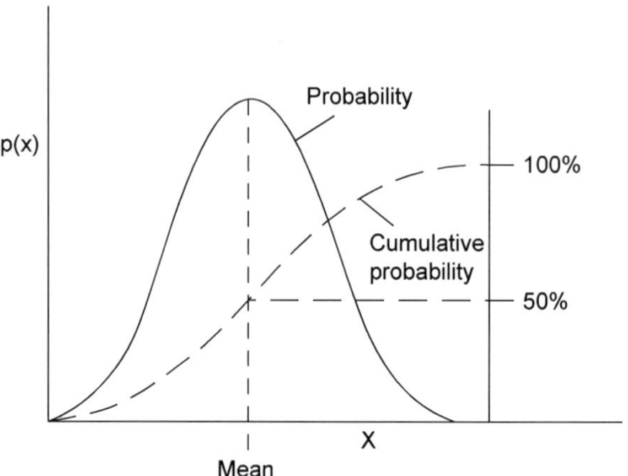

FIGURE A.1 Probability and cumulative probability curves.

For any set of data:

- the *mean* is the sum of the values divided by the number of results. Correspondingly $2a$ is found by squaring the values, summing them and dividing by the number of results;
- the *variation* is found by subtracting the mean from each value, squaring the result and then finding the mean of the values;
- the *standard deviation* is the square root of the variation.

The area under the curve for an increment dx of the variable represents the probability that the variable will have a value within dx. Integrating the area under the curve up to a value X of the variable represents the probability that the variable will be less than X. Plotting these areas against the variable gives the *cumulative probability function.* See Figure A.1 for a normal distribution curve and the corresponding cumulative probability curve.

For more information on these and other probability distributions the reader should refer to a textbook on statistics.

Appendix B

The Displacement Sheet and Hydrostatics

Chapter 4 introduced the concepts of Bonjean curves, a displacement sheet and hydrostatic curves. Rather than place a lot of related numerical work in the main text an example of how they can be derived is placed in this appendix. For an expanded version of this Appendix please visit http://booksite.elsevier.com/9780080982373.

Table B.1 is a displacement sheet, using Microsoft Excel, for a vessel in which the waterplanes are 2 m apart and the sections 14.1 m apart. The actual half ordinates defining the underwater form of a body are shown in bold. For greater definition in way of the turn of bilge an intermediate waterplane has been introduced between waterplanes 5 and 6, the Simpson's multipliers being adjusted accordingly. To simplify the arithmetic the appendages which would usually be found below number 6 waterplane and aft of ordinate 11, have been ignored.

The figures in Row 6 are obtained from multiplying the half ordinates in Row 5 by the corresponding Simpson's multipliers in Row 3. Thus cell M6 is the product of the contents of cells M3 and M5. Cell R6 is the sum of the cells in Row 6 and represents the area of the section at ordinate 1 up to the summer waterline (SWL). The figures in Column S are the result of multiplying the figures in Column R by the Simpson's multipliers in Column B. Cell S28 is the sum of the figures in Column S and represents the volume of the immersed body. The figures in Column U are the products of Columns S and T. Cell U28 is the sum of the figures in Column U and represents the moment of the buoyancy force about amidships.

Correspondingly, the figure in Cell D7 is the product of the figure in cell C7 and the Simpson's multiplier in cell B7. Then cell D28 is the sum of the figures in Column D and represents the area of waterplane 6. The figures in Row 30 are the result of multiplying the figures in Row 28 by the Simpson's multipliers in Row 3, noting that the SM for Column D is that appearing in Column C, and so on. Cell R30 is the sum of the figures in Row 30 and represents the immersed volume of the body. *Note*: As a check on the accuracy of the arithmetic the figures in Cells S28 and R30 are the same at 3903.66. The figures in Row 33 are obtained by multiplying the figures in Row 30 by the

TABLE B.1 Displacement Sheet

	A	B	C	D	E	F	G	H	I	J	K	L	M	N	O	P	R	S	T	U
1	Waterline		6		5.5		5		4		3		2		SWL					
2																				
3	Ordinate	SM	0.5		2		1.5		4		2		4		1		F(*A*)	F(*A*) × SM	Lever aft	F(*M*)
4																				
5	1	1	**0.000**	0.000	**0.000**	0.000	**0.000**	0.000	**0.100**	0.100	**0.100**	0.100	**0.100**	0.100	**0.100**	0.100				
6			0.000		0.000		0.000		0.400		0.200		0.400		0.100		1.100	1.1	−5	−5.5
7	2	4	**0.500**	2.000	**1.060**	4.240	**1.640**	6.560	**2.760**	11.040	**3.640**	14.560	**4.440**	17.760	**5.200**	20.800				
8			0.250		2.120		2.460		11.040		7.280		17.760		5.200		46.110	184.44	−4	−737.76
9	3	2	**1.900**	3.800	**3.280**	6.560	**4.700**	9.400	**6.660**	13.320	**8.100**	16.200	**9.100**	18.200	**9.840**	19.680				
10			0.950		6.560		7.050		26.640		16.200		36.400		9.840		103.640	207.28	−3	−621.84
11	4	4	**4.500**	18.000	**6.800**	27.200	**8.600**	34.400	**10.640**	42.560	**11.800**	47.200	**12.480**	49.920	**12.800**	51.200				
12			2.250		13.600		12.900		42.560		23.600		49.920		12.800		157.630	630.52	−2	−1261.04
13	5	2	**7.240**	14.480	**10.100**	20.200	**11.800**	23.600	**13.300**	26.600	**13.800**	27.600	**14.000**	28.000	**14.040**	28.080				

14			3.620		20.200		17.700		53.200		27.600		56.000		14.040		192.360	384.72	−1	−384.72
15	6	4	**9.000**	36.000	**11.900**	47.600	**13.240**	52.960	**14.200**	56.800	**14.500**	58.000	**14.500**	58.000	**14.400**	57.600				
16			4.500		23.800		19.860		56.800		29.000		58.000		14.400		206.360	825.44	0	0
17	7	2	**7.900**	15.800	**10.700**	21.400	**12.400**	24.800	**13.640**	27.280	**14.080**	28.160	**14.220**	28.440	**14.200**	28.400				
18			3.950		21.400		18.600		54.560		28.160		56.880		14.200		197.750	395.5	1	395.5
19	8	4	**5.500**	22.000	**8.000**	32.000	**9.700**	38.800	**11.900**	47.600	**13.020**	52.080	**13.540**	54.160	**13.700**	54.800				0
20			2.750		16.000		14.550		47.600		26.040		54.160		13.700		174.800	699.2	2	1398.4
21	9	2	**3.000**	6.000	**4.500**	9.000	**6.040**	12.080	**8.640**	17.280	**10.700**	21.400	**12.020**	24.040	**12.600**	25.200				
22			1.500		9.000		9.060		34.560		21.400		48.080		12.600		136.200	272.4	3	817.2
23	10	4	**0.940**	3.760	**1.560**	6.240	**2.160**	8.640	**3.700**	14.800	**5.700**	22.800	**8.000**	32.000	**10.060**	40.240				
24			0.470		3.120		3.240		14.800		11.400		32.000		10.060		75.090	300.36	4	1201.44
25	11	1	**0.100**	0.100	**0.100**	0.100	**0.100**	0.100	**0.100**	0.100	**0.100**	0.100	**0.100**	0.100	**1.300**	1.300				
26			0.050		0.200		0.150		0.400		0.200		0.400		1.300		2.700	2.7	5	13.5
27																				
28				121.940		174.540		211.340		257.480		288.200		310.720		327.400		3903.66		815.18

(*Continued*)

TABLE B.1 Displacement Sheet—(cont.)

	A	B	C	D	E	F	G	H	I	J	K	L	M	N	O	P	R	S	T	U
29																				
30				60.97		349.08		317.01		1029.92		576.40		1242.88		327.40	3903.66			
31	Lever			5		4.5		4		3		2		1		0				
32																				
33	Moment			304.85		1570.86		1268.04		3089.76		1152.8		1242.88		0	8629.19			
34																				
35	Volume	24,463																		
36	Mass Disp	25,075																		
37	LCB aft	2.94																		
38	VCB	5.58																		
39																				
40																				

corresponding levers in Row 31. Cell R33 is the sum of the figures in Row 33 and represents the moment of buoyancy about the SWL.

Since the ordinates are for half the hull the total hull volume is given by:

Volume $= 2 \times (2/3) \times (14.1/3) \times 3903.66 = 24{,}463$ m^3.
Displacement, in tonnes $= 24{,}463\ (1.025) = 25{,}075$ tonnes in sea water.
The centre of buoyancy of the hull from amidships
$= 14.1(815.18)/(3903.66) = 2.94$ m aft.
The centre of buoyancy below the SWL $= 2(8629.19)/(3903.66) = 4.42$ m.

If wished these figures can be calculated within Excel and placed in designated cells in the table.

Once a template has been created for the calculations it can be used repeatedly with new sets of ordinates. If one figure has to be changed in the table the computer will automatically correct all the related figures. The computer calculates figures to the full number of decimal places. The number printed out can be controlled by the relevant command. This should not lead the reader to suppose that the volume and centre of buoyancy position have been calculated this accurately. The ordinates used will have limited accuracy and this will be compounded by the use of approximate integration methods. As a check on the latter, Table B.2 compares the vertical centre of buoyancy (VCB) position by taking moments about the SWL and the keel. It will be noted that the two figures added together correspond very closely to the draught of 10 m.

TABLE B.2 Comparison of VCB

Waterline	Area	SM	F(*V*)	Lever	F(*M*)1	Lever	F(*M*)^2
SWL	3078	1	3078	0	0	10	30,780
2	2921	4	11,684	2	23,368	8	93,472
3	2709	2	5418	4	21,672	6	32,508
4	2420	4	9680	6	58,080	4	38,720
5	1987	1.5	2980.5	8	23,844	2	5961
5.5	1641	2	3282	9	29,538	1	3282
6	1146	0.5	573	10	5730	0	0
			36,695.5		162,232		204,723
Volume			24,463.67				
CB below SWL					4.421033		
CB above keel							5.57967

WATERPLANE AND SECTION AREAS

Embedded in Table B.1 are figures that can be used to derive the area of each waterplane and the area of each section up to the SWL. The former are in the second column of figures under each waterline (Columns D, F, H, etc.); the latter in the second row against each ordinate (Rows 6, 8, 10, etc.). These could be calculated within the main table, but for clarity of presentation they are here presented in Table B.3. The tonnes per cm immersion are calculated for each waterplane.

Tables can be produced for each waterplane, similar to Table 3.2 in Chapter 3, to give the area, centroid position and the longitudinal and transverse moments of inertia. That for the SWL is presented as Table B.4.

Note: In the row against *I*(long) the first figure is the moment of inertia about amidships and the second is the inertia about the centre of flotation.

If desired, displacement sheets can be produced for waterplanes other than the SWL.

A convenient way of calculating the volume of displacement and VCB position for waterlines 2 and 4 (as well as the SWL) is to plot the waterplane areas to obtain the figures for intermediate waterplanes as shown in Table B.5.

BONJEAN CURVES

The Bonjean curves can be calculated, for any section, by integration up to each waterline in turn. The Simpson's rule chosen in each case, and hence the multiplying factor to be used, will depend upon the number of ordinates. Table B.6 derives the section areas up to waterline 5.5 and between SWL and 2WL using Simpson's 5, 8, 1 Rule. Table B.7 derives the section areas between the SWL and 4WL and 5WL using the 1, 3, 3, 1 Rule and between SWL and 5WL using the 1, 4, 1 Rule. Then a table of cross sectional areas can be drawn up as in Table B.8 from which the Bonjean curves can be drawn. Figure B.1 uses the data in Table B.8 to show the Bonjean curves for ordinates 2, 3, 4 and 5.

VOLUMES AND LONGITUDINAL CENTRES OF BUOYANCY

The section areas can be used to calculate the volumes of displacement up to each waterline and the corresponding longitudinal centres of buoyancy as in Table B.9.

METACENTRIC DIAGRAM

The metacentric diagram shows how the VCB and metacentre positions vary with draught. The VCB values have been found above. BM is given by I/V and values are derived in Table B.10 using I and V figures from the other

TABLE B.3 Waterplane and Section Areas

Waterplane Areas											
Waterplane	6	5.5	5	4	3	2	SWL				
F(*A*)	121.94	174.54	211.34	257.48	288.20	310.72	327.40				
Area = F(*A*) × 2 × (14.1/3)	1146.24	1640.68	1986.60	2420.31	2709.08	2920.77	3077.56				
TPC(tonnes/CM) = Area(1.025/100)	11.75	16.82	20.36	24.81	27.77	29.94	31.54				
Section areas up to SWL											
Ordinate	1	2	3	4	5	6	7	8	9	10	11
F(*A*)	1.10	46.11	103.64	157.63	192.36	206.36	197.75	174.80	136.20	75.09	2.70
Area up to SWL = F(*A*) × 2 × (2/3)	1.47	61.48	138.19	210.17	256.48	275.15	263.67	233.07	181.60	100.12	3.60

TABLE B.4 SWL

Station	Half Ord, *y*	SM	F(*A*)	Lever	F(*M*)	Lever	F(*I*) long	yyy	F(*I*) trans
1	0.10	1	0.10	5	0.50	5	2.50	0	0
2	5.20	4	20.80	4	83.20	4	332.80	141	562
3	9.84	2	19.68	3	59.04	3	177.12	953	1906
4	12.80	4	51.20	2	102.40	2	204.80	2097	8389
5	14.04	2	28.08	1	28.08	1	28.08	2768	5535
6	14.40	4	57.60	0	0.00	0	0.00	2986	11,944
7	14.20	2	28.40	−1	− 28.40	−1	28.40	2863	5727
8	13.70	4	54.80	−2	− 109.60	−2	219.20	2571	10,285
9	12.60	2	25.20	−3	− 75.60	−3	226.80	2000	4001
10	10.06	4	40.24	−4	− 160.96	−4	643.84	1018	4072
11	1.30	1	1.30	−5	− 6.50	−5	32.50	2	2
Totals			327.40		− 107.84		1896.04		52,423
Area = (2/3) × 14.1 × F(*A*) =					3077.56				
Moment					− 14,293.1				
Centre of flotation (CF)					− 4.6443				
I(long)					3,543,346		3,476,965		
I(trans)					164,258.9				

TABLE B.5 Volumes and VCBs for Intermediate Waterplanes

Waterline	Area	SM	F(*V*)	Lever	F(*M*)[1]	Lever	F(*M*)[2]	SM	F(*V*)	Lever	F(*M*)	SM	F(*V*)	Lever	F(*M*)
SWL	3078	1	3078	0	0	10	30,780								
1.5	3015	4	12,060	1	12,060	9	108,540								
2	2921	2	5842	2	11,684	8	46,736	1	2921	8	23,368				
2.5	2820	4	11,280	3	33,840	7	78,960	4	11,280	7	78,960				
3	2709	2	5418	4	21,672	6	32,508	2	5418	6	32,508	1	2709	6	16,254
3.5	2570	4	10,280	5	51,400	5	51,400	4	10,280	5	51,400	4	10,280	5	51,400
4	2420	2	4840	6	29,040	4	19,360	2	4840	4	19,360	2	4840	4	19,360
4.5	2230	4	8920	7	62,440	3	26,760	4	8920	3	26,760	4	8920	3	26,760
5	1987	2	3974	8	31,792	2	7948	2	3974	2	7948	2	3974	2	7948
5.5	1641	4	6564	9	59,076	1	6564	4	6564	1	6564	4	6564	1	6564
6	1146	1	1146	10	11,460	0	0	1	1146	0	0	1	1146	0	0
			73,402		324,464		409,556		55,343		246,868		38,433		128,286
Volume			24,467.33						18,447.67				12,811		
CB below SWL					4.42037										
CB above keel							5.57963				4.460691				3.337913

TABLE B.6 Calculations for Bonjean Curves

Ordinate	Area Between Keel and 5.5WL						F(*A*)	Area	F(*V*)	Areas Between SWL and 2WL						F(*A*)	Area	F(*V*)
	SM	*5*		*8*		*−1*				*−1*		*8*		*5*				
1	1	**0.000**	0.000	**0.000**	0.000	**0.000**				**0.100**	0.100	**0.100**	0.100	**0.100**	0.100			
		0.000	0.000	0.000		0.000	0.00	0.00	0.00	−0.100		0.800		0.500		1.20	0.40	0.4
2	4	**0.500**	2.000	**1.060**	4.240	**1.640**				**3.640**	14.560	**4.440**	17.760	**5.200**	20.800			
		2.500		8.480		−1.640	9.34	1.56	6.23	−3.640		35.520		26.000		57.88	19.29	77.17333
3	2	**1.900**	3.800	**3.280**	6.560	**4.700**				**8.100**	16.200	**9.100**	18.200	**9.840**	19.680			
		9.500		26.240		−4.700	31.04	5.17	10.35	−8.100		72.800		49.200		113.90	37.97	75.93333
4	4	**4.500**	18.000	**6.800**	27.200	**8.600**				**11.800**	47.200	**12.480**	49.920	**12.800**	51.200			
		22.500		54.400		−8.600	68.30	11.38	45.53	−11.800		99.840		64.000		152.04	50.68	202.72
5	2	**7.240**	14.480	**10.100**	20.200	**11.800**				**13.800**	27.600	**14.000**	28.000	**14.040**	28.080			
		36.200		80.800		−11.800	105.20	17.53	35.07	−13.800		112.000		70.200		168.40	56.13	112.2667
6	4	**9.000**	36.000	**11.900**	47.600	**13.240**				**14.500**	58.000	**14.500**	58.000	**14.400**	57.600			
		45.000		95.200		−13.240	126.96	21.16	84.64	−14.500		116.000		72.000		173.50	57.83	231.3333

7	2	**7.900**	15.800	**10.700**	21.400	**12.400**				**14.080**	28.160	**14.220**	28.440	**14.200**	28.400			
		39.500		85.600		−12.400	112.70	18.78	37.57	−14.080		113.760		71.000		170.68	56.89	113.7867
8	4	**5.500**	22.000	**8.000**	32.000	**9.700**				**13.020**	52.080	**13.540**	54.160	**13.700**	54.800			
		27.500		64.000		−9.700	81.80	13.63	54.53	−13.020		108.320		68.500		163.80	54.60	218.4
9	2	**3.000**	6.000	**4.500**	9.000	**6.040**				**10.700**	21.400	**12.020**	24.040	**12.600**	25.200			
		15.000		36.000		−6.040	44.96	7.49	14.99	−10.700		96.160		63.000		148.46	49.49	98.97333
10	4	**0.940**	3.760	**1.560**	6.240	**2.160**				**5.700**	22.800	**8.000**	32.000	**10.060**	40.240			
		4.700		12.480		−2.160	15.02	2.50	10.01	−5.700		64.000		50.300		108.60	36.20	144.8
11	1	**0.100**	0.100	**0.100**	0.100	**0.100**				**0.100**	0.100	**0.100**	0.100	**1.300**	1.300			
		0.500		0.800		−0.100	1.20	0.20	0.20	−0.100		0.800		6.500		7.20	2.40	2.4
			F(V)						299.11									1278.187
			Volume						1405.83									6007.48

TABLE B.7 Section Areas Between SWL and 4WL and 5WL

		5		4		3		2		SWL					
WL	*SM*	*1*		*4*		*2*		*4*		*1*	*F(A)*		*F(A)*		
Ordinate	SM			1		3		3		1	(4WL)		(5WL)	Area(5)	Area(4)
1	1	**0.00**	0.00	**0.10**	0.10	**0.10**	0.10	**0.10**	0.10	**0.10**	0.10				
		0.00		0.40		0.20		0.40		0.10			1.10	1.47	
				0.10		0.30		0.30		0.10		0.80			1.20
2	4	**1.64**	6.56	**2.76**	11.04	**3.64**	14.56	**4.44**		**5.20**	20.80				
		1.64		11.04		7.28		17.76	17.76	5.20			42.92	57.23	
				2.76		10.92		13.32		5.20		32.20			48.30
3	2	**4.70**	9.40	**6.66**	13.32	**8.10**	16.20	**9.10**		**9.84**	19.68				
		4.70		26.64		16.20		36.40	18.20	9.84			93.78	125.04	
				6.66		24.30		27.30		9.84		68.10			102.15
4	4	**8.60**	34.40	**10.64**	42.56	**11.80**	47.20	**12.48**		**12.80**	51.20				
		8.60		42.56		23.60		49.92	49.92	12.80			137.48	183.31	
				10.64		35.40		37.44		12.80		96.28			144.42
5	2	**11.80**	23.60	**13.30**	26.60	**13.80**	27.60	**14.00**		**14.04**	28.08				
		11.80		53.20		27.60		56.00	28.00	14.04			162.64	216.85	
				13.30		41.40		42.00		14.04		110.74			166.11

6	4	**13.24**	52.96	**14.20**	56.80	**14.50**	58.00	**14.50**		**14.40**	57.60				
		13.24		56.80		29.00		58.00	58.00	14.40			171.44	228.59	
				14.20		43.50		43.50		14.40		115.60			173.40
7	2	**12.40**	24.80	**13.64**	27.28	**14.08**	28.16	**14.22**		**14.20**	28.40				
		12.40		54.56		28.16		56.88	28.44	14.20			166.20	221.60	
				13.64		42.24		42.66		14.20		112.74			169.11
8	4	**9.70**	38.80	**11.90**	47.60	**13.02**	52.08	**13.54**		**13.70**	54.80				
		9.70		47.60		26.04		54.16	54.16	13.70			151.20	201.60	
				11.90		39.06		40.62		13.70		105.28			157.92
9	2	**6.04**	12.08	**8.64**	17.28	**10.70**	21.40	**12.02**		**12.60**	25.20				
		6.04		34.56		21.40		48.08	24.04	12.60			122.68	163.57	
				8.64		32.10		36.06		12.60		89.40			134.10
10	4	**2.16**	8.64	**3.70**	14.80	**5.70**	22.80	**8.00**		**10.06**	40.24				
		2.16		14.80		11.40		32.00	32.00	10.06			70.42	93.89	
				3.70		17.10		24.00		10.06		54.86			82.29
11	1	**0.10**	0.10	**0.10**	0.10	**0.10**		**0.10**		**1.30**					
		0.10		0.40		0.20	0.10	0.40		1.30	1.30		2.60	3.47	
				0.10		0.30		0.30		0.10	1.30		2.00		3.00

Area to 5WL = F(*A*)(5WL) × 2 × 2/3 = (4/3) F(*A*)(5WL);
Area to 4WL = F(*A*)(4WL) × 2 × 2 × (3/8) = 1.5 F(*A*)(4WL).

TABLE B.8 Table of Section Areas

Ord	*Area to*						
	6WL	*5.5WL*	*5WL*	*4WL*	*3WL*	*2WL*	*SWL*
1	0.00	0.00	0.00	0.27	0.67	1.07	1.47
2	0.00	1.56	4.25	13.18	26.01	42.19	61.48
3	0.00	5.17	13.15	36.04	65.73	100.22	138.19
4	0.00	11.38	26.87	65.75	110.81	159.49	210.17
5	0.00	17.53	39.63	90.37	144.69	200.35	256.48
6	0.00	21.16	46.56	101.75	159.28	217.32	275.15
7	0.00	18.78	42.07	94.56	150.12	206.78	263.67
8	0.00	13.63	31.47	75.15	125.23	178.47	233.07
9	0.00	7.49	18.03	47.50	86.43	132.11	181.60
10	0.00	2.50	6.23	17.83	36.44	63.92	100.12
11	0.00	0.20	0.40	0.60	1.20	1.60	3.60

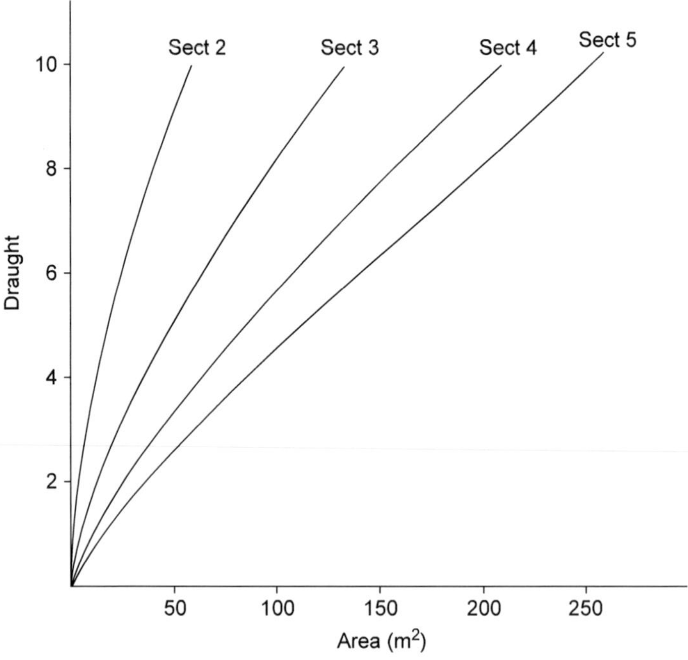

FIGURE B.1 Bonjean curves.

TABLE B.9 Volumes and LCBs

Ord	*SM*	*Lever*	*Area to*					
			5.5WL			*5WL*		
			Area	*F(V)*	*F(M)*	*Area*	*F(V)*	*F(M)*
1	1	5	0.00	0.00	0.00	0.00	0.00	0.00
2	4	4	1.56	6.24	24.96	4.25	17.00	68.00
3	2	3	5.17	10.34	31.02	13.15	26.30	78.90
4	4	2	11.38	45.52	91.04	26.87	107.48	214.96
5	2	1	17.53	35.06	35.06	39.63	79.26	79.26
6	4	0	21.16	84.64	0.00	46.56	186.24	0.00
7	2	−1	18.78	37.56	−37.56	42.07	84.14	−84.14
8	4	−2	13.63	54.52	−109.04	31.47	125.88	−251.76
9	2	−3	7.49	14.98	−44.94	17.53	35.06	−105.18
10	4	−4	2.50	10.00	−40.00	6.23	24.92	−99.68
11	1	−5	0.20	0.20	−1.00	0.40	0.40	−2.00
Total			99.40	299.06	−50.46		686.68	−101.64
Volume	CB aft			1405.582	−2.379074		3227.396	−2.087033
Mass Disp				1440.722			3308.081	

(*Continued*)

TABLE B.9 Volumes and LCBs—(cont.)

Ord	*SM*	*Lever*	*4WL*			*3WL*		
			Area	*F(V)*	*F(M)*	*Area*	*F(V)*	*F(M)*
1	1	5	0.27	0.27	1.35	0.67	0.67	3.35
2	4	4	13.18	52.72	210.88	26.01	104.04	416.16
3	2	3	36.04	72.08	216.24	65.73	131.46	394.38
4	4	2	65.75	263.00	526.00	110.81	443.24	886.48
5	2	1	90.37	180.74	180.74	144.69	289.38	289.38
6	4	0	101.75	407.00	0.00	159.28	637.12	0.00
7	2	−1	94.56	189.12	−189.12	150.12	300.24	−300.24
8	4	−2	75.15	300.60	−601.20	125.23	500.92	−1001.84
9	2	−3	47.50	95.00	−285.00	86.43	172.86	−518.58
10	4	−4	17.83	71.32	−285.28	36.44	145.76	−583.04
11	1	−5	0.60	0.60	−3.00	1.20	1.20	−6.00
Total				1632.45	−228.39		2726.89	−419.95
Volume	CB aft			7672.515	−1.972678		12,816.38	−2.171446
Mass Disp				7864.328			13,136.79	

Ord	SM	Lever	2WL			SWL		
			Area	F(V)	F(M)	Area	F(V)	F(M)
1	1	5	1.07	1.07	5.35	1.47	1.47	7.35
2	4	4	42.19	168.76	675.04	61.48	245.92	983.68
3	2	3	100.22	200.44	601.32	138.19	276.38	829.14
4	4	2	159.49	637.96	1275.92	210.17	840.68	1681.36
5	2	1	200.35	400.70	400.70	256.48	512.96	512.96
6	4	0	217.32	869.28	0.00	275.15	1100.60	0.00
7	2	−1	206.78	413.56	−413.56	263.67	527.34	−527.34
8	4	−2	178.47	713.88	−1427.76	233.07	932.28	−1864.56
9	2	−3	132.11	264.22	−792.66	181.60	363.20	−1089.60
10	4	−4	63.92	255.68	−1022.72	100.12	400.48	−1601.92
11	1	−5	1.60	1.60	−8.00	3.60	3.60	−18.00
Total				3927.15	−706.37		5204.91	−1086.93
Volume	CB aft			18,457.61	−2.536144		24,463.08	−2.944472
Mass Disp				18,919.05			25,074.65	

TABLE B.10 KB and KM Values

WL	6	5.5	5	4	3	2	SWL
Trans *I*	18,056	46,467	72,957	109,542	134,359	152,017	164,259
Volume, V	0	1406	3232	7505	12,816	18,470	24,463
Trans BM		33.05	22.57	14.60	10.48	8.23	6.71
KB	0	0.53	1.09	2.22	3.34	4.46	5.58
Trans KM		33.58	23.66	16.82	13.82	12.69	12.29
Long *I*	688,900	1,092,400	1,451,700	2,061,000	258,700	3,049,300	3,477,000
Long BM		777	449	275	202	165	142
Long KM		777	450	277	205	170	148
Additional data for plotting hydrostatic curves							
Mass Disp	0	1441	3313	7693	13,136	18,932	25,075
TPC	11.75	16.82	20.36	24.81	27.77	29.94	31.54
CF aft	2.71	2.15	1.79	2.06	2.88	3.78	4.64
LCB aft		2.38	2.09	1.97	2.17	2.54	2.94
Further if KG = 11 m							
Long GM		766	439	266	194	159	137
MCT		7830	10,320	14,510	18,070	21,350	24,360

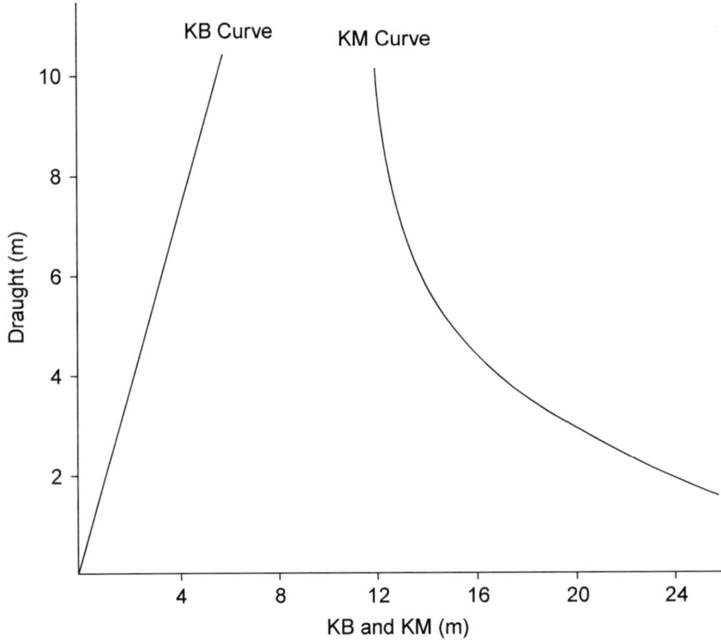

FIGURE B.2 Metacentric diagram.

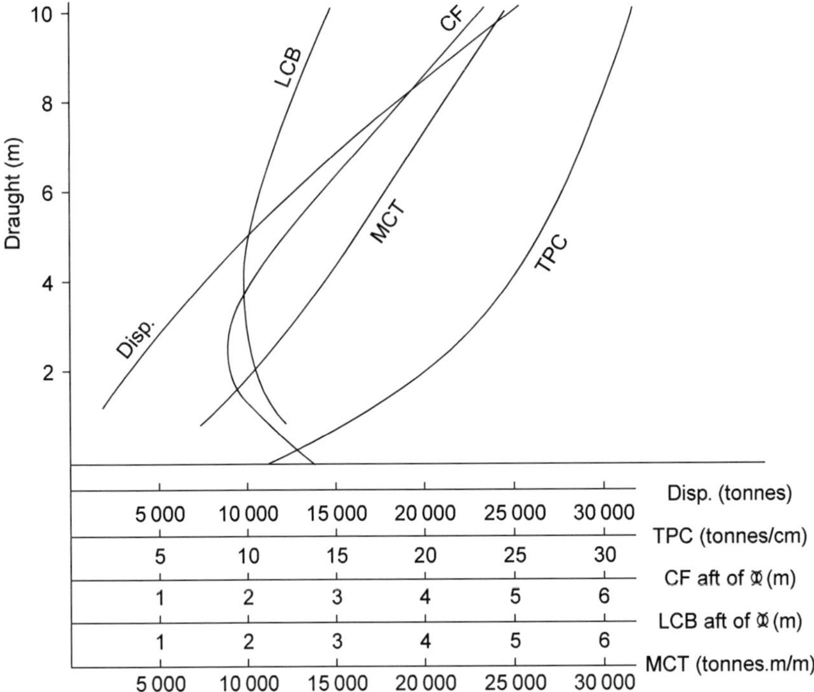

FIGURE B.3 Hydrostatic curves.

tables. KM is KB + BM and thus a metacentric diagram can be produced for the body which is the subject of this appendix. Table B.10 also includes a calculation of the longitudinal metacentre position which is needed to obtain the hydrostatic curves. The metacentric diagram is plotted in Figure B.2.

HYDROSTATIC CURVES

Chapter 4 introduced the concept of hydrostatic curves. The information in Table B.10, assuming a KG of 11 m, is used to plot Figure B.3.

Note: The data used in the Bonjean and hydrostatic curves can be derived in a number of different ways. Those above have been selected to give the reader an idea of the procedures involved.

Appendix C

Basic Ship Dynamics

GENERAL

For many calculations a ship is treated as a rigid body and is considered to be static or moving slowly between positions of equilibrium. Whilst unrealistic, such assumptions are necessary for the simplest studies of flotation and stability. In reality the ship is a flexible structure subject to many fluctuating forces – both internal and external. The responses of the ship to these forces include the motions of a ship as a rigid body but also its distortion as an elastic structure. This appendix considers the basic responses which provide the foundation for studies of ship motions, vibrations, noise and structural responses.

DYNAMIC RESPONSES

The Basic Oscillatory Responses

It is useful to set the scene by describing briefly the fundamental response of an elastic system to applied forces.

Simple Oscillations

The simplest oscillatory motion is one in which the restoring force acting is proportional to the body's displacement from a position of stable equilibrium. A mass on a spring exhibits such a motion and this is the building block from which the response of complex structures, regarded as the combination of many masses and springs, can be assessed. In the absence of damping the body, once disturbed, would oscillate indefinitely. Its distance from the equilibrium position would vary sinusoidally. Such motions are said to be *simple harmonic*. The presence of *damping*, due say to friction or viscous effects, causes the motion to die down with time. The greater the damping, the more quickly the motion dies down. The motion is also affected by *added mass* effects due to the moving body interacting with the fluid around it. These are not usually significant for a body vibrating in air but in water they can be important. There are many standard texts to which the reader can refer for a mathematical treatment of these motions. Only the basic findings are summarised here.

The motion is characterised by its amplitude, A, and period, T. For undamped motions the displacement at any time, t, is given by:

$$A \sin\left[\left(\frac{k}{M}\right)^{0.5} t + \delta\right]$$

where

M is the mass of the body,
k is the force acting per unit displacement and
δ is a phase angle.

The *period* of this motion is $T = 2\pi(M/k)^{0.5}$, and its frequency is $n = 1/T$. These are said to be the system's *natural period* and *frequency*.

Damping

All systems are subject to some damping, the simplest case being when the damping is proportional to the velocity. This modifies the period of the motion and causes the amplitude to diminish with time.

The period becomes $T_d = 2\pi/[(k/M) - (\mu/2M)^2]^{0.5}$, the frequency being $1/T_d$, where μ is a damping coefficient such that damping force equals μ (velocity).

Successive amplitudes decay according to the equation:

$$A \exp[-(\mu/2M)\, t].$$

As the damping increases the number of oscillations about the mean position will reduce until finally the body does not overshoot the equilibrium position at all. The system is then said to be a *dead beat*.

When damping is not proportional to the angular or linear velocity the differential equation is not capable of easy solution. For more background on these types of motion reference should be made to standard textbooks.

Regular Forced Oscillations

Free oscillations can occur when, for instance, a structural member is struck an instantaneous blow. More generally the disturbing force will continue to be applied to the system for a longish period (relative to the period of motion) and will itself fluctuate in amplitude and frequency. The simplest disturbing force to assume for analysis purposes is one with constant amplitude varying sinusoidally with time. This would be the case where the ship is in a regular wave system. The differential equation of motion, taking x as the displacement at time t, becomes:

$$M\frac{dx^2}{dt^2} + \mu\frac{dx}{dt} + kx = F_0 \sin \omega t$$

The solution of this equation for x is the sum of two parts. The first part is the solution of the equation with no forcing function. That is, it is the solution of the damped oscillation previously considered. The second part is an oscillation at the frequency of the applied force. It is $x = B\sin(\omega t - \gamma)$.

After a time the first part will die away leaving the oscillation in the frequency of the forcing function. This is called a *forced oscillation*. It is important to know its amplitude, B, and the phase angle, γ. These can be shown to be:

$$B = \frac{F_0}{k} \times \frac{1}{[(1-\Lambda^2)^2 + (\mu^2\Lambda^2/Mk)]^{0.5}}$$

and

$$\tan\gamma = \frac{\mu\Lambda}{(Mk)^{0.5}} \times \frac{1}{(1-\Lambda^2)}$$

In these expressions Λ is called the *tuning factor* and is equal to $\omega/(k/M)^{0.5}$. That is the tuning factor is the ratio of the frequency of the applied force to the natural frequency of the system. Since k represents the stiffness of the system, F_0/k is the displacement which would be caused by a static force F_0. The ratio of the amplitude of the dynamic displacement to the static displacement is termed the *magnification factor*, Q, given by:

$$Q = \left[(1-\Lambda^2)^2 + \frac{\mu^2\Lambda^2}{Mk}\right]^{-0.5}$$

Curves of magnification factor can be plotted against tuning factor for a range of damping coefficients as in Figure C.1. At small values of Λ, Q tends to unity and at very large values it tends to zero. In between these extremes the response builds up to a maximum value which is higher the

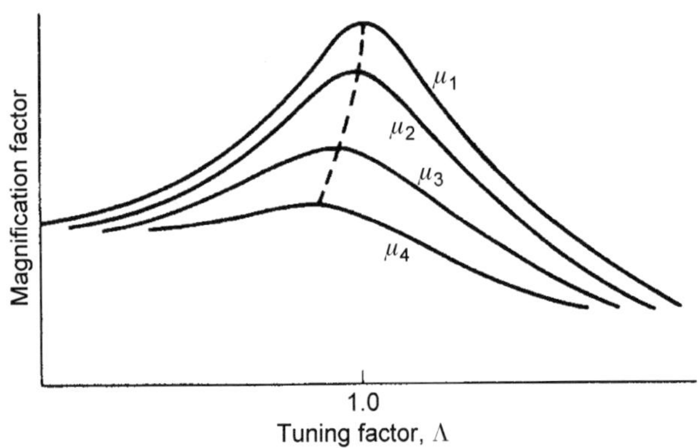

FIGURE C.1 Tuning factor.

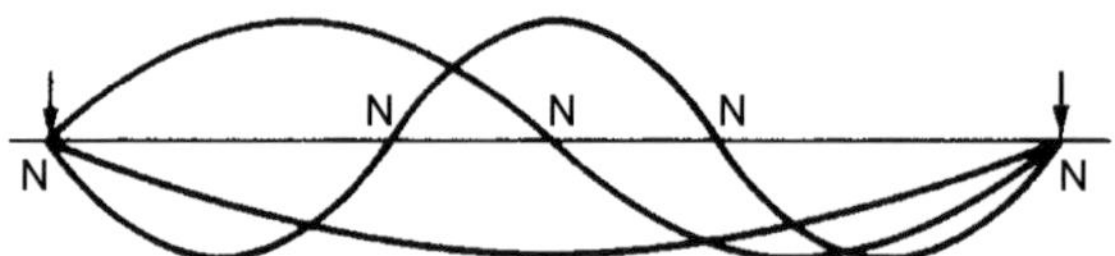

FIGURE C.2 Vibrating beam.

lower the damping coefficient. If the damping were zero the response would be infinite. For lightly damped systems the maximum displacement occurs very close to the system's natural frequency and the tuning factor can be taken as unity. Where the frequency of the applied force is equal to the system's natural frequency it is said that there is *resonance*. It is necessary to keep the forcing frequency and natural frequency well separated if large amplitude oscillations are to be avoided. At resonance the expression for the phase angle gives $\gamma = \tan^{-1}\infty$, giving a phase lag of 90°.

In endeavouring to avoid resonance it is important to remember that many systems have several natural frequencies associated with different deflection profiles or *modes* of vibration. An example is a vibrating beam, fixed at the ends, that has many modes, the first three of which are shown in Figure C.2. All these modes will be excited and the overall response may show more than one resonance peak. The points, labelled N, at which there is no displacement are called *nodes*.

Irregular Forcing Function

In the above the forcing function was assumed sinusoidal and of constant amplitude. The more general case would be a force varying in an irregular way. In this case the force can be analysed to obtain its constituent regular components as was done for the waves in an irregular sea. The vibratory response of the system to the irregular force can then be taken as the sum of its responses to all the regular components.

Appendix D

The Froude Notation

For presenting resistance and propulsion data, Froude introduced a special notation which was commonly called the *constant notation* or the *circular notation*. Although it has not been used for many years there is still a lot of data in the notation. Also the student needs knowledge of it if it is desired to read up the early work, much of which is fundamental to the subject. The rather curious name arose because the key characters were surrounded by circles.

Froude took as a characteristic length the cube root of the volume of displacement, and denoted this by U. He then defined the ship's geometry with the following:

$$Ⓜ = \text{length constant} = \frac{\text{wetted length}}{U}$$

$$Ⓑ = \text{breadth constant} = \frac{\text{wetted breadth}}{U}$$

$$Ⓓ = \text{draught constant} = \frac{\text{draught at largest section}}{U}$$

$$Ⓢ = \text{wetted surface constant} = \frac{\text{wetted surface area}}{U^2}$$

$$Ⓐ = \text{section area constant} = \frac{\text{section area}}{U^2}$$

In verbal debate Ⓜ and Ⓑ are referred to as 'circular M' and 'circular B' and so on.

To cover the ship's performance Froude introduced:

$$Ⓚ = \frac{\text{speed of ship}}{\text{speed of wave of length } U/2}$$

$$Ⓛ = \frac{\text{speed of ship}}{\text{speed of wave of length } L/2}$$

$$Ⓒ = \frac{1000\,(\text{resistance})}{\Delta Ⓚ^2}$$

with subscripts to denote total, frictional or residuary resistance as necessary.

Although looking a little odd at first sight it was a truly non-dimensional form of presentation. Whereas Schoenherr and the International Towing

Tank Conference (ITTC) used Reynolds' number as the basis for scaling the frictional resistance, Froude based his method on measurements of the resistance of planks extrapolated to ship-like lengths. He tried fitting the results with a formula such as:

$$R_f = fSV^n$$

where f and n were empirical constants.

He found that both f and n depended upon the nature of the surface. For very rough surfaces n tended towards 2. The value of f reduced with increasing length. For smooth surfaces, at least, n tended to decline with increasing length.

Later his son proposed:

$$R_f = fSV^{1.825}$$

in conjunction with f values as in Table D.1.

The f values in Table D.1 apply to a wax surface for a model and a freshly painted surface for a full-scale ship.

TABLE D.1 R. E. Froude's Skin Friction Constants

Length (m)	*f*	Length (m)	*f*	Length (m)	*f*
2	1.966	18	1.526	70	1.441
3	1.867	20	1.515	80	1.437
4	1.791	22	1.506	90	1.432
5	1.736	24	1.499	100	1.428
6	1.696	26	1.492	120	1.421
7	1.667	28	1.487	140	1.415
8	1.643	30	1.482	160	1.410
9	1.622	35	1.472	180	1.404
10	1.604	40	1.464	200	1.399
12	1.577	45	1.459	250	1.389
14	1.556	50	1.454	300	1.380
16	1.539	60	1.447	350	1.373

Note: f values (metric units): frictional resistance = $fSV^{1.825}$, newtons; wetted surface, S, in square metres; ship speed, V, in m/s. Values are for salt water. Values in fresh water may be obtained by multiplying by 0.975.

Within the limits of experimental error, the values of f in the above formula, can be replaced by:

$$f = 0.00871 + \frac{0.053}{8.8 + L}$$

where R_f is in lbf, l in ft, S in ft^2 and V in knots, or:

$$f = 1.365 + \frac{2.530}{2.68 + L}$$

where R_f is in newtons, l in m, S in m^2 and V in m/s.

FROUDE METHOD OF CALCULATING RESISTANCE

Elements of Form Diagram

This diagram was used by Froude to present data from model resistance tests. Resistance is plotted as ©–® curves, corrected to a standard 16 ft model. Separate curves are drawn for each ship condition used in the tests. Superimposed on these are curves of skin friction correction needed when passing from the 16 ft model to geometrically similar ships of varying length. The complete elements of form diagram includes, in addition, the principal dimensions and form coefficients, and non-dimensional plottings of the curve of areas, waterline and midship section.

Using Froude's '*circular*' or '*constant*' *notation*:

$$©_f = \frac{1000\,(\text{frictional resistance})}{\Delta ®^2}$$

$$= \frac{\dfrac{1000}{\rho g U^3} f S V^{1.825}}{4\pi V^2 / gU}$$

$$= OSL^{-0.175}$$

where:

$$O = \frac{1000f}{4\pi\rho\left(\dfrac{gL}{4\pi}\right)^{0.0875}} = \text{'Circular } O\text{'}$$

From which:

$$[©_t]_{ship} = [©_t]_{model} - [O_m - O_s]\ Ⓢ Ⓛ^{-0.175}$$

A selection of O and f values are presented in Table D.2. These apply to a standard temperature of 15°C (59°F). The $©_f$ value is increased or

TABLE D.2 R. E. Froude's Frictional Data

Length (ft)	*O*	*f*	Length (ft)	*O*	*f*
5	0.15485	0.012585	80	0.08987	0.009309
10	0.13409	0.011579	90	0.08840	0.009252
15	0.12210	0.010925	100	0.08716	0.009207
20	0.11470	0.010524	200	0.08012	0.008992
25	0.10976	0.010269	300	0.07655	0.008902
30	0.10590	0.010068	400	0.07406	0.008832
35	0.10282	0.009908	500	0.07217	0.008776
40	0.10043	0.009791	600	0.07062	0.008726
45	0.09839	0.009691	700	0.06931	0.008680
50	0.09664	0.009607	800	0.06818	0.008639
60	0.09380	0.009475	1000	0.06636	0.008574
70	0.09164	0.009382	1200	0.06493	0.008524

Note: *O* and *f* values. Frictional resistance = $fSV^{1.825}$ lbf. Values are for a standard temperature of 15°C. *f* values are for salt water with *S* in ft^2, *V* in knots. Values of *f* in fresh water may be obtained by multiplying by 0.975.

TABLE D.3 Ⓛ$^{-0.175}$ Values

Ⓛ	Ⓛ$^{-0.175}$	Ⓛ	Ⓛ$^{-0.175}$
0.05	1.6892	1.00	1.0000
0.10	1.4962	1.20	0.9686
0.15	1.3937	1.40	0.9428
0.20	1.3253	1.60	0.9210
0.30	1.2345	1.80	0.9023
0.40	1.1739	2.00	0.8858
0.50	1.1290	2.20	0.8711
0.60	1.0935	2.40	0.8580
0.70	1.0644	2.60	0.8460
0.80	1.0398	2.80	0.8351
0.90	1.0186	3.00	0.8251

decreased by 4.3% for every 10°C (2.4% for every 10°F) the temperature is below or above this value Table D.3 gives Ⓛ$^{-0.175}$ for a range of Ⓛ values.

Froude Applied to Example 7.1

Assume a Froude coefficient, $f = 0.42134$
Midship area coefficient = 0.98
Wetted surface area = 3300 m^2

Solution

Speed of model = $15(4.9/140)^{0.5} = 2.81$ knots
= 1.44 m/s
Wetted surface of model = $3300(4.9/140)^2 = 4.04$ m^2
Froude coefficient for model (fresh water) = 1.698
Frictional resistance of model = $1.698 \times 4.04 \times 2.81^{1.825} = 13.35$ N
Wave resistance of model = 19 − 13.35 = 5.65 N
Wave resistance of ship = $5.65 \times (1025/1000) \times (140/4.9)^3$
= 135,060 N
Speed of ship = $15 \times 0.5148 = 7.72$ m/s
Frictional resistance of ship = $1.415 \times 3300 \times (7.72)^{1.825}$
= 194,600 N
Total resistance of ship = 329,660 N
It would then follow that the effective power
= $(329{,}660 \times 15 \times 1852)/(1000 \times 3600)$
= 2544 kW

This page intentionally left blank

Appendix E

Questions

The following questions are presented for the benefit of students and lecturers using *Introduction to Naval Architecture*. Solutions to Questions can be freely accessed online by visiting http://booksite.elsevier.com/9780080982373.

Although grouped by chapters, additional matter relevant to the answer may be found in other sections of the book. Students are encouraged to carry out additional research to provide more up-to-date, or more complete, information than that which can be contained in a single volume. In this research use can be made of the websites given under references.

CHAPTER 2. DEFINITION AND REGULATION

Question 1. How is the overall geometry of a ship defined? How are the coefficients of fineness related?

Question 2. Discuss the various displacements and tonnages used to define the overall size of a ship. Why are freeboard and reserve of buoyancy important?

Question 3. Discuss the various international and national bodies linked to shipping and their roles.

Question 4. A ship floating freely in salt water is 150 m long, 22 m beam and 9 m draught and has a mass displacement of 24,000 tonnes. The midships wetted area is 180 m^2. Find the block, prismatic and midship area coefficients.

Question 5. A circular cylinder 100 m long and 7 m diameter floats with its axis in the waterline. Find its mass in tonnes in salt water. How much must the mass be reduced to float with its axis in the waterline in fresh water?

CHAPTER 3. SHIP FORM CALCULATIONS

Question 1. What do you understand by approximate integration? Discuss the various rules used in simple naval architectural calculations and how they are used.

Question 2. Show that Simpson's 1, 4, 1 Rule is accurate for defining the area under a curve defined by the equation $y = a + bx + cx^2 + dx^3$, between $x = -h$ *to* $x = h$.

Question 3. Find the area of the waterplane defined by the following half-ordinates (m) which are 16 m apart:

2.50, 11.00, 20.50, 27.00, 29.50, 29.00, 25.00, 18.00, 7.00

Question 4. The half-ordinates (m) of a section of a ship at waterlines which are 1.5 m apart are, reading from the design waterplane down:

33.45, 33.30, 32.55, 30.90, 25.80, 12.00

At the turn of bilge, midway between the last two half-ordinates, an extra half-ordinate is 19.80 m.

Find the area of the section up to the design waterplane and the height of the centroid of area above the keel.

Question 5. The half-ordinates (m) of a waterplane, which are 6 m apart, are given by:

11.16, 24.84, 39.42, 47.52, 40.23, 26.46, 13.23

Calculate, and compare, the areas of the waterplane as given by the 1, 4, 1 Rule, the 1, 3, 3, 1 Rule and the trapezoidal rule.

Question 6. A three-dimensional body 54 m long has regularly spaced sectional areas (m^2) of:

1.13, 2.50, 4.49, 6.40, 9.92, 14.52, 20.35, 26.18, 29.62, 29.59, 27.07, 21.48, 15.19, 10.08, 6.83, 4.48, 3.13, 2.47, 0.00

Calculate the volume of the body and the distance of its centroid of volume from the centre of length.

Question 7. Find the area, tonnes per cm, centre of flotation and the transverse inertia of the waterplane defined by the following half-ordinates (m) which are 15 m apart:

0.26, 2.99, 8.32, 12.87, 16.38, 17.55, 17.94, 17.81, 16.64, 13.78, 8.32, 2.47, 0.26

If the displacement is 70,000 tonnes what is the value of BM?

Question 8. A homogeneous log of square cross section, side 1 m, and 6 m long is floating in a position of stable equilibrium. The log's density is half that of the water in which it is floating. Find the longitudinal metacentric height.

CHAPTER 4. FLOTATION

Question 1. What do you understand by equilibrium? For a ship being heeled to small angles what do you understand by stability? What do you understand by the term metacentre? What is the significance of the positions of the centre of gravity and the metacentre in relation to stability?

Question 2. A rectangular box of length, L, beam, B, and depth, D, floats at a uniform draught, T. Deduce expressions for KB, BM and KM in terms of the principal dimensions. If the beam is 9 m what must be the draught for the metacentre to lie in the waterplane?

Question 3. A uniform body of regular triangular cross section of length, L, base, B, and height, H, floats apex down. If its density is half that of the water in which it is floating, find expressions for KB, BM and KM in terms of B and H.

Question 4. Describe an easy way to establish whether a complex three-dimensional fitting is made of brass, lead or steel.

Question 5. A small craft is floating in an enclosed dock. Then:

1. Two bulks of timber floating in the dock are lifted on to the craft.
2. A large stone in the craft is dumped into the dock.

State whether the water depth in the dock will increase, decrease or remain the same in each case. Does your answer depend upon the density of the water?

Question 6. A ship of 100 m length floats in sea water at a draught of 4 m forward and 4.73 m aft. Data for the ship is:

Tonnes per cm = 12.

Centre of flotation is 4.1 m aft of amidships.

Moment to change trim 1 m = 3700 tonnes m/m.

Where should a weight, of 50 tonnes, be added to bring the ship to a level keel? What is the new level draught?

CHAPTER 5. STABILITY

Question 1. When and why is an inclining experiment carried out? Discuss how it is carried out and the steps taken to ensure accurate results.

Question 2. A ship of 11,500 tonnes is inclined using four groups of weights, each group of 20 tonnes separated by 12 m across the upper deck. The weights are moved in sequence, leading to the following deflections of a pendulum 6 m long. Each weight movement is 12 m transversely.

Weight Moved (tonnes)	Movement	Pendulum Reading (m)
20	P to S	0.13 to S
20	P to S	0.25 to S
40	S to P	0.01 to S
20	S to P	0.11 to P
20	S to P	0.23 to P
40	P to S	0.01 to S

Comment upon the readings and deduce the metacentric height as inclined.

Question 3. Show how the position of the metacentre for a right cylinder of circular cross section, floating with its axis horizontal, will be at the centre of the circular cross section whatever the draught. Use this fact to show that the centroid of a semicircle of diameter d is $2d/3\pi$ from the diameter.

Question 4. A waterplane is defined by the following half-ordinates, spaced 6 m apart:

0.12, 2.25, 4.35, 6.30, 7.98, 9.27, 10.20, 10.80, 11.10, 11.19, 11.19, 11.19, 11.19, 11.16, 11.13, 11.04, 10.74, 10.02, 8.64, 6.51, 3.45

If the ship's displacement is 14,540 tonnes in salt water, find:

1. The waterplane area.
2. The longitudinal position of the centre of flotation.
3. BM_T
4. BM_L

Question 5. The half-ordinates defining a ship's waterplane, reading from forward, are:

0.12, 6.36, 12.96, 18.00, 21.12, 22.20, 22.08, 21.00, 18.36, 12.96, 4.56

The ordinates are 30 m apart and there is an additional (total) area aft of the last ordinate of 78 m^2 with its centroid 4.8 m aft of the last ordinate.

Find:

1. The total waterplane area.
2. The position of the centre of flotation (CF).
3. The longitudinal inertia about the CF.

Question 6. How is a ship's stability at large angles measured? Why are *cross curves used*? What do you understand by an angle of loll?

Question 7. What effects do liquid free surfaces in a ship have on its stability? Can other cargoes present similar effects?

Question 8. Why does a ship need stability? What factors are usually considered in setting standards?

Question 9. A ship with vertical sides in way of the waterline is said to be wall-sided. Show for such a vessel that for inclinations, φ, within the range of wall-sidedness:

$$GZ = \sin\phi[GM + 0.5BM\tan^2\phi]$$

Question 10. Show that a wall-sided ship with an initial metacentric height, GM, which is negative will loll to an angle, φ, such that:

$$\tan\phi = +[2GM/BM]^{0.5} \text{ or } -[2GM/BM]^{0.5}$$

Question 11. A ship of 10,000 tonnes has cross curves of stability which give, for that displacement, SZ values of:

φ(°)	15	30	45	60	75	90
SZ (m)	0.85	1.84	2.82	2.80	2.06	1.14

If the pole S is 0.5 m below G:

1. find the corresponding values of GZ,
2. plot the GZ curve,
3. find the angle and value of the maximum GZ,
4. find the angle of vanishing stability.

Question 12. A ship of 12,000 tonnes has cross curves of stability which give, for that displacement, SZ values of:

φ(°)	15	30	45	60	75	90
SZ (m)	0.68	1.64	2.58	2.86	2.72	2.20

If the pole S is 0.65 m below G:

1. find the corresponding values of GZ,
2. plot the GZ curve,
3. find the angle and value of the maximum GZ,
4. find the angle of vanishing stability,
5. find the dynamical stability up to 60°.

Question 13. Outline the latest International Maritime Organisation (IMO) probabilistic method for assessing damage stability.

Question 14. Discuss the special problems of Ro-Ro vessels with large vehicle decks.

CHAPTER 6. LAUNCHING, DOCKING AND GROUNDING

Question 1. Discuss the means of transferring ships from dry land to the sea on first build. Describe the slipway and the means of ensuring a successful conventional launch.

Question 2. Discuss the calculations carried out to ensure a successful end on launch. What safety precautions are taken?

Question 3. Discuss the docking process and the precautions taken to ensure a successful operation.

CHAPTER 7. RESISTANCE

Question 1. Name two key relationships between the physical quantities that are important in a study of the resistance of ships. Why is each important?

Question 2. Discuss the various types of resistance encountered by a ship. How can they be reduced? What is their relative importance at different speed regimes?

Question 3. Discuss how ship model data can be used to determine ship resistance.

Question 4. How are the following used?

1. The International Towing Tank Conference (ITTC) correlation line.
2. Methodical series.

Question 5. What do you understand by roughness? How does it come about? How is it measured? How can it be reduced?

Question 6. Describe a typical ship tank in which resistance experiments are carried out. How can model results be compared with full-scale data? Why are special ship trials needed?

CHAPTER 8. PROPULSION

Question 1. Discuss what you understand by the term effective power. What are the factors affecting the overall propulsive efficiency?
Question 2. Outline the simple momentum theory for a propulsor. Show that the ideal efficiency is related to the axial and rotational inflow factors.
Question 3. Discuss the physical features of a screw propeller and the blade section. How are the thrust and torque estimated?
Question 4. What do you understand by the term cavitation? Why is it important, how is it studied and how can it be reduced?
Question 5. Discuss the various types of propulsive device used to propel marine vehicles.
Question 6. Describe the use of a measured distance to determine a ship's speed accurately. What precautions are taken to achieve an accurate result?

CHAPTER 9. THE SHIP ENVIRONMENTS

Question 1. Discuss the features of the environment in which a ship operates that affect its design.
Question 2. Discuss the two 'standard' waves used in ship design. The sea surface usually looks completely irregular. How does the naval architect define such seas?
Question 3. Discuss the international rules governing the pollution of the environment by ships. What steps can be taken by the naval architect to reduce or eliminate pollution?
Question 4. Discuss the important features of a ship's internal environment. Why are they important and how are they controlled?
Question 5. Discuss how noise is measured. How does it vary with distance from the source? How does it arise in a ship and why should it be reduced as much as possible? How can noise levels be reduced in critical areas?

CHAPTER 10. SEAKEEPING

Question 1. What do you understand by the terms *simple harmonic motion, added mass, damping, tuning factor* and *magnification factor*?
Question 2. What are the main causes of vibration in ships? How can the levels of vibration be reduced?
Question 3. Show that in the case of undamped, small amplitude motion, a ship when heeled in still water, and released, will roll in simple harmonic motion. Deduce the period of roll. Repeat this for a ship heaving.

Question 4. Discuss how ship motion data can be presented, including the concepts of response amplitude operators and an energy spectrum.
Question 5. Discuss the various factors that can affect a ship's performance in waves. How can some of these affects be minimised?
Question 6. Discuss how the overall seakeeping performance of two different designs can be compared.
Question 7. Discuss the various ways in which seakeeping data can be acquired.
Question 8. Discuss how a ship can be stabilised to reduce roll motions.
Question 9. Discuss, in general terms, the effect of ship form on seakeeping performance.
Question 10. Discuss the reasons why a ship may experience very large rolling angles.

CHAPTER 11. VIBRATION, NOICE AND SHOCK

Question 1. Discuss the ways in which a ship's structure may flex and vibrate. How can the levels of vibration experienced be reduced?
Question 2. Discuss the shock experienced by a ship due to an underwater explosion. How can the effects be mitigated?

CHAPTER 12. MANOEUVRING

Question 1. Discuss what you understand by directional stability and manoeuvring. How are these attributes provided in a ship?
Question 2. Describe a number of measures that can be used to define a ship's manoeuvring characteristics.
Question 3. Describe, and sketch, a number of different types of rudder. How are rudder forces and torques on a conventional rudder calculated?

CHAPTER 13. STRUCTURES

Question 1. Discuss how a ship structure may fail in service.
Question 2. Outline the simple standard method of assessing a ship's longitudinal strength. How are bending moments translated into hull stresses?
Question 3. Discuss superstructures and their contribution to longitudinal strength.
Question 4. Discuss the transverse strength of a ship, the loading and methods of calculation.
Question 5. Discuss fatigue, cracking and stress concentrations in relation to ship type structures.
Question 6. Discuss buckling and load shortening curves in relation to ships' structures.

CHAPTER 14. SHIP DESIGN

Question 1. How would you expect a good set of ship requirements to be set out? What would you expect them to cover?
Question 2. What factors should be taken into account in assessing the cost of a ship? How can different design solutions be compared?
Question 3. Describe the design process and discuss the various phases of design?
Question 4. Discuss the impact of computers upon the design, build and operation of ships.
Question 5. Describe how a formal safety assessment can be conducted.
Question 6. How can ship design be approached in a methodical manner?

CHAPTER 15. SHIP TYPES

Question 1. Discuss the various types of merchant ships. Discuss how container ships were evolved and the advantages they possess over general cargo carriers.
Question 2. Discuss the various types of fast craft with their advantages and disadvantages. Indicate typical uses to which each type is put. How would you go about comparing the relative merits of several types?
Question 3. Discuss the various types of tug and their main design features.
Question 4. Discuss the types of passenger vessel now in use.
Question 5. Discuss a number of warship types. Give their main functions. Discuss in more detail the design and use of destroyers and frigates.
Question 6. What are the main features of submarines that distinguish them from surface ships?

References and Further Reading

American Bureau of Shipping (ABS). Documents can be accessed online. www.eagle.org Guidelines include:

98. Passenger ships (2011)

134. Safehull finite element analysis of hull structures (2003)

145. Vessel manoeuverability (2006)

147. Ship vibration (2006)

151. Vessels operating in low temperature environments

Aldwinkle, D.S., Pomeroy, R.V., 1982. A rational assessment of ship reliability and safety. TRINA.

Allan, R.G., 1997. From Kort to escort: the evolution of tug design 1947–1997. Ship and Boat International, RINA.

Andrews, D., 1992. The management of warship design. The MOD warship project manager's perspective. TRINA.

Andrews, D., 2004. A creative approach to ship architecture. TRINA.

Andrews, D., Burger, D., Zhang, J., 2005. Design for production using the building block approach. TRINA.

Andrews, D.J., Pawling, R.G., 2008. A case study in preliminary ship design. TRINA.

Andrews, D.J., 2010. 150 years of ship design. TRINA.

Andrews, D., 2011a. Marine requirements elucidation and the nature of preliminary design. TRINA.

Andrews, D., 2011b. Art and science in the design of physically large and complex systems. Proc. Royal Soc.

Barrass, C.B., 2001. Ship Stability, Notes and Examples, third ed. Butterworth–Heinemann, Oxford.

Barrass, C.B., 2004. Ship Design and Performance for Masters and Mates. Butterworth–Heinemann, Oxford.

Barrass, C.B., Derrett, D.R., 2012. Ship Stability for Masters and Mates, seventh ed. Butterworth–Heinemann, Oxford.

Bateman, W., 2010. The formation of freak waves. Lloyds Regist. Technol. Days[(1)].

Bazari, Z., McStay, P., 2011. Energy Efficient Design Index (EEDI) for marine greenhouse gas emissions control. Lloyds Regist. Technol. Days[(1)].

Bertram, V., 2012. Practical Ship Hydrodynamics. Butterworth–Heinemann, Oxford.

Biran, A., 2003. Ship Hydrostatics and Stability. Butterworth–Heinemann, Oxford.

Bittner, A.C., Guignard, J.C., 1985. Human factors engineering principles for minimising adverse ship motion effects: theory and practice. Naval Engineering Journal 97 (4).

Blyth, A.G., 2005. An ISO standard for stability and buoyancy of small craft. TRINA.

Brook, A.K., 1992. Improving the safety of tankers and bulk carriers through a hull condition monitoring scheme. RINA Conference on Tankers and Bulk Carriers – the Way Ahead).

Brown, D.K., 1985. The value of reducing ship motions. Naval Engineers Journal.

Brown, D.K., Tupper, E.C., 1989. The naval architecture of surface warships. TRINA.

Brown, D.K., Moore, G., 2003. Rebuilding the Royal Navy. Chatham Publishing, London.

Bruce, G., Eyres, D., 2012. Ship Construction, seventh ed. Butterworth–Heinemann, Oxford.

BS 5555: 1995; ISO 1000: 1992. SI Units and recommendations for the use of their multiples and certain other units.

BS 5760. Reliability of system equipment.

BS 6634: 1985; ISO 6954: 1984 (Rev 2000). Mechanical vibration and shock. Guidelines for the overall evaluation of vibration in merchant ships.

BS 6841: 1987. Guide to measurement and evaluation of human exposure to whole-body mechanical vibration and repeated shock.

BS 7608. Code of practice for fatigue design and assessment of steel structures.

Burcher, R.K., 1991. The prediction of the manoeuvring characteristics of vessels. The Dynamics of Ships. The Royal Society, London.

Burrill, L.C., 1934–1935. Ship vibration: simple methods of estimating critical frequencies. TNECI.

Burrill, L.C., Emerson, A., 1962–1963. Propeller cavitation: further tests on 16 in propeller models in the King's College cavitation tunnel. TNECI.

Canham, H.J.S., 1974. Resistance, propulsion and wake tests with HMS Penelope. TRINA.

Carlton, J.S., Vlasic, D., 2005. Ship vibration and noise: Some topical aspects.[1] First International Ship Noise and Vibration Conference, London.

Carlton, J.S., 2012. Marine Propellers and Propulsion. Butterworth–Heinemann, Oxford.

Chislett, H.W.J., 1972. Replenishment at sea. TRINA.

Clancy, T., 1993. Submarine. Harper Collins, London.

Clark, I.C., 2008. Stability, Trim and Strength for Merchant Ships and Fishing Vessels. The Nautical Institute, London.

Clark, I.C., 2005. Ship Dynamics for Mariners. The Nautical Institute, London.

Conn, J.F.C., Lackenby, H., Walker, W.B., 1953. BSRA resistance experiments on the Lucy Ashton. Part II: the ship-model correlation for the naked hull condition. TINA.

Corlett, E.C.B., Colman, J.C., Hendy, N.R., 1988. Kurdistan – the anatomy of a marine disaster. TRINA.

Dand, I.W., 1977. The physical causes of interaction and its effects. Nautical Institute Conference on Ship Handling.

Dand, I.W., 1981. On ship–bank interaction. TRINA.

Dand, I.W., Barnes, J., Austen, S., 2008. The speed of fast inflatable lifeboats. TRINA.

Dand, I.W., Ferguson, A.M., 1973. The squat of full ships in shallow water. TRINA.

Dawson, P., 2000. Cruise Ships: An Evolution in Design. Conway Maritime Press, London.

Dawson, P., 2005. The Liner: Retrospective & Renaissance. Conway, London.

Denny, Sir Maurice, E., 1951. BSRA resistance experiments on the Lucy Ashton. Part I: full scale measurements. TINA.

Department of Transport, 1988. A Report into the Circumstances Attending the Loss of MV Derbyshire. Appendix 7: Examination of Fractured Deck Plate of MV Tyne Bridge.

Det Norske Veritas documents on line. www.dnv.com Guidelines and Classification Notes (CN):

CN 20, 2000. Recommended Practice. Corrosion protection of ships.

CN 20.1, 2011. Stability documentation for approval.

CN 30.6, 1992. Structural reliability analysis.

CN 30.7, 2011. Fatigue assessment of ship structures.

CN 31.1 to 31.9 Strength analysis – various ship types.

CN 32,1, 1989. Strength analysis of rudder arrangements.

CN 41.5, 2012. Calculation of marine propeller.

Dieudonne, J., 1959. Vibration in ships. TINA.

Dobie, T.G., 2003. Critical significance of human factors in ship design. Proceedings of the 2003 RVOC Meeting, University of Minnesota.

Dodman, J., 2010. SOLAS 2009 Stability requirements implementation. Lloyds Regist. Technol. Days[1].

Dokkum, K. van, 2003. Ship Knowledge. A Modern Encyclopedia. Dokmar, Enkhuizen, The Netherlands.

Dorey, A.L., 1989. High speed small craft. The 54th Parsons Memorial Lecture. TRINA.

DPTAC, 2000. The design of large passenger ships and passenger infrastructure: guidance on meeting the needs of disabled people. Marine Guidance Note 31(M).

Eyres, D.J., 2007. Ship Construction, sixth ed. Butterworth–Heinemann, Oxford.

Ewing, J.A., Goodrich, G.J., 1967. The influence on ship motions of different wave spectra and of ship length. TRINA.

Faulkner, D., 2003. Freak waves – what can we do about them? Lecture to the Honourable Company of Master Mariners and the Royal Institute of Navigation, London.

Ferreiro, L.D., Stonehouse, M.H., 1994. A comparative study of US and UK frigate design. TRINA.

Ferreiro, L.D., 2007. Ships and science: The birth of naval architecture in the scientific revolution, 1600–1800. The MIT Press, Cambridge, Massachusetts.

Final Act and Recommendations of the International Conference on Tonnage Measurement of Ships, 1969; International Convention on Tonnage Measurement of Ships, 1969. HMSO Publication, Miscellaneous No. 6 (1970) Cmmd. 4332.

Fischer, J.O., Holbach, G., 2011. Cost Management in Shipbuilding. GKP Publishing, Cologne.

Francescutto, A., 2002. Intact ship stability – the way ahead. Sixth International Ship Stability Workshop, Webb Institute.

Fredriksen, A., 2000. Classification of naval craft. Warship 2000. RINA.

Froude, R.E., 1905. Model experiments on hollow versus straight lines in still water and among artificial waves. TINA.

Froude, W., 1874. On experiments with HMS Greyhound. TINA.

Froude, W., 1877. On experiments upon the effect produced on the wavemaking resistance of ships by length of parallel middle body. TINA.

Froude, R.E., 1888. On the 'constant' system of notation of results of experiments on models used at the Admiralty Experiment Works. TINA.

Garzke, W.H., Yoerger, D.R., Harris, S., Dulin, R.O., Brown, D.K., 1993. Deep underwater exploration vessels – past, present and future. SNAME Centennial Meeting.

Gawn, R.W., 1943. Steering experiments, Part 1. TINA.

Gawn, R.W., 1953. Effect of pitch and blade width on propeller performance. TINA.

Germanischer Lloyd (GL), 2009. Preliminary guidelines for safe return to port capability of passenger ships.

Gertler, M., 1954. A Re-analysis of the Original Test Data for the Taylor Standard Series. Navy Department, Washington, DC.

Gibbons, G.E., James, P., 2001. The design, construction and maintenance of naval ships to classification society rules. Warship 2001. RINA.

Gibbons, G., 2003. Fatigue in ship structure. NA, January.

Gibbons, G., 2003. Fracture in ship structure. NA, February.

Glover, E.J., 1966–1967. Contra rotating propellers, for high speed cargo vessels. TNECI.

Greenhorn, J., 1989. The assessment of surface ship vulnerability to underwater attack. TRINA.

Gullaksen, J., 2011. A practical guide to damage stability assessment – regulation on damage stability. RINA International Conference, The Damaged Ship.

Hadler, J.B., 1958. Coefficients for International Towing Tank Conference 1957 Model-Ship Correlation Line. DTMB, Report 1185.

Havelock, T.H., 1956. The damping of heave and pitch: a comparison of two-dimensional and three-dimensional calculations. TINA.

Hogben, N., 1995. Increases in wave heights over the North Atlantic: a review of the evidence and some implications for the naval architect. TRINA.

Hogben, N., Lumb, F.E., 1967. Ocean Wave Statistics. HMSO, London.

Hogben, N., Dacunha, N.M.C., Oliver, G.F., 1986. Global Wave Statistics. British Maritime Technology Ltd, London.

Honnor, A.F., Andrews, D.J., 1981. HMS Invincible: the first of a new genus of aircraft carrying ships. TRINA.

Hudson, D.A., Price, W.G., Temeral, P., Turnock, S.R., 2010. A brief discussion of the development of mathematical models in ship motion theory. The Royal Institution of Naval Architects 1860–2010. RINA.

Hutchinson, K.W., Scott, A.L., Wright, P.N.H., Woodward, M.D., Downes, J., 2011. Consideration of damage to ships from conceptual design to operation: the implications of recent and potential future regulations regarding application, impact and education. RINA International Conference, The Damaged Ship.

International Association of Classification Societies (IACS) has produced a number of Unified requirements (UR), Common Structural Rules (CSR) and Recommendations (REC) that will be found on their web site.

CSR.01, 2010. Common Structural Rules for Double Hull Oil Tankers.

CSR.02, 2010. Common Structural Rules for Bulk Carriers.

UR F, 2011. Fire protection.

UR I, 2011. Polar Class.

UR L, 2011. Load line.

UR S, 2011. Strength of ships.

REC 031, 2004. Inclining test unified procedures.

REC 034, 2001. Standard wave data.

REC 046, 1997. Bulk carriers. Guidance and information on loading and discharging to reduce the likelihood of over stressing the hull structure.

REC 056, 1999. Fatigue assessment of ship structures.

REC 076, 2007. IACS Guidelines for survey, assessment and repair of hull structure – bulk carrier.

REC 082, 2003. Surveyor's glossary of hull terms and hull surveying terms.

REC 084, 2005. Container ships – Guidelines for survey, assessment and repair of hull structure.

REC 085, 2005. Recommendations on voyage data recorder.

REC 096, 2007. Double hull oil tankers - Guidelines for survey, assessment and repair of hull structure.

REC 110, 2010. Guidelines for scope of damage stability verification on new oil tankers, chemical carriers and gas carriers.

REC 111, 2010. Passenger ships – Guidelines for preparation of hull structural surveys.

Improving Ship Operational Designs, 1999. The Nautical Institute, London.

Isherwood, J.W., 1908. A new system of ship construction. TINA.

ISO 2633. Guide for the evaluation of human exposure to whole body vibration.

ISO 6954, 2000. Guidelines for permissible mechanical vibration on board seagoing vessels.

ISO 12217, 2002. Small craft – stability and buoyancy assessment and categorization.

ITTC Recommended Procedures and Guidelines. The papers are available on http://ittc.sname.org.

ITTC 7.5-02-02-01, 2011. Resistance Tests.

ITTC 7.5-02-01-03, 2011. Fresh Water and Seawater Properties.
ITTC 7.5-02-03-01.4, 2011. 1978 ITTC Performance Prediction Method.
ITTC 7.5-02-05-04.1, 2002. Excerpt of ISO 2631, Seasickness and fatigue.
ITTC 7.5-02-06-01, 2008. Manoeuvrability. Free Running Model Test Procedure
ITTC 7.5-02-06-02, 2008. Maneuvrability Captive Model Test Procedure.
ITTC 7.5-02-07-02.1, 2011. Seakeeping Experiments.
ITTC 7.5-02-07-2.6, 2011. Global Loads: Seakeeping Procedure.
ITTC 7.5-02-07-04.1, 2008. Model Tests on Intact stability.
ITTC 7.5-02-07-04.2, 2005. Model Tests on Damage Stability in Waves.
ITTC 7.5-03-01-03, 1999. CFD User's Guide.
ITTC 7.5-04-01-01.1, 2005. Full Scale Measurements. Speed and Power Trials. Preparation and Conduct of Speed/Power Trials Data.
ITTC 7.5-04-01-01.2, 2005. Full Scale Measurements. Speed and Power Trials. Analysis of Speed/Power Trials.

Jurgen, D., Moltrecht, T., 2001. Cycloidal rudder and screw propeller for a very manoeuvrable combatant. RINA International Symposium. Warship 2001.

Karaminas, L., 1999. International rules and regulations and ship design – a healthy relationship? RINA (London Branch).

Kristiansen, S., 2005. Maritime Transportation. Elsevier, Butterworth–Heinemann, Oxford.

Kuo, C., 1996. Defining the safety case concept. NA.

Lackenby, H., 1955. BSRA resistance experiments on the Lucy Ashton. Part III: the ship-model correlation for the shaft appendage conditions. TINA.

Lackenby, H., 1966. The BSRA methodical series. An overall presentation. Variation of resistance with breadth/draught ratio and length/displacement ratio. TRINA.

Lackenby, H., Milton, P., 1972. DTMB Standard Series 60. A new presentation of the resistance data for block coefficient, LCB, breadth/draught ratio and length/breadth ratio variations. TRINA.

Lamb, H., 1993. Hydrodynamics. Cambridge University Press, Cambridge.

Landweber, L., deMacagno, M.C., 1957. Added mass of two dimensional forms oscillating in a free surface. J. Ship Res. SNAME.

Leaper, R.C., Renilson, M.R., 2012. A review of practical methods for reducing underwater noise pollution from large commercial vessels. TRINA.

Lewis, F.M., 1929. The inertia of the water surrounding a vibrating ship. TSNAME.

Liu, D., et al., 1992. Dynamic load approach in tanker design. TSNAME.

Livingstone Smith, S., 1955. BSRA resistance experiments on the Lucy Ashton. Part IV: miscellaneous investigations and general appraisal. TINA.

Lloyd, A.J.R.M., 1998. Seakeeping. Ship Behaviour in Rough Weather. Ellis Horwood Series in Marine Technology.

Lloyd, A.J.R.M., Andrew, R.N., 1977. Criteria for ship speed in rough weather. 18th American Towing Tank Conference.

McCallum, J., 1974. The strength of fast cargo ships. TRINA.

Marshall, S., 2011. Tolerable safety of damaged naval ships. RINA International Conference, The Damaged Ship.

Meek, M., et al., 1972. The structural design of the OCL container ships. TRINA.

Meek, M., 2003. There Go the Ships. The Memoir Club, County Durham.

Miles, A., Wellicome, J.F., Molland, A.F., 1993. The technical and commercial development of self pitching propellers. TRINA.

Ministry of Defence defence Standard 02-109 (NES109) Stability Standards for Surface Ships. Part 1. Conventional Ships.

Molland, A.F., Turnock, S.R., 2007. Marine Rudders and Control Surfaces. Elsevier, Butterworth–Heinemann, Oxford.

Molland., A.F. (Ed.), 2008. The Maritime Engineering Reference Book. Elsevier, Butterworth-Heinemann, Oxford.

Molland, A.F., Wilson, P., 2010. The Development of Hydrodynamics 1860–2010. The Royal Institution of Naval Architects 1860–2010. RINA.

Molland, A.F., Turnock, S.R., Hudson, D.A., 2011. Ship Resistance and Propulsion. Cambridge University Press, Cambridge.

Moor, D.I., Parker, M.N., Pattullo, R.N.M., 1961. The BSRA methodical series. An overall presentation. Geometry of forms and variation of resistance with block coefficient and longitudinal centre of buoyancy. TRINA.

Morrow, R.T., 1989. Noise reduction methods for ships. TRINA.

Muckle, W., 1954. The buoyancy curve in longitudinal strength calculations. Shipbuilder Mar Engine Builder February.

Murray, J.M., 1965. Notes on the longitudinal strength of tankers. TNEC.

Newton, R.N., 1960. Some notes on interaction effects between ships close aboard in deep water. First Symposium on Ship Manoeuvrability, David Taylor Model Basin.

Nishida, S., 1994. Failure Analysis in Engineering Applications. Butterworth–Heinemann, Oxford.

Nonweiler, T.R.F., 1961. The stability and control of deeply submerged submarines. TRINA.

O'Hanlon, J.F., McCauley, M.E., 1973. Motion sickness incidence as a function of the frequency and acceleration of vertical sinusoidal motion. Human Factors Res Inc, Tech Memo 1733–1.

Paik, J., Kim, D., Park, D., Kim, H., 2012. A new methodology for assessing the safety of damaged ships. TRINA.

Parker, T.J., 1998. Innovative marine vessels. The 1998 Lloyd's Register Lecture. The Royal Academy of Engineering, London.

Pattison, D.R., Zhang, J.W., 1994. Trimaran ships. TRINA.

Payne, S.M., 1992. From Tropicale to Fantasy: a decade of cruise ship development. TRINA.

Perrault, D.E., Hughes, T., Marshall, S., 2010. Naval ship stability guideline: developing a shared vision for naval stability assessment. TRINA.

Petershagen, H., 1986. Fatigue problems in ship structures. Advances in Marine Structure. Elsevier Applied Science Publishers, Oxford.

Raper, R.G., 1970–1971. Designing warships for a cost-effective life. TIMarE.

Rawson, K., 2006. Ever the Apprentice. The Memoir Club, County Durham.

Rawson, K.J., Tupper, E.C., 2001. Basic Ship Theory, fifth ed. Butterworth–Heinemann, Oxford.

RINA International Conference with the Nautical Institute, 1993. Escort tugs; design, construction and handling – the way ahead.

RINA Conference, 1997. Design and operation for abnormal conditions. Glasgow.

Sarchin, T.H., Goldberg, L.L., 1962. Stability and buoyancy criteria for US naval surface warships. TSNAME.

Schlick, O., 1884. Vibration of steam vessels. TINA.

Schoenherr, K.E., 1932. Resistance of flat surfaces moving through a fluid. TSNAME.

Shearer, K.D.A., Lynn, W.M., 1959–1960. Wind tunnel tests on models of merchant ships. TNECI.

Significant ships. Annual publication of the RINA.

Significant small ships. Annual publication of the RINA.

Smith, C.S., Chalmers, D.W., 1987. Design of ship superstructures in fibre reinforced plastic. NA, May.

Smith, C.S., Anderson, N., Chapman, J.C., Davidson, P.C., Dowling, P.J., 1992. Strength of stiffened plating under combined compression and lateral pressure. TRINA.

SOLAS 2009 takes ro-ros to new limits. NA, September 2010.

Somerville, W.L., Swan, J.W., Clarke, J.D., 1977. Measurements of residual stresses and distortions in stiffened panels. J. Strain Anal. 12 (2).

Spouge, J.R., 2003. The safety of general cargo ships. TRINA.

Standard procedure for resistance and propulsion experiments with ship models. National Physical Laboratory Ship Division Report No. 10.

Sumpter, J.D.G., 1986. Design against fracture in welded structures. Advances in Marine Structure. Elsevier Applied Science Publishers, Oxford.

Sumpter, J.D.G., Bird, J., Clarke, J.D., Caudrey, A.J., 1989. Fracture toughness of ship steels. TRINA.

Taylor, D.W, 1989. Speed and Power of Ships. United States Shipping Board, Revised.

Taylor, J.L., 1924–1925. The theory of longitudinal bending of ships. TNECI.

Taylor, J.L., 1927–1928. Ship vibration periods. TNECI.

Taylor, J.L., 1930. Vibration of ships. TINA.

The Nautical Institute, 1998. Improving ship operational design.

Thomas, T.R., Easton, M.S., 1991. The Type 23 Duke Class frigate. TRINA.

Thornton, A.T., 1992. Design visualisation of yacht interiors. TRINA.

Todd, F.H., 1961. Ship Hull Vibrations, Arnold, London.

Townsin, R.L., 1969. Virtual mass reduction factors: J values for ship vibration calculations derived from tests with beams including ellipsoids and ship models. TRINA.

Troost, L., 1950–1951. Open water test series with modern propeller forms. TNECI.

van Lammeren, W.P.A., van Manen, J.D., Oosterveld, M.W.C., 1969. The Wageningen B-screw series. TSNAME.

Violette, F.L.M., 1994. The effect of corrosion on structural detail design. RINA International Conference on Marine Corrosion Prevention.

Walker, F.M., 2010. Ships and Shipbuilders. Seaforth Publishing, Barnsley.

Ward, G., Willshare, G.T., 1975. Propeller excited vibration with particular reference to full scale measurements. TRINA.

Warnaka, G.E., 1982. Active attenuation of noise – the state of the art. Noise Control Eng.

Watson, D.G.M., 1998. Practical Ship Design. Elsevier Ocean Engineering Book Series, Oxford.

Weitsendorf, E.-A, Friesch, J., Song, C.S.S., 1987. Considerations for the new hydrodynamics and cavitation tunnel (HYCAT) of the Hamburg Ship Model Basin (HSVA). ASME International Symposium on Cavitation Research Facilities and Techniques, Boston.

Yuille, I.M., Wilson, L.B., 1960. Transverse strength of single hulled ships. TRINA.

Notes

1. These papers indicated by superscript (1) can be found on the Lloyds Register under Lloyd's Register Technology Days. Web site: www.lr.org.
2. Where the papers quoted are contained in the transactions of a learned society they are referred to as, for instance, TRINA, TSNAME, etc., plus year of publication. For references to articles in *The Naval Architect* (the Journal of the RINA) the abbreviation NA is used. Two other institutions which no longer exist but whose papers are quoted are the Institution of Engineers and Shipbuilders of Scotland TIESS and the North East Coast Institution of Engineers and Shipbuilders (TNECI). The David Taylor Model Basin is abbreviated to DTMB.

The Internet

A lot of information is available on the Internet. Some useful web sites are listed below and others will be found in the technical press. The web sites associated with the regulatory bodies are useful in indicating the latest versions of rules and regulations. The web sites of the learned societies include details of membership, papers and other publications. That of the Royal Institution of Naval Architects includes a Maritime Directory which provides links to the web sites of other professional, academic, industrial, governmental and international organisations.

Some useful web sites

International and Government Organisations

The following organisations are referred to in the main text:

International Maritime Organisation (IMO)	www.imo.org
Department of the Environment, Transport and the Regions (DETR)	www.detr.gov.uk
UK Maritime and Coastguard Agency (MCA)	www.hmcoastguard.co.uk
US Coastguard	www.uscg.mil
Defence Evaluation and Research Agency (DERA) (UK)	www.dera.gov.uk
MARIN (Netherlands)	www.marin.nl
David Taylor Model Basin (USA)	www50.dt.navy.mil
British Standards Institution	www.bsi-global.com
European Maritime Safety Agency	www.emsa.europa.eu

Note that in 2001 DERA was divided into two organisations. They are:

1. QinetiQ. An independent science and technology company: www.qinetiq.com;
2. The Defence Science and Technology Laboratory, an agency of the Ministry of Defence: www.dstl.gov.uk.

Classification societies

Most classification societies belong to IACS and details are on the web site.

The International Association of Classification Societies (IACS)	www.iacs.org.uk
American Bureau of Shipping	www.eagle.org
Bureau Veritas	www.veristar.com
China Classification Society	www.ccs.org.cn
Det Norske Veritas	www.dnv.com
Germanischer Lloyd	www.GermanLloyd.org
Korean Register of Shipping	www.krs.co.kr
Lloyds Register of Shipping	www.lr.org
Nippon Kaiji Kyokai	www.classnk.or.jp
Registro Italiano Navale	www.rina.it
Russian Maritime Register of Shipping	www.rs-head.spb.ru

Learned societies

Membership of a learned society is recommended as a means of keeping abreast of the many new developments in the discipline as they occur and for the opportunity to meet like-minded people. Additionally, there is the status associated with membership of a recognised body with the possibility, in the UK, of being able to register as a member of the Engineering Council.

Corresponding bodies exist in other countries. It is likely that such membership will become more important in future as an aid to obtaining employment in industry. Many of the societies allow students to join, and gain the benefits of membership, free or at low cost.

The principal societies, with their abbreviated titles and web sites, whose papers or publications have been cited in this book are as follows:

The Royal Institution of Naval Architects: RINA (INA before 1960) www.rina.org.uk.

The Society of Naval Architects and Marine Engineers: SNAME www.sname.org.

The Japan Society of Naval Architects and Ocean Engineers JASNAOE www.jasnaoe.or.jp.

The Institute of Marine Engineering, Science and Technology: IMarEST (Formerly IMarE) www.imarest.org.uk.

The Nautical Institute: NI www.nautinst.org.

This page intentionally left blank

Index

C

D

E

F

G

H

I

Q

R

S

T

Printed in Great Britain
by Amazon